P9-DTP-785

WITHDRAWN

MAY 14 2010

WDC LIB - STACKS

The Economic
Analysis Of
Technological
Change

The Economic Analysis of Technological Change

PAUL STONEMAN

OXFORD UNIVERSITY PRESS

1983

Oxford University Press, Walton Street, Oxford OX2 6DP

London Glasgow New York Toronto
Delhi Bombay Calcutta Madras Karachi
Kuala Lumpur Singapore Hong Kong Tokyo
Nairobi Dar es Salaam Cape Town
Melbourne Auckland

and associated companies in
Beirut Berlin Ibadan Mexico City Nicosia

Oxford is a trade mark of Oxford University Press

Published in the United States
by Oxford University Press, New York

© Paul Stoneman 1983

All rights reserved. No part of this publication may be reproduced,
stored in a retrieval system, or transmitted, in any form or by any means,
electronic, mechanical, photocopying, recording, or otherwise, without
the prior permission of Oxford University Press

British Library Cataloguing in Publication Data
Stoneman, Paul
The economic analysis of technological change.
1. Technological innovation – Economic aspects
I. Title
338'.06   HC79.T4
ISBN 0-19-877194-0
ISBN 0-19-877193-2 Pbk

Library of Congress Cataloging in Publication Data
Stoneman, Paul.
The economic analysis of technological change.
Bibliography: p.
Includes indexes.
1. Technological innovations – Economic aspects.
2. Diffusion of innovations. I. Title.
HD45.S84 1983   338'.06   83-8311
ISBN 0-19-877194-0 (U.S.)
ISBN 0-19-877193-2 (U.S. : pbk.)

Set by Hope Services, Abingdon
Printed in Great Britain
at the University Press, Oxford
by Eric Buckley
Printer to the University

Preface

The economics of technological change has, for a topic in a young science, a long history. Despite this, recent advances in the subject are not readily available in one source for the interested student or researcher. One aim of this book is to make them available.

My biases should be recognized. As an economist I consider technological change from the economic point of view. The perspectives provided by sociologists, technologists, industrial relations researchers and others are not considered in this volume. This is not to say that such literature should be ignored, but rather that I claim no expertise in expositing the material. Even as an economist, my bias is towards the analytical aspects of the subject rather than its empirical manifestations. For this reason the book reflects a view that empirics must be preceded by theory.

The material in the book is related to material used in lectures given at Warwick University to undergraduate and postgraduate students of industrial economics and macroeconomics. It also reflects material used in a course, taught jointly with Maxine Berg, on the Technology and Industrial Development of the UK Economy since 1800. The degree of mathematical sophistication reflects the form in which this material has been presented to the students on these courses. The mathematical content of the volume is reasonably high, but in general the mathematics is simple. For the most part, it should not really be beyond any undergraduate who has followed a standard course in mathematical analysis. In addition to assuming such a knowledge, much of the book assumes a basic knowledge of macro- and microeconomic theory. This I hope suggests that much of the material will be comprehensible to many second- and third-year UK undergraduates. Other, more sophisticated, material assumes a background appropriate to a graduate student. However, the book is not written exclusively as a text for students; it is also aimed at the researcher in the field, and thus only on occasion is simplification introduced solely to ease exposition.

An attempt has been made to make the book reasonably comprehensive as to the current state of the art. This does not imply however that it even pretends to be a survey of all the literature in the area. The selection is governed by objectives, space and limitations on my own knowledge of the literature. I have attempted to present what I see as the main analytical material, but I have not made a conscious effort to be exhaustive in the empirical area. The book should thus be considered complementary to more descriptive views of the process of technological change. It is also orientated more towards the spread of new technology and the impact of new technology than to the sources of new technology. This reflects my own conception of the relative importance of the different topics and the extent to which material on spread and impact are available in the existing literature. We have not ignored the sources; one cannot do that. We have, however, devoted more space to spread and impact.

The biases I have as an author will mean that others, also economists, will find that certain topics are not covered in this book. Thus, for example, the social consequences of technological change, defined as effects on social structures, the environment, conditions of work, etc., have not really been treated. The economics of education have also been largely ignored. Again, the excuses are a lack of expertise and a shortage of space. It is however hoped that the material that is presented will provide a useful core for any study of technological change considered more widely than we have chosen to consider it here.

In writing this book I have been helped and encouraged by a number of people, not all of whom I can mention here. I owe particular thanks however to my colleagues at Warwick, especially Keith Cowling, who acted as a most useful guinea-pig on whom I could test an earlier version. I am also grateful to Paul Geroski and Chris Freeman for their comments on an earlier draft. I believe that by reacting to their comments the book has been much improved. I must also thank the various typists who have struggled with my handwriting, but especially Liz Cross, who has borne the largest part of this burden. In the usual way, however, any errors or omissions that still remain are my sole responsibility.

University of Warwick
December 1982

Contents

Part III   THE IMPACT OF TECHNOLOGICAL CHANGE

Chapter 1

A Preamble and Some Basic Concepts

1.1 The importance of technological change

Technological advance has probably been the major influence on the nature of
the lives that we lead relative to the lives that our forebears had and our children
and grandchildren will have. In a world that has over the last half-century
acquired the ability to undertake space travel, destroy itself through nuclear
explosions, transmit messages around the globe by satellite, and travel at twice
the speed of sound, it is difficult to conceive of any other factor that has so
dramatically shaped our lives. However, we also live in a world where the benefits
of new technology do not extend to all, where many of the world's children
are starving, and where people die through a lack of appropriate medical care.
We live in a world where technological advance introduces fear as well as improved living standards — the fear of nuclear holocaust or the fear of mistakes
in genetic engineering may dominate the pleasure we derive from cheaper power
or new improved strains of wheat. We see, by looking at our own lives, or by
comparing the lives of our children with our own childhood, how much technology
has influenced our lives over a relatively short period. Our children take for
granted the power of micro-computers and the novelty of video games. They
regard the modern motor car as a basic component of life and yawn at the latest
space shot. Our children view powerful modern drugs as commonplace. By
comparing our own childhood with those of our own children, we can attempt
to bring into focus the power of technology. By watching a child and its grand-
parent together, the strength of technological advance is perhaps even more
vividly illustrated. Watching a three-year-old show a seventy-year-old how to
operate a video recorder is an illuminating experience.

Do we, however, feel that our own children will have a better life than
ourselves? As we observe the environmental pollution that new technology has
caused and the changes in social habits that television and other forms of communication have introduced, can the new really be considered to be better than
the old? As one hears of the latest advances in weapon warfare can one really
believe that advances in peaceful technology are sufficent to compensate for
this? Moreover, do we really see that the technological advances of the developed
world will improve, or even lengthen, the lives of many children in the poorer
nations of the world?

Technological change is often two-edged: it changes many things, often for
the better, but these improvements are often achieved at a cost. Why is it that
technology keeps advancing, and what is it that leads it to advance in the directions that it does advance in? As it advances, what are the benefits and costs of
this advance? These are the questions with which this book is concerned. It

would be misleading to even pretend, however, that we have provided answers to the global questions that condition our view towards new technology. Our aims are much more modest. First, we have restricted ourselves largely to the analysis of modern capitalist economies. Second, we have constrained ourselves largely to that part of those economies that might be called the commercial sector. Finally, we have constrained ourselves to the consideration of the economist's contribution to the analysis of the relevant issues. We thus do not squarely face why the world has developed nuclear weapons or undertaken space travel — the decision to proceed with such projects was probably not based on economic reasoning at all. At the same time, we do not discuss the very wide issues of how new technology has impacted on social structures, social mores of behaviour, or society in general. This is not to dispute the importance of such issues, for they are important. My own limitations and the limitations of time and space preclude them. Thus to a large extent the perspectives provided by sociologists, technologists, industrial relations researchers and others are not considered in this volume.

The economist's viewpoint is of course concerned with social welfare and the impact of technology on this, but practicalities largely limit the concepts of social welfare discussed. Our discussions of the impact of technological change are largely centred on output, employment, growth, and income distribution. This is justified on the grounds that these factors impinge on social welfare, and although they may not give us the whole picture, economic wellbeing for which they are a proxy is a major component of social welfare.

The justification for presenting the economic point of view is two-fold.

1   Unlike many social sciences, economics has a number of well defined conceptual and theoretical frameworks that enable one to structure questions and provide answers to relevant questions. These answers may not be precise and they may not be definitive, but they represent an important contribution to an important debate.
2   Over the last few years the economic contribution to the analysis of technological change has grown apace, and it appears to be a particularly appropriate time to bring together these contributions.

The time is appropriate on a number of grounds. From a European perspective, we are at present going through one of the periodic explosions of interest in the impact of particular new technologies. For the last five years a major topic of concern within much of Europe has been the impact that new micro-electronic or information technology will have on national economies. In fact, 1982, the year in which most of this volume was written, was Information Technology Year in the UK. Such periodic bouts of interest at the national level with regard to particular new technologies seem to appear every twenty years or so. The last such occasion concerned the then new computer technology. In the present discussion three main issues have been at the centre of the debate:

1   What impact will the new technology have on employment?

2   What impact will the new technology have on the international division of labour and wealth?

3   What opportunities does the new technology offer to improve upon or maintain previous levels of economic performance?

The first of these topics is particularly relevant when unemployment (in the UK) is at record high levels. In the UK also, after having undergone a long period of relative industrial decline, the fear that the new technology will accelerate this decline, plus the belief that new technology could reverse it, means that a particularly heavy emphasis is being placed on the importance of new technology in economic development.

In such discussions the majority of the emphasis is naturally on the impacts of new technology. However, the impact will be at the end of a long chain of economic decision-making involving the generation and implementation of new technology. To consider only the impacts would be misleading. To consider only the impacts would suggest that new technology arrives *deus ex machina*, whereas the processes of research and development and selection and rejection of technologies will shape the nature of new technologies coming forth, the rate at which they arise, and the origin of the new technologies. Moreover, the time it takes for a new technology to have an impact may be long: it can take up to twenty years for it to be used extensively, and if one is to gain any insight into what impacts a technology is going to have, one must investigate the process by which it spreads. A concern about impacts thus necessitates an investigation of the whole process of technological change.

It is the grand vision of this whole process that we try to represent in this volume. In Part I we look at the generation of new technology and in Part II we consider its spread. In Part III we discuss the impact of new technology. To fill in this division with greater detail, it is useful first to explore some of the basic representations of technological change that have been used in the literature, which will also provide building blocks for later work. Having done so, we can then proceed to explain more fully the structure of the book.

1.2 Some basic concepts

We do not have a neat, completely general definition of the concept of technological change; thus we turn to some of the representations of technological change that have been used in the literature. In the neoclassical literature one defines a production function, $Q = F(K;L;t)$ where K and L are capital and labour inputs, Q is output, and t is time; technological change is the process justifying t in this function. On a more specific level, technological change is a change in the economy's information set detailing the relationships between inputs and outputs in the economy. To be even more concrete, technological change is the process by which economies change over time in respect of the products they produce and the processes used to produce them. To continue such representations is not, however, particularly instructive in itself. What is more interesting is to consider some of the distinctions between types of technological change that have been advanced, and thereby to illuminate the whole issue.

Consider first the neoclassical concepts as summarized by t in the production function. If technology is so represented, then a technological advance (as opposed to merely a change) enables the economy to obtain greater outputs from the same inputs as time proceeds. The major distinctions that have been derived from this approach are two fold: the first is the distinction between embodied and disembodied technical change; the second concerns the bias or direction of technological change.

Technological advance is disembodied if, 'independent of any changes in the factor inputs, the isoquant contours of the production function shift towards the origin as time passes' (Burmeister and Dobell, 1970, p. 66). This is often referred to as technological change of the 'manna from heaven' type. Embodied technological change comprises improvements that can be introduced only by investment in new equipment and skills, with old equipment left unenhanced. The new technology 'is built into or embodied in new capital equipment or newly trained or retrained labour' (Burmeister and Dobell, 1970, p. 66). Disembodied change is automatic; embodied change is not. The rate of embodied change and its direction will be the result of numerous forces in the economy, to the analysis of which a large part of this book is directed.

To illustrate the concept of bias, consider disembodied change. Write the production function as

$$Q = F(K, L; t). \tag{1.1}$$

Allow that the production function has positive first- and negative second-order derivatives with respect to K and L for a given t, and also has the property that it can be rewritten in per capita form (i.e., is homogeneous of degree one) as

$$q = f(k; t) \tag{1.2}$$

where $q = Q/L$ and $k = K/L$.

We now consider that (1.1) can be written as

$$Q = F(K, L; t) = G\{b(t)K, a(t)L\}, \tag{1.3}$$

in which case technological change is known as the factor-augmenting type. Using the dot convention for representation of a derivative with respect to time, if $\dot{b}(t) > 0$ and $\dot{a}(t) = 0$, the change is purely capital-augmenting, and if $\dot{a}(t) > 0$, $\dot{b}(t) = 0$, the change is purely labour-augmenting. If $\dot{a}(t) = \dot{b}(t)$, then the change is equally labour- and capital-augmenting.

To define the bias of technological change we must first define neutrality. A neutral technical change is one that shifts the production function in such a way as to leave undisturbed the balance between capital and labour in current production. The most common definition of this balance refers to factor shares in national income. A change is neutral if factor shares are not affected. However, the development of factor shares will also depend on how the economy is developing over time. We thus have different definitions of neutrality depending on how the economy is developing.

1   If we consider an economy developing such that the capital–output ratio remains constant, then 'Harrod neutrality' exists if factor shares are not affected by technological change.
2   If we consider an economy developing such that the labour–output ratio is constant, then 'Solow neutrality' exists if factor shares do not change with technological change.
3   If we consider an economy developing such that the capital–labour ratio is constant, then 'Hicks neutrality' exists if factor shares do not change with technological change.

Thus, Harrod neutrality requires that, for a given capital–output ratio, the shift in the production function is such as to keep the marginal product of capital constant. To show this we define capital's share in national income as Π. Then, assuming factors are paid their marginal products,

$$\Pi = \frac{\partial f}{\partial k} \cdot \frac{k}{q}$$

If $d(k/q)/dt = 0$ and Harrod neutrality holds, then $\partial f/\partial k$ *is constant and* $\dot{\Pi} = 0$. It can be shown (Burmeister and Dobell, 1969) that Harrod neutrality requires the production function to be of the form of (1.4); i.e., $b(t) = 1$, $\dot{b}(t) = 0$:

$$Q = G\{K, a(t)L\}. \tag{1.4}$$

For obvious reasons Harrod-neutral progress is called labour-augmenting.

By similar arguments we may show that Solow neutrality will occur if, for a given output–labour ratio, the marginal product of labour does not change with a change in technology. In this case the production function can be written as

$$Q = G\{b(t)K, L\} \tag{1.5}$$

and technical change is purely capital-augmenting.

In the Hicksian case, technical change is Hicks-neutral if, for a given capital–labour ratio, the marginal rate of substitution between capital and labour does not change with a change in technology. In this case the production function can be written as (1.6), where $c(t) = a(t) = b(t)$:

$$Q = c(t) G(K, L). \tag{1.6}$$

Expression (1.6) represents equal capital and labour augmentation and is often called product-augmenting technical change.

These three types of neutrality thus refer to pure labour-augmenting, pure capital-augmenting, and product-augmenting technical change. One should not confuse these concepts with technological change brought about by using better labour, better machines, or by making better products. A change could be pure labour-augmenting although it is realized by introducing better machines. All one is doing is characterizing the shift in the isoquant or production function.

With these definitions of neutrality we can now consider bias. A technological change is labour-saving if the relative share of labour falls along the path with respect to which neutrality is defined. A technological change is capital-saving if the relative share of capital falls. The logic of this classification is that a labour-saving change with a falling labour share implies that a given labour force can be employed only at lower wages after the change and thus must be in less demand. Now, although these definitions are essentially macroeconomic-based, shifts in micro-production functions can be classified using the same definitions of neutrality. Moreover, one should note that what is labour- or capital-saving under one definition of neutrality may not be under another.

We have then three definitions of neutrality. We now state the following two properties.

1   Only under Harrod neutrality can a steady-state growth path in an economy be maintained.
2   Two types of technical progress are compatible if there exists a production function that represents both concepts.

The three concepts above are compatible and the production function within which they are compatible is the Cobb–Douglas. Let technical progress proceed at a given proportional rate λ, and write the Cobb–Douglas as

$$Q = e^{\lambda t} K^{\alpha} L^{1-\alpha}. \tag{1.7}$$

If technical progress is Harrod-neutral at rate m, then

$$Q = K^{\alpha} (Le^{mt})^{1-\alpha} = e^{m(1-\alpha)t} K^{\alpha} L^{1-\alpha} \tag{1.8}$$

and $\lambda = m(1 - \alpha)$.

If technical progress is Solow-neutral at rate m, then

$$Q = (Ke^{mt})^{\alpha} L^{1-\alpha} = e^{m\alpha t} K^{\alpha} L^{1-\alpha} \tag{1.9}$$

and $\lambda = m\alpha$.

If technical progress is Hicks-neutral at rate m, then

$$Q = (Ke^{mt})^{\alpha} (Le^{mt})^{1-\alpha} = e^{mt} K^{\alpha} L^{1-\alpha} \tag{1.10}$$

and $\lambda = m$.

Thus (1.7) is at the same time Harrod-, Hicks-, and Solow-neutral. Only the Cobb–Douglas has this property (Uzawa, 1961).

We therefore have three concepts of neutrality. There are however others in the literature. To the three above we can add the following (Gehrig, 1980).

1   For a constant capital–output ratio the marginal product of labour is constant. This requires a production function of the form

$$F(K, L; t) = G\{K, L + b(t)K\} \tag{1.11}$$

and is known as labour-combining.

2   For a constant output–labour ratio the marginal product of capital is constant. This requires

$$F(K, L; t) = G\{K + a(t)L, L\} \tag{1.12}$$

and is known as capital combining.

3   At a constant capital–labour ratio the marginal product of labour is constant, yielding

$$F(K, L; t) = b(t)K + G(K, L), \tag{1.13}$$

known as capital-additive.

4   At a constant capital–labour ratio the marginal product of capital is constant, yielding

$$F(K, L; t) = a(t)L + G(K, L) \tag{1.14}$$

and known as labour-additive.

These variants have not however received the attention in the literature given to the original three.

We have defined these concepts of bias for *disembodied* technical change. Consider now that technical change is *embodied*. We identify a machine by a number v, denoting its date of manufacture or vintage. Machines of different vintages are assumed to be operated independently. We then can define a vintage production function that relates output to inputs for machines of vintage v. Thus, if we let $Q(v, t)$ be output on vintage v at time t, $K_v(t)$ be capital stock of vintage v in time t, and $L_v(t)$ be labour employed on vintage v at time t, we write the production function as

$$Q(v, t) = F\{K_v(t), L_v(t), v\}. \tag{1.15}$$

This relation stresses the point that it is date of manufacture that affects productivity and not time. Definitions of neutrality follow as above, but now are taken with respect to different vintages rather than times.

This vintage approach with its concept of embodiment does take us away from the automaticity of the disembodied approach; however, it is still open to two basic objections.

1   No distinction is made between new technology that changes processes and that which changes products. The product/process distinction can be an important one. Of course, in an interrelated economy one industry's new products can be another's new process, but especially when we consider technological change at the micro-level the product/process distinction can be important.

2   The production function approach does not tell us anything of the sources of technological change. Why, for example, is vintage v equipment better than that of vintage $v - 1$? To investigate this further we turn to the Schumpeter trilogy.

Schumpeter defined three phases in the process of technological change: invention, innovation, and diffusion. Using Freeman's (1974) definitions, invention is 'an idea, a sketch or a model for a new improved device, product process or system. Such inventions may often (not always) be patented but they do not necessarily lead to technical innovation. . . . An innovation, in the economic sense, is accomplished only with the first commercial transaction involving the new product, process, system or device.' Unfortunately, innovation is often used to describe either the whole process of technological change or an act that is original to the decision-maker under discussion but not to the economy as a whole. In general, we shall use the term as defined above. The third part of the trilogy is diffusion, which occurs after the invention and innovation stages and refers to the process by which the innovation spreads across the market.

In this trilogy, patenting comes in at the invention stage and inventions will be produced partly through expenditure on research and development (R & D). However, the step from invention to innovation also depends on R & D. In the UK, R & D expenditures have split on average in the postwar period in the following proportions: basic research, 3 per cent; applied research, 22 per cent; development, 75 per cent. Although it is not a completely accurate distinction, development expenditures could be associated with innovation, suggesting that most R & D is an input to the innovation process. However, unless the user of the innovation is the same firm (individual) as the producer, we would expect there be two parties to an innovation – a commercial transaction has two sides. Only one of these sides – the producer – may be a major spender of R & D funds.

The innovation process is sometimes referred to as additions to the economy's book of blueprints; however to confuse matters further, this blueprint terminology is sometimes used to refer to invention, and then only certain blueprints are selected for development to the innovation stage. However, once we have innovations, these start to have an impact on the economy as they are used or produced. This is the process of diffusion. As diffusion proceeds we may think of the disembodied production function shifting. Alternatively, we may think of the embodied production function being different for different vintages because of innovation, and the process of diffusion concerns how machines of vintage v take over from those of other vintages. We do not wish to carry this parallel too far, however, for diffusion has also been analysed outside the context of the vintage model.

1.3 An outline

We can now proceed to outline the structure of the book. Part I, on the generation of new technology, is divided into three chapters. The first looks at the patent system, patenting activity as a proxy for invention, and innovation. The second considers the determination of R & D expenditures, and the third is concerned with the determination of the capital or labour saving bias in technological change. The three chapters as a whole represent an attempt to look at those forces that will affect the rate at which new technology will come on to the market and the directions that it will take.

In Part II the diffusion process is studied. Chapter 5 is an introduction to the whole concept, and Chapters 6-9 consider different approaches that have been taken in the analysis of diffusion processes. The objective is to isolate those forces that will determine the rate at which new technology is taken up. Chapter 10 is a report on two case studies in diffusion.

In Part III we come to the impacts of technological change. The relationship between technological change, output, and the demand for inputs is covered in Chapters 11-13. In these chapters the question of the impact of technological change on employment is addressed. Chapter 13 is an attempt to bring to the fore the (seemingly hitherto neglected) role of technological change in the determination of investment expenditure. Chapter 14 is a study of the impact of technological change on the factoral distribution of income. In Chapter 15 the welfare consequences of new technology are addressed directly. Chapter 16 is concerned with the impact of new technology on market structure, and Chapter 17 discusses a number of issues relevant to the open economy. Conclusions are provided in Chapter 18.

The greater part of this book is concerned with economic *analysis* rather than with the empirical detail of technological change. Such detail can be found in, for example, Freeman's (1974) excellent volume. However, the analysis itself is often supported by empirical evidence. In many cases this evidence refers to relatively recent phenomena; however, in other places we have made a conscious effort to introduce the work of the economic historians. The process of technological change is not new; in fact, one of the constants of economic development is that change will occur. It seems most appropriate, therefore, that such an historical aspect should be reflected in the work. The introduction of historical material also reflects the view that the historians have made considerable contributions to our knowledge of the subject. The emphasis on analysis, however, reflects a view that one needs theory before empirical investigation, and that it is largely in the theoretical field that the major advances in the economics of technological change have recently been made.

References

Burmeister, E. and Dobell, R. (1969), 'Disembodied Technical Change with Several Factors', *Journal of Economic Theory*, 1, 1-8.
Burmeister, E. and Dobell, R. (1970). *Mathematical Theories of Economic Growth*, Macmillan, London.
Freeman, C. (1974), *The Economics of Industrial Innovation*, Penguin, Harmondsworth.
Gehrig, W. (1980), 'On Certain Concepts of Neutral Technical Progress: Definitions, Implications and Compatability', in T. Puu and S. Wibe (eds), *The Economics of Technological Progress*, Macmillan, London.
Uzawa, H. (1961), 'Neutral Inventions and the Stability of Growth Equilibrium', *Review of Economic Studies*, 28, 117-24.

Part I
The Generation of New Technology

Chapter 2

Invention and Innovation I:
Output Measures

2.1 Introduction

In the following three chapters we explore the process that generates changes in technology. In terms of the Schumpeter trilogy, we are interested in invention and innovation. Diffusion is to be considered in Part II of the book; our prime concern here is to analyse both the rate and direction of inventive and innovative activity.

The majority of the work in this field is concerned with the Schumpeterian hypothesis, which has been interpreted broadly as stating that bigness and few-ness encourage technological advance. Thus a considerable part of the literature is concerned with the role of large firms and monopoly power in the process of technological change. We must right at the start however issue a warning. It has been argued by Fisher and Temin (1973) that (1) the true Schumpeter hypothesis is not as we broadly stated it above; (2) the majority of the tests that have been made of the broad Schumpeter hypothesis do not correctly test even that; and (3) the true Schumpeter hypothesis is not sufficient to ensure that bigness and fewness encourage technological advance.

This work raises some important issues, so it is worthy of early consideration. Following Fisher and Temin, and using their notation, let N be the number of workers in a firm, S the non-R & D workforce, and R the number engaged in R & D. Let $F(R, N)$ be the value of per worker output from R & D. Now the first point is that, if we consider the question of the effect of firm size on R & D output, we can look at how the value of output increases relative to firm size; in particular, we can see whether

$$\Omega \equiv \frac{S}{RF} \frac{d(RF)}{dS} > 1$$

However,

$$\Omega \equiv \frac{S}{R} \frac{dR}{dS} + \frac{S}{F} \frac{dF}{dS};$$

thus whether the value of the output of technological advances will increase more than proportionately with firm size depends not only on whether R & D inputs increase but also on whether output per unit input does not decrease with firm size.

The second point made by Fisher and Temin is that the true Schumpeter hypothesis is that $\partial F/\partial R > 0$ and $\partial F/\partial N > 0$; i.e., large R & D departments are more productive than small (given firm size), and that a given R & D spending yields a greater return per worker to large firms than to small ones. However, they argue that these conditions do not guarantee that $\Omega > 1$. Thus, if one accepts Fisher and Temin's view of the Schumpeterian case, then the broad Schumpeterian hypothesis is a misleading derivative of Schumpeter's views. In a series of papers these arguments have been challenged and refined (Rodriguez, 1979; Fisher and Temin, 1979; Kohn and Scott, 1982). Kohn and Scott's view is that it is correct to state that basic assumptions about increasing returns to scale in R & D do not imply (1) that the output from the R & D process increases more than proportionately with firm size; (2) that the elasticity of R & D spending with respect to firm size is greater than unity; or (3) that the *value* of R & D output increases more than proportionately with firm size. However, they argue that if $S/R \ dR/dS > 1$, then the elasticity of R & D output with respect to R & D will be greater than unity, and thus a test of whether (dR/dS) $(S/R) > 1$ will be sufficient to test the relationships between output (technological advance) and firm size. If this is the case one need look at only R & D spending and not at the output of the process.

To detail their argument, let us concentrate on the elasticity of the output from the R & D process (Q) and not its value (as used by Fisher and Temin). Kohn and Scott state that, if $(dR/dS) \ (S/R) > 1$, then $(dQ/dS) \ (S/Q) > 1$. Define

$$Q = G(R)$$

as the R & D production function. Specify that $G'(R) > 0, G''(R) > 0$. Now,

$$\frac{dR}{dS} \frac{S}{R} \equiv \frac{(dQ/dS) \ (S/Q)}{(dQ/dR) \ (R/Q)}.$$

If $(dR/dS) \ (S/R) > 1$, then $(dQ/dS) \ (S/Q) > 1$ if $(dQ/dR) \ (R/Q) > 1$. If $G'(R) > 0$ and $G''(R) > 0$, then $G'(R) > Q/R$ and thus $(dQ/dR) \ (R/Q) > 1$. Thus, if $G'(R) > 0$ and $G''(R) > 0$, it is the case that a test of whether (dQ/dS) $(S/Q) > 1$ can be undertaken by looking at $(dR/dS) \ (S/R)$. However, by assuming $G''(R) > 0$, one is assuming that $(dQ/dR) \ (R/Q) > 1$, and thus that output from R & D per unit input increases more than proportionately with firm size. In terms of our expression for Ω, one is saying that we can be sure that $\Omega > 1$ when $(S/R) \ (dR/dS) > 1$ if $(S/F) \ (dF/dS) > 1$, and this we knew anyway. The Kohn and Scott position does not therefore invalidate a statement that one cannot draw conclusions about the relationship between technological advance and firm size just by looking at inputs. One also needs to study the input–output relationships. In this chapter, therefore, we study output indicators; in the next we look at inputs.

The work in this area is concerned with both the rate of technological advance and with its direction. The models will indicate (because of differences in technological opportunities, elasticities, etc.) that there will be inter-industry differences in the rate of technological advance, and therefore give some indication of direction. In Chapter 4 we consider further approaches that have been explicitly constructed to yield results on direction.

2.2 Output measures

The output of the process of advancing technology is, by tautology, techno-logical advance. Exactly how to measure such advances is a basic problem in the field. In this chapter we consider the analysis of two basic measures: (1) patenting activity and (2) innovative activity. One may if one wishes consider these as synonymous with invention and innovation, but such definitions are not so clear in practice.

If for the moment we consider inventions and innovations separately, it should be clear that not all inventions will lead to innovations. One can think of inventions as additions to the pages of a book of blueprints, only certain pages of which will be fully developed into innovations. We should stress therefore that, by looking at inventions and innovations, we are looking at two separate stages in the process of technological advance.[1]

We turn first to an analysis of patenting activity. This discussion however raises a number of issues with respect to the patent system itself, and so prior to analysing patents as a measure of technological advance we will consider the patent system.

2.3 The patent system: advantages and drawbacks

A patent is a certificate of ownership which confers a monopoly right for a limited period for a given piece of intellectual property with an industrial application. The patent is given in recognition that the inventor has produced and disclosed knowledge of a kind that can be used to produce a substantially improved product or process. (Bosworth, 1980a)

In the UK a patent has a maximum twenty-year term, but continuation does depend on payments of continuation fees and abstinence from abuse of the monopoly rights. Patents 'may be granted for any invention which is new, involves an inventive step and is susceptible of industrial application'. An inven-tion is to be considered as new 'if it does not form part of the state of the art'. The state of the art is defined as 'everything made available to the public by means of a written or oral description by use or in any other way before the date of the application' (Bosworth, 1980a).

The existence of a patent system is justified on the grounds that, by guaran-teeing a return to the inventor for his effort in the inventive process, it stimulates the search for inventions. Scherer (1980) argues that the patent system also encour-ages the development and commercial utilization of inventions and encourages the disclosure of information, although the stimulation of invention is the most important desired objective.

If the patent system does encourage invention, we can ask what the costs of operating the system actually are. Arrow argues that the invention process

---

[1] To the extent that it is considered, the selection of blueprints to be developed is considered in Chapter 4.

produces information, a commodity with a very low transmission cost:

Any information obtained should, from a welfare point of view, be available free of charge (apart from the cost of transmitting information). This ensures optimal utilization of the information but of course provides no incentive for investment in research . . . In a free enterprise economy, inventive activity is supported by using the invention to create property rights: precisely to the extent that it is successful, there is an underutilization of the investment. The property rights may be in the information itself, through patents and similar legal devices, or in the intangible assets of the firm if the information is retained by the firm and used only to increase its profits. The first problem, then, is that in a free enterprise economy the profitability of invention requires a non-optimal allocation of resources. (Arrow, 1962)

The argument therefore is that a patent system will encourage invention, but by granting the inventor monopoly power over the invention it will lead to a diffusion process that is too slow. Given that one cannot have both incentives for invention and the free transmission of information, one could look for a second-best optimum by seeking that length of patent life that will maximize social returns to invention. As patent life varies, the incentive to the inventor varies (and thus the number of inventions produced vary); but also, as patent life varies the social losses through costly information will vary. In the light of the results below on the effectiveness of the patent system, we do not pursue this line of enquiry; for a survey of the relevant literature see, for example, Scherer (1980). What is also worth stressing, however, is the implication of Arrow's statement that patents are not the only way to maintain property rights in an invention. Secrecy is another way; also, learning economies may yield benefit to an inventor, as may brand loyalty, other natural reaction lags, or significant entry barriers. All such 'imperfections' can allow the inventor to capitalize on his invention before imitators can reduce his monopoly rents to zero.

The conflict between encouraging invention and discouraging diffusion is perhaps best considered at an empirical level. We can try to answer two questions. First, has the patent system worked effectively as an incentive to invention? Second, has the existence of property rights in invention significantly affected the spread of new technology? We consider the second question in Chapters 9 and 10, where it is suggested that monopoly power in the hands of the inventor can or will slow down the use of new technology to some degree. The first question we consider here.

Taylor and Silberston (1973) and Scherer (1977) investigate the extent to which, in a sample of research-intensive firms, R & D expenditure depended on the incentive provided by patents. In both cases patent protection was not considered to be of vital importance (except in pharmaceuticals). The obvious implication is that protection, and thus returns, can be obtained by other means.

In Mansfield *et al.* (1981) a study of imitation costs and times in the chemical, drug, electronics, and machinery industries is undertaken. A number of innovations of major new products are studied. Of the forty-eight innovations, 70 per cent were patented, and in only one case was a licence granted to an imitator.

It was found that within four years of their introduction 60 per cent of the patented successful innovations were imitated. However, imitation costs were increased by the existence of a patent by approximately 11 per cent (an average, masking an increase of 30 per cent in ethical drugs, 10 per cent in chemicals but only 7 per cent in electronics and machinery). It was also found that imitation was more costly for innovations that had a high ratio of applied to basic research costs, were in the field of ethical drugs, or did not involve the use of an existing material. It was further found that imitation was more likely if imitation costs were small. As to the importance of patents, Mansfield's sample results are that about a half of the patented innovations in the sample would not have been introduced without patent protection — the bulk of these in the drug industry. Outside the drug industry, less than 25 per cent would have been affected. This result confirms that of Taylor and Silberston.

An indication of the importance of patents can perhaps be gauged by looking at patent royalties. In his study Wilson (1977) argues that firms were likely to seek licences for major technical innovations but would simply 'design around' minor ones. This suggests that patents may be important only for major innovations. Bosworth (1980b), however, finds that Wilson's model does not perform well on UK data, and needs extension.

Finally on this issue, we make reference to Mansfield's findings (Mansfield *et al.*, 1971) on returns to innovations (which is discussed more fully in Chapter 15). It is found that, in the sample of seventeen innovations, private returns were significantly less than social returns, and in many cases the private return would not in hindsight have justified the R & D programme. The patent system does not therefore guarantee a return to the inventor for a socially desirable innovation.

From this we can argue that patents possibly play only a minor role in stimulating invention and in most cases do not prevent imitation — and thus may not slow down diffusion to any *great* extent. We now move to an analysis of patenting activity as an indicator of inventive activity.

2.4 The determinants of patenting activity

If we are to use patents as a measure of the rate of technological advance, we wish to know how closely patenting activity reflects such advances. It is usually argued that a head count of patents issued is a poor measure because of the following.

1   Patents differ in quality. Bosworth (1980a) suggests that a way round this is to weight a patent by the number of years it remains extant on the grounds that a minor patent will not be worth paying renewal fees for.
2   There are international differences in what is patentable; thus national patent statistics cannot be compared. Here Soete (1981) suggests using foreign patents issued in the USA to correct for such biases.
3   Patent laws have changed over time, making time-series analysis suspect.

4   Not all advances will be patented. Comanor and Scherer (1969) have found
     that in the USA the propensities to patent vary
     (a) over time, with a secular decline in the propensity;
     (b) by product and process inventions, with products more likely to be
          patented;
     (c) by firm size, large firms being more likely to patent;
     (d) by industry, defence industries for example being less likely to patent.

Despite these reservations, however, one of the most influential contributions
to the literature on technological change has been Schmookler's (1966) analysis
of patenting activity. Schmookler's major result is that, measuring inventions by
a head count of patents, 'capital goods invention tends to be distributed among
industries in proportion to the prevailing distribution of investment. The central
reason for this evidently is that investing is heavily influenced by considerations
of profitability.'

The logic of Schmookler's argument is that, defining Π_i as the expected
profit to be derived from a capital goods invention, i, we have

$$\Pi_i = p_i x_i - c_i x_i - E_i \qquad (2.1)$$

where $x_i = s_i S/p_i$ = output of the capital good embodying the invention;
       s_i = expected market share to be derived from an invention;
       S = size of the market;
       c_i = cost of manufacturing one machine;
       E_i = expected cost of inventing;
       p_i = price of capital good i.
We may now write

$$\Pi_i = s_i S - \frac{c_i s_i S}{p_i} - E_i. \qquad (2.2)$$

We assume prices are set by a cost plus mark-up principle, so $c_i/p_i = k$, a constant.
Then from (2.2) we have

$$\Pi_i = s_i S (1-k) - E_i. \qquad (2.3)$$

It is now argued that the inventor will

1   tend to invent machines for which $(1 - k) s_i S > E_i$;
2   given S and E_i, invent those machines that will capture the largest market
     shares;
3   given S and s_i, invent those machines for which E_i is at a minimum.

However, in his empirical work Schmookler does not pursue these lines of
enquiry. Instead he pursues

4   given E and s_i, one would tend to invent those machines for which S is
     greatest.

Given that one is discussing capital goods inventions, the size of the market will be represented by the level of investment activity. Schmookler looks extensively at patenting activity in US railroads and finds support for his hypothesis. He also undertakes some cross-section analysis yielding the following results:

$$\log P^j_{1940-2} = 1.174 + 0.927 \log I^j_{1939} \quad R^2 = 0.918 \tag{2.4}$$
$$(0.080) \quad (0.070)$$

$$\log P^j_{1948-50} = 0.598 + 0.940 \log I^j_{1947} \quad R^2 = 0.905 \tag{2.5}$$
$$(0.116) \quad (0.070)$$

where $P^j_{t-(t+2)}$ = total number of patents issued in industry j between t and $t + 2$;

I^j_t = investment in industry j in time t.

Standard errors are in parentheses.

The lags in (2.4) and (2.5) are important, for they indicate that investment affects patenting activity and not the reverse. The key result of this work of Schmookler is that invention will be biased towards investing industries.

The Schmookler hypothesis has not met with wholehearted support. Rosenberg (1974) in particular stresses that, in his empirical work, Schmookler ignores the effect of cost differences on patenting activity (E_i in (2.3)). If cost differences are an important determinant of patenting activity, then the purely demand-led view can no longer be supported. Scherer (1965) found some support for the contention that inter-market differences in technological opportunity are major influences on the rate of patenting activity (a finding further reinforced in Scherer, 1982). In Stoneman (1979) further support is provided by an analysis of UK data. However, in this study a revised theoretical framework is used. The Schmookler framework does not allow for any explicit maximizing behaviour on the part of the firm. The intention of this analysis is partly to remedy that situation. The analysis starts from a proposition that the inventor is a profit-maximizer. Thus the profit function (equivalent to (2.1)) is formally maximized. Consider firm i in industry j ($i = 1 \ldots N_j, j = 1 \ldots M$) that undertakes invention-producing activities. Let the number of patents produced by firm i be P_{ij}, and define the revenue from the kth invention as R_{ijk} ($k = 1 \ldots P_{ij}$). The determinants of R_{ijk} may be considered in three categories; (1) invention-specific, (2) firm-specific, and (3) industry-specific. It may well be argued that the return to an invention will be determined by the characteristics of the invention itself. However, it is assumed that a count of the number of patents is a good indicator of the rate of technological change, and this suggests that R_{ijk} is independent of k.

A particular firm-specific factor affecting revenue may be considered the potential captive market represented by the firm's own output. However, given that in principle the patent can be licensed for use by all firms in an industry, the size of the inventing firm's existing output would not seem to be an important factor. In fact, there seems no good reason why the return to an invention should vary with the identity of the firm producing it, and thus it is assumed that R_{ijk} is independent of i as well as k, and will thus be written as R_j.

Consider, now, the industry-specific factors. Schmookler suggests that the level of investment expenditure of the industry that will use the jth industry's patents (with some lag structure) is a good indicator of the expected revenue from an invention. Here three prime determinants of expected revenue are suggested. The first is the size of the potential market for an invention (which would seem to be best indicated by the output level of the user industry, especially in the case of a process innovation). Second, in a world with a fixed patent life, the speed with which an invention is adopted is a crucial determinant of the return to the inventor. A slow diffusion means lower returns than a fast diffusion. As an indicator of the expected speed of diffusion for a given industry, the ratio of investment to net output, I/Y, with some lag structure is used. The reasoning is that process innovations, and most probably product innovations, require positive gross investment. The faster the acceptance rate, the higher the level of gross investment. Thus the ratio of gross investment to output would seem a good indicator of diffusion speed and thus expected return. The third factor considered important is the number of patents, or inventions, being produced for use in a given industry. One would expect that, the greater is the flow of inventions, the lower will be the return to any one, either because of technological overtaking or pure competitive pressure.

Thus, if one lets

$$P_j = \sum_{i=1}^{N_j} P_{ij} = \text{the number of inventions produced by industry } j$$

I_j = investment in the industry using the inventions of industry j
Y_j = output in the industry using the inventions of industry j
J_j = a constant

then it is postulated that[2] (ignoring lag structures for the moment)

$$R_{ijk} = R_j = J_j \left(\frac{I_j}{Y_j}\right)^{\xi} Y_j^{\gamma} (P_j)^{\eta_{RP}^j} \tag{2.6}$$

where η_{RP}^j is the elasticity of expected revenue from a patent in industry j with respect to the number of patents produced in industry j. It is assumed that firms are profit-maximizers and that their invention-producing activities can be separated from their other activities. Defining Π_{ijk} as the profit derived by firm i in industry j from its kth invention, and C_{ijk} as the total cost to this firm of making invention k, it is then assumed that the firm maximizes

$$\sum_{k=1}^{P_{ij}} \Pi_{ijk} = \Pi_{ij} = \sum_{k=1}^{P_{ij}} R_{ijk} - \sum_{k=1}^{P_{ij}} C_{ijk} \tag{2.7}$$

---

[2] One could allow ξ and γ to be industry-specific, but this would have the effect of further complicating the notation below without providing any noticeable benefit.

which, using (2.6), one may write as

$$\Pi_{ij} = P_{ij} R_j - \sum_{k=1}^{P_{ij}} C_{ijk}. \tag{2.8}$$

Just as it is argued that R_{ijk} cannot for practical reasons (among others) be considered as a function of k, it is argued that C_{ijk} can be treated as similarly independent of k. Then hypothesize that C_{ij}, as for any normal cost curve, is a function of P_{ij}; i.e., $C_{ij} = C_{ij}(P_{ij})$; then one can write (2.8) as

$$\Pi_{ij} = P_{ij} R_j - C_{ij}(P_{ij}). \tag{2.9}$$

If one allows for conjectural variations such that (2.10) holds,

$$\frac{dP_j}{dP_{ij}} \frac{P_{ij}}{P_j} = \rho_{ij}, \tag{2.10}$$

then the first-order conditions for profit maximization imply that

$$R_j (1 + \rho_{ij} \eta^i_{RP}) = \frac{dC_{ij}}{dP_{ij}} \tag{2.11}$$

which is the condition that holds for firm i. To aggregate to the industry level, assume (i) $\rho_{ij} = \rho_j$ for all i; (ii) $dC_{ij}/dP_{ij} = \bar{C}_j$ for all i; then from (2.11) and (2.6) one obtains

$$\log P_j = \frac{1}{\eta^j_{RP}} \{ \log \bar{C}_j - \log (1 + \rho_j \eta^j_{RP}) - \xi \log (I_j/Y_j) \tag{2.12}$$

$$- \gamma \log Y_j - \log J_j \} ;$$

i.e., as one would expect, the equilibrium output of inventions is a function of both cost and demand parameters (both of which one would expect to vary across industries), and both have to be considered when one comes to estimation. Schmookler in essence ignores this 'cost of invention' variable in his empirical work.

Consider then the term, $\bar{C}$, cost per patent produced. Given that $\eta^j_{RP} < 0$ (i.e., that revenue per invention reduces as the number of inventions increases), we expect from (2.12) that higher levels of cost per patent, *ceteris parabus*, will generate lower levels of patenting activity. It would seem logical to argue that an industry with extensive inventive opportunities would have lower costs per patent produced than one with limited opportunities. One can interpret this factor, therefore, as reflecting the supply side of the inventing industry, i.e., the element not considered empirically by Schmookler, and for which he has been criticized by, among others, Rosenberg (1974). Where attempts have been made elsewhere to measure this concept, qualitative measures have been used (see Scherer, 1965, or Wilson 1977), but these are always subject to personal judgement.

In the empirical work, despite rather severe data problems, and using R & D costs per patent as a proxy for $\bar{C}$, it is shown that the cost of producing an invention as measured by this proxy significantly influenced the level of patenting activity. The demand proxies were also significant. One may say therefore that it is necessary to modify Schmookler's view to one in which both demand and technological opportunity determine the level of patenting activity. In addition, it is argued that a preferred measure of the demand effect ought to consider both market size and diffusion speed.

This work ignored the influence of firm size and market structure on inventive activity. The relationship between bigness and fewness and technological change is however a key one in the literature. Both concentration and firm size have been considered as explanatory factors for patenting activity.

Scherer (1965) investigated the relationship between patenting activity and firm size and found that, in general, patents issued did not increase proportionately with firm size (also, concentration, liquidity, and profitability were not significant influences on patenting activity). Scherer also looked to see if diversification of the inventing firm was important (on the grounds that diversification implies that there are more markets in which unexpected inventions can be marketed by the firm). He found no significant effect.

The relationship being estimated by Scherer related patents issued to firm sales. Mansfield (1968) has also attempted to estimate a similar relationship. Thus he estimates

$$P_i = (R\&D)_i \, \{a + b \, (R\&D)_i + c X_i\} + \epsilon_i \qquad (2.13)$$

where X_i = firm size;
P_i = number of inventions of firm i;
$(R\&D)_i$ = firm R & D expenditure;
ϵ_i = an error term.

In this work P_i is related to patenting activity rather than just being a head count of patents. Mansfield's results for the chemicals industry are

$$P_i = (R\&D)_i \, \{2.38 + 0.404 \, (R\&D)_i - 0.247 \, X_i\}, \quad R^2 = 0.98$$
$$ (0.04) \quad (0.075) \qquad\quad (0.005) \qquad\qquad\qquad (2.14)$$

(standard errors in parentheses), suggesting that firm size has a significant negative effect on inventive activity. Mansfield also has results for petroleum and steel. Bosworth and Wilson (1980) perform a similar exercise for the world chemical industry. They measure inventive activity by patents issued generating, for example, on data for 1967, the results in (2.15) suggesting a positive effect of sales on patenting activity:

$$P_i = 3.54 - 0.0814 \, (R\&D)_i^2 + 0.0062 \, X_i \, (R\&D)_i, \quad R^2 = 0.82$$
$$ (3.02) \quad (-4.58) \qquad\quad (3.54) \qquad\qquad\qquad (2.15)$$

(t-statistics in parentheses).

The important point about these studies, however, is that both (2.14) and

(2.15) are considered to be production relationships. They are supposed to reflect the production process in invention. They incorporate the two Schumpeter hypotheses that large R & D departments are more efficient than small and that large firms get greater returns from given R & D inputs.[3] (See Fisher and Temin, 1973 and Section 2.1 above.)

One can cast considerable doubt on whether (2.14) or (2.15) will reflect a production relationship, however. Consider for example that we have a profit-maximizing firm, the trading profit of which can be written as $\Pi = \Pi(X, P)$, where P is the number of inventions and X is firm size (total sales) and the production relationship for inventions is given by

$$P = a + b\,(R\&D) + cX \qquad (2.16)$$

The firm chooses X and $R\&D$ to maximize profits defined as $G = \Pi - (R\&D)$. Then we have

$$\frac{\partial \Pi}{\partial P} = \frac{1}{b} \qquad (2.17)$$

and

$$\frac{\partial \Pi}{\partial X} = -\frac{c}{b}. \qquad (2.18)$$

If $\Pi(X, P)$ is homogeneous of degree θ we may write

$$\theta \Pi = X\frac{\partial \Pi}{\partial X} + P\frac{\partial \Pi}{\partial P} \qquad (2.19)$$

which with (2.17) and (2.18) yields

$$P = \theta bG + \theta b\,(R\&D) + cX. \qquad (2.20)$$

Estimating an equation relating P to $(R\&D)$ and X will thus yield estimates that reflect not only production relationships but also demand relationships. Moreover in (2.20) no causality is implied — P, G, $R\&D$ and X are all endogenous to the system, and (2.20) could just as easily have been written with $R\&D$ on the left-hand side. The point is that one cannot just write down a relationship between P, $R\&D$, and X and then call estimates of its coefficients production coefficients; for P, $R\&D$ and X are all endogenous. (This line of reasoning is further developed in Chapter 3, when we discuss the determination of R & D spending. If there exists a production relationship between the output of inventions and $R\&D$, then the determination of $R\&D$ will determine the output of inventions.)

One of the problems of the above analyses of patenting activities is that they generally do not distinguish between patents issued to domestic and foreign firms or individuals. If a large proportion of patents are the property of foreign

---

[3] In his latest paper (Scherer, 1983), Scherer found that patenting activity most commonly rises roughly proportionately with R & D, shedding some doubt on the Schumpeter hypotheses.

firms, domestic level explanatory variables may not be relevant. Moreover, if domestic firms can patent overseas, their domestic characteristics may not be fair explanatory factors. Pursuing similar arguments, Bosworth (1980b) has studied empirically flows of international patenting activity. Let P_{ij} be patent applications from country i in country j and A_{ji} be applications of country j in country i. He then suggests that

$$P_{ij} = A Y^{\alpha}_j H_j^{\beta} X_{ij}^{\gamma} S_{ij}^{\delta} \epsilon_{ij}$$

$$A_{ji} = B T_{ij}^{a} M_{ij}^{b} D_j^{c} \hat{\epsilon}_{ji}$$

where Y_j = GDP of recipient country j;
       H_j = income per head of recipient country j;
       X_{ij} = exports by country i to recipient j;
       S_{ij} = size of multinational operations in recipient j;
       T_{ij} = size of foreign country (j)'s multinationals' interest in country i;
       M_{ij} = country i's imports from j;
       D_j = supply of domestically produced inventions in j;
       $\epsilon_{ij}, \hat{\epsilon}_{ji}$ = error terms.

Taking country i as the UK, it is found that α, β, γ, and δ are significantly different from zero (and all positive), confirming the importance of foreign markets. Also, a and c are both positive and significantly different from zero. This realization of the fact that the world rather than the nation is the appropriate market for patents is an important one.

To summarize this section, we have argued that a head count of patents may not be a good measure of technological advance, but it can at least be measured. We have investigated the determinants of patenting activity. We argued that both technological opportunity (or the cost of invention) and demand factors were important in determining such activity. The role of firm size and monopoly is not quite so clear, in that the beneficial effect of bigness and fewness is not proven. We also throw some doubt on closed-economy analysis, suggesting that, for the UK at least, foreign patenting activity and foreign markets will significantly influence the process of technological change.

2.5 The sources of invention

We could now turn to discuss rates of change of productivity as an indication of inventive activity, but it is felt this has more to do with the whole invention–innovation–diffusion trilogy than just the generation of new technology. Instead, therefore, we shall concentrate in this section on the 'sources of invention' literature. The aim of this literature is to track the development of major technological advances and by doing so learn about the inventing process.

Jewkes *et al.* (1969) look at 64 major twentieth-century innovations. Of these they find that 40 can be attributed to individual inventors, and only 24 to corporate R & D. Hamberg (1966) reports that only 7 inventions out of 27 in his sample came from R & D units of firms. Peck (1962) shows that only 17 out of 149 major inventions in the aluminium industry (1946-57) came from major firms. Freeman (1974), however, in his most informative study of a number of twentieth-century *innovations*, gives a much larger role to corporate R & D departments and 'professional science'. This is not however completely at odds with the above results. Thus, Jewkes *et al.* (1969) find that, of the 40 external inventions, at least half needed a major corporate R & D effort in development before innovation occurred. Thus, even if a large part of invention is exogenous, innovation is not. This is the first important result to be derived from this literature. The second result is an implication of this. If much invention is external, then it may not be realistic to analyse patenting activity in the sort of models we have discussed in the previous section (unless patents reflect innovation rather than invention).

2.6 Innovations

An invention we consider to be a new idea, sketch, or model for a new or improved device, product, process, or system. An innovation is accomplished only with the first commercial transaction involving the new product process, etc. Mansfield's (1968) data suggest an average lag between invention and innovation of ten to fifteen years. As we have argued above, Jewkes *et al.* illustrate that, even if invention has not been the prerogative of corporate organizations, innovation has. (Moreover, innovation requires the input of invention, which, allied with R & D effort, leads to the commerical transaction.) However, a commercial transaction has two parties to it, a buyer and a seller. The study of innovation must discuss both. The purpose of this section is to analyse the literature that uses measures of innovative activity.

Mansfield (1968) presents an analysis of the role of firm size and market structure in innovative activity. He considers three basic industries; iron and steel, petroleum refining, and bituminous coal. He investigates the nature of innovative activity by seeing which firms first introduced each innovation. Some innovations represent new processes, others new products; however, in each case only one innovator is discussed. To the extent that the same firm produces and uses the innovation, this is fine. However, when the two are different one would expect to find two parties to any innovation. The sample covers 25 innovations in the 1919-38 period and 25 for 1939-58 in iron and steel, 30 and 36 in petroleum refining, and 11 and 16 in bituminous coal.

It is argued that, if Z is defined as the proportion of all innovations (given type, product/process, industry, and period) introduced by the four largest firms, then one can test for the importance of large firms in the innovative process by estimating

$$Z = \frac{4}{L\,(M)} + 4\,\alpha_1\,(\bar{S}_4 - \bar{S}_M) + 4\,\alpha_2\,\bar{I}\,\frac{\bar{S}_4 - \bar{S}_M}{\bar{S}_M} + \epsilon \qquad (2.21)$$

where $L\,(M)$ = number of firms with assets greater than or equal to M;

M = threshold firm size above which innovations are profitable;

$\bar{S}_4$ = average assets of the four largest firms;

$\bar{S}_M$ = average assets of the firms with assets $\geqslant M$;

$\bar{I}$ = average minimum investment required to introduce the innovation;

ϵ = random error term.

Obtaining rough estimates for $\bar{I}$, M, $L\,(M)$, $\bar{S}_4$, and $\bar{S}_M$ for the innovations of each type, for three industries (excluding coal), Mansfield estimates

$$Z - 4/L\,(M) = \underset{(0.00007)}{0.00014\ (\bar{S}_4 - \bar{S}_M)} + \underset{(0.0063)}{0.0289\ \bar{I}\,(\bar{S}_4 - \bar{S}_M)/\bar{S}_M} \qquad (2.22)$$

(standard errors in parentheses). This predicts that $\alpha_2 > 0$ and statistically significant, suggesting that large firms account for more than their share of innovations.

Analysis of data on individual industries suggests, however, that in petroleum refining, coal, and railroads, large firms accounted for a greater share of innovations than of the market, but that in steel they did not. The advantage of large firms was particularly apparent for innovations with high threshold firm size, high investment requirements, and where the largest four firms were particularly larger than other firms. There is also evidence that small firms were less important in the postwar than the prewar innovation process.

In further work Mansfield (1968) analyses data on the timing of innovation. On basically a similar sample he fits

$$n_t = \beta_0 + \beta_1\,u_t + \beta_2\,u_t{}^2 + e_t \qquad (2.23)$$

where n_t = number of innovations occurring in year t;

u_t = average capacity utilization;

e_t = random error term.

He finds that 'the rate of process innovation varied substantially over the business cycle and it reached a maximum when about 75% of the industry's capacity was utilized. The rate of product innovation did not vary significantly' (Mansfield, 1968, p. 119).

Freeman (1974) presents a considerable amount of evidence on innovative activity, for the detail of which the reader is referred to the source. Essentially, Freeman's argument is that small firms should not be ignored as innovators and that their potential impact on reducing market power through innovation is important. However, in certain industries the small firm was an almost irrelevant source of innovations. This is as one would expect; in industries such as passenger aircraft, computers, nuclear reactors, etc., the R & D costs of innovations are so high that no small firm can be expected to be able to carry them.

Freeman goes further, however, by investigating success in innovation. Given the findings of Mansfield (and others) that successful innovation is a major contributory factor to the growth of firms, we ask what makes for success.

Project SAPPHO (Freeman, 1974) suggests that success is associated with

1   strong in-house professional R & D;
2   careful attention to the potential market and substantial efforts to involve, educate, and assist users;
3   entrepreneurship that is strong enough effectively to co-ordinate R & D, production and marketing;
4   good communications with the outside scientific world and customers;
5   the use of patents to gain protection and to bargain with customers.

This last finding throws some light on our above discussion of the importance of patents. It suggests that the patent system may be important to *successful* innovators.

The evidence from studies of innovative activity is again mixed, although it seems fair to conclude that, in certain industries especially, innovation is the prerogative of large firms. However, this does not preclude small firms from having an important role in other industries.

2.7 Conclusions

We started this chapter by distinguishing between the outputs from the technological change process and the inputs to that process, arguing that one cannot analyse the rate of advance purely by looking at inputs, for the relationship between inputs and outputs may vary according to the environment in which the process was taking place. However, one could not solely consider output, for this is very difficult to quantify. In this chapter therefore we have concentrated on output measures, the analysis of inputs being postponed to the next chapter.

Our discussion of the patent system concerned the role that patents play in promoting technological advance; the suggestion was that the existence of such a system was probably neither always necessary nor sufficient to guarantee a return to the inventor. It was also shown that there are significant differences between firms, industries, and time periods in the propensity to patent. This latter observation, plus the quality problem, makes the use of patenting data as a proxy for the rate of technological advance rather suspect; however, the literature that uses such a proxy was considered in some detail. The outcome of this analysis was to suggest that there is some evidence that patenting activity is positively related to expected profitability, this in turn being dependent on available returns and technological opportunity. This implies that technology will advance in the direction yielding greatest returns and in which advance is cheapest.

The work on the sources of innovation suggested, however, that much inventing and thus patenting activity was exogenous to the commerical sector, which again throws some doubt on whether the analysis of patenting activity as an economic activity is completely relevant. These sources of invention literature do, however, also indicate that innovation as opposed to invention is very much the prerogative of the commercial sector. But not all inventions will be turned into innovations. The selection process will very much influence the direction of technological advance. We have little insight into this process, though what we do find with innovations is that, although small firms still have some role to play, it is to a large extent the province of large firms, especially in research-intensive industries. This leads us to argue that, although small firms and research in the non-corporate sector may be important sources of new ideas, the next stage in the technological change process may well be encouraged by large firms in more concentrated industries.

References

Arrow, K. (1962), 'Economic Welfare and the Allocation of Resources for Inventions', in R.R. Nelson (ed.), *The Rate and Direction of Inventive Activity*, Princeton University Press.

Bosworth, D. (1980a), *Statistics of Technology (Invention and Innovation)*, Review no. 28 (provisional), Reviews of UK Statistical Sources, Heinemann Educational, London.

Bosworth, D. (1980b), 'The Movement of Technological Knowhow To and From the UK', Discussion Paper no. 40, Department of Economics, Loughborough University of Technology.

Bosworth, D. and Wilson, R. (1980), 'Returns to Scale in R & D: Empirical Evidence from the World Chemical Industry', Discussion Paper no. 80, CIEBR, University of Warwick, Coventry.

Comanor, W.S. and Scherer, F.M. (1969), 'Patent Statistics as a Measure of Technical Change', *Journal of Political Economy*, 77, 392–8.

Fisher, F.M. and Temin, P. (1973), 'Returns to Scale in Research and Development: What Does the Schumpeterian Hypothesis Imply? *Journal of Political Economy*, 81, 56–70.

Fisher, F.M. and Temin, P. (1979), 'The Schumpeterian Hypothesis: Reply', *Journal of Political Economy*, 87, 386–9.

Freeman, C. (1974), *The Economics of Industrial Innovation*, Penguin, Harmondsworth.

Hamberg, D. (1966), *R & D: Essays on the Economics of Research and Development*, Random House, New York.

Jewkes, J., Sawers, D. and Stillerman, R. (1969), *The Sources of Invention*, W.W. Norton, New York.

Kamien, M. and Schwartz, N. (1980), *Market Structure and Innovation*, Cambridge University Press.

Kohn, M. and Scott, J.T. (1982), 'Scale Economies in Research and Development', *Journal of Industrial Economics*, 30(3), 239–50.

Mansfield, E. (1968), *Industrial Research and Technological Innovation*, W.W. Norton, New York.

Mansfield, E., *et al.* (1971), 'Social and Private Rates of Return from Industrial Innovations', *Quarterly Journal of Economics*, 91, 221–40.

Mansfield, E., *et al.* (1981), 'Imitation Costs and Patents: An Empirical Study', *Economic Journal*, 91, 907–18.

Peck, M.J. (1962), 'Inventions in the Post-War American Aluminium Industry', in R.R. Nelson (ed.), *The Rate and Direction of Inventive Activity, Economic and Social Factors*, Princeton University Press, 279–98.

Rodriguez, C.A. (1979), 'A Comment on Fisher and Temin on the Schumpeterian

Hypothesis', *Journal of Political Economy*, 87, 383–5.

Rosenberg, N. (1974), 'Science, Invention and Economic Growth', *Economic Journal*, 84, 90–108.

Scherer, F.M. (1965), 'Firm Size, Market Structure, Opportunity and the Output of Patented Inventions', *American Economic Review*, 55, 1097–1125.

Scherer, F.M. (1977), *The Economic Effect of Compulsory Patent Licensing*, New York University, Monograph Series in Finance and Economics.

Scherer, F.M. (1980), *Industrial Market Structure and Economic Performance* (2nd ed.), Rand McNally, Chicago.

Scherer, F.M. (1982), 'Demand-Pull and Technological Invention: Schmookler Revisited', *Journal of Industrial Economics*, 30(3), 225–38.

Scherer, F.M. (1983), 'The Propensity to Patent', *International Journal of Industrial Organisation*, 1, 107–28.

Schmookler, J. (1966), *Invention and Economic Growth*, Harvard University Press, Cambridge, Mass.

Soete, L. (1981), 'A General Test of Technological Trade Gap Theory', *Weltwirtschaftliches Archiv*, 117, 638–59.

Stoneman, P. (1979), 'Patenting Activity: A Re-evaluation of the Influence of Demand Pressures', *Journal of Industrial Economics*, 27, 385–401.

Taylor, C., and Silberston, Z.A. (1973), *The Economic Impact of the Patent System*, Cambridge University Press.

Wilson, R.W. (1977), 'The Effect of Technological Environment and Product Rivalry on R & D Effort and Licensing of Innovations', *Review of Economics and Statistics*, 59, 171–8.

Chapter 3

Invention and Innovation II: Research and Development

3.1 Introduction

Research and development spending may be considered as the monetary equivalent of inputs to the process producing technological advances. In the last chapter we considered the output of the process. For reasons discussed there we now consider the input measure.

R & D spending is an input to both invention and innovation. However, to the extent that R & D tends to be directed more towards development than research, we might argue that an analysis of R & D is concerned more with innovation than with invention. This is not necessarily reflected in the models discussed below, however.

In the literature the basic underlying assumption is that research and development is undertaken to obtain profits. The research expenditure of the firm may yield process or product improvements. The profits that the firm can obtain from its research process will depend upon the cost of making the advances and the gains in revenue that can be derived from the use or marketing of the improvements. The former may depend on technological opportunity, efficiency, and the speed of development. The latter will depend upon how quickly other firms can copy or match the advances made. Revenue from innovations can be derived either by licensing them to those who have not developed the technology themselves, or by the innovating firm incorporating advances in its own products and gaining market share or introducing new processes into its production methods and using the cost advantage to follow a profit-maximizing strategy on the product market. The extent to which the firm will be able to capitalize on any advances will depend upon the degree to which it is able to gain property rights on its inventions. If inventions can be easily copied, the firm is likely to get a lower return. The returns to R & D will thus depend on the effectiveness of the patent system (on which we cast some doubt in Chapter 2) and on the importance of other mechanisms that can protect technology (e.g., secrecy). The costs and returns involved in R & D, moreover, may well be uncertain. The process of R & D is essentially generating something previously unknown, and thus uncertainty is almost by definition a part of the environment. Both the costs of development and the potential returns cannot be known with certainty.

The theoretical and empirical literature in this area is concerned primarily with investigations of interrelationships between R & D spending, firm size, and market structure in an attempt to provide a theoretical grounding for Schumpeterian hypotheses and empirical support for these hypotheses. In this chapter we first consider theory (section 3.2) and then (section 3.3) we consider the empirical work in this area. As will become apparent, this empirical work is somewhat divorced from the analytical models in section 3.2; however, as it is so extensive it cannot be ignored.

3.2 Models of R & D determination

3.2.1 The Arrow model

In Arrow (1962) the incentives to invent for monopolistic and competitive markets are compared with each other and with those that would exist in a socially managed economy.

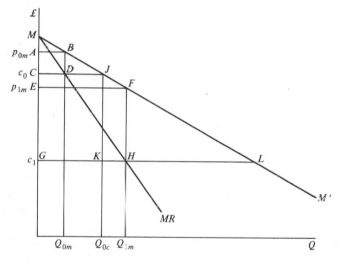

Fig. 3.1

Consider an industry facing a demand curve MM' with pre-invention costs, c_0, as shown in Fig. 3.1. With costs c_0, a competitive profit-maximizing industry would generate a price c_0, a monopoly industry, a price p_{0m} (i.e., where marginal revenue equals marginal cost). The monopoly industry would have profits $ABDC$.

Allow now that a new technology is discovered that enables unit costs of production to fall to c_1. With cost c_1, a monopoly would charge price p_{1m}. If $p_{1m} < c_0$, we term the innovation large. If $p_{1m} > c_0$ (but of course $p_{1m} < p_{0m}$), then the innovation is considered small.

Arrow now considers that the knowledge incorporated in the new technology is monopolized by a patent holder. In the monopoly industry case this is the same monopolist that monopolizes the product. In this case, with costs c_1 the monopolist can make profits $EFHG$, and thus the gain in profits from the use of the new technology is $EFGH - ABCD \equiv \Pi_m$.

In the competitive case, the new technology is controlled by an inventor who allows firms to use it by charging a royalty (r). The royalty is set to maximize the return to the inventor. Given that firms charge a price equal to $c_1 + r$, the returns are maximized for a large innovation if $r = p_{1m} - c_1$, yielding profits $\Pi_c = EFHG > \Pi_m$, or with a small innovation if $r = c_0 - c_1$, yielding profits $CGJK$, which Arrow shows is also greater than Π_m. Given that $\Pi_c > \Pi_m$, it is argued that the pre-invention monopoly power acts as a strong disincentive to further innovation. The socially managed market would generate, with price equal to c_1, social gains, Π_s, equal to $CJLG$. It is therefore considered that $\Pi_s > \Pi_c > \Pi_m$; thus, competition encourages innovation and invention, but in general, the incentives to both competitive and monopoly industries are less than the potential social benefit, which suggests that there will be an under-investment in research.

This model of Arrow's has been criticized on a number of grounds.

1   Demsetz (1969) argues that Arrow's result will not hold if pre-invention outputs are the same for both monopolized and competitive industries. To argue this, however, the industries have to face different demand or cost curves, and thus in this context cannot be the same industry.
2   The model applies only to process innovation and does not consider product innovation, although such an extension is provided by Usher (1964).
3   The model assumes that both industries take up the new technology equally fast and thus ignores diffusion.

We treat point 2 to some extent below, and in Chapter 7 point 3 is discussed to see if there are any differences in diffusion speed between different market structures (the evidence is mixed).

One might also question whether the inventor is really able to appropriate the profit as Arrow has argued. In Arrow the technology is controlled by a monopolist — in the monopoly industry case by the monopoly producer, in the competitive case by an outside inventor. We might ask whether a comparison with an outside inventor in both cases would yield the Arrow result. One might argue that, in the monopoly case, with an outside inventor one would have a bilateral monopoly situation and so Π_m would be the maximum the inventor could appropriate. However, this would only reinforce the conclusions that $\Pi_c > \Pi_m$.

The major problem with this Arrow approach, however, is that the analysis begins with one invention having already been made and takes no account of the process leading to invention. Arrow's conclusions on the relative effectiveness of different market structures are thus based on a comparison of gross returns from this one invention. If one considers that inputs are necessary to produce inventions, then the market structure that is most conducive to innovation should be related to comparisons of net returns — in other words, gross returns minus R & D costs. Moreover, as Arrow only has one invention being marketed, there is no competition in the sale of inventions. It is quite possible that two inventions appear on a market at the same time, both directed towards the same end but based on different technologies (for example, in the consumer case, VHS and Betamax video recorders). In such a situation the assumed monopoly power of the inventor is no longer apparent. These arguments suggest to us the following points:

1   that if we are to consider net returns to R & D, then we must model the determination of R & D expenditures *per se* as well as the revenue increases to be obtained from technological advance; and
2   that if firms are competing to make technological advances, then the problem requires an analysis of R & D determination at the industry level.

To proceed to the industry level, we must start by looking at the firm's decision-making process, and later consider the competitive game. We thus start with the firm.

3.2.2 Invention and innovation at the firm level

The process of developing new technology can be considered from two directions: either the firm spends on research and development to develop one particular advance, or else there is a continuum of advances that can be made, and the amount spent on R & D determines the extent of the advance produced. We consider the latter approach first and start with the work of Dasgupta and Stiglitz (1980a).

The firm is assumed to be able to generate reductions in unit costs by spending on R & D. It is allowed that, if R & D spending is x, then unit cost of production c, is related to x by

$$c = c(x) \qquad c'(x) < 0 \qquad (3.1)$$

where $c'(x) = dc/dx$. One might particularly note here that the R & D spending itself yields reductions in unit costs. It is assumed that, once the technological advance is made, no further costs are involved in introducing and using that technology. It might even be argued that the advance represents a completely disembodied innovation. Product innovations are not considered.

The new technology generated is exploited on the market not by the sale of rights to use and the collection of royalties, but by production using the new technology. Consider firm i in an oligopoly industry; then it is argued that the firm will choose R & D spending x_i and its output Q_i so as to maximize profits Π_i, where

$$\Pi_i = \{p(Q_i + Q_r) - c(x_i)\} Q_i - x_i \qquad (3.2)$$

where Q_r is the output of rivals, and industry output $Q \equiv Q_i + Q_r$. The choice of output and R & D by the firm will depend upon its expectations of rivals reactions. Assume Cournot conjectures, so that firm i does not expect other firms to react to its own decisions, and then first-order profit maximization conditions yield

$$p \left\{ 1 - \frac{\epsilon (Q) Q_i}{Q} \right\} = c (x_i) \tag{3.3}$$

and

$$- Q_i c' (x_i) = 1 \tag{3.4}$$

where $\epsilon (Q)$ is $- (\partial p / \partial Q) (Q/p)$, i.e., the inverse elasticity of demand. From (3.3) and (3.4) we may generate

$$\frac{x_i}{pQ_i} = \alpha (x_i) \left\{ 1 - \frac{Q_i}{Q} \epsilon (Q) \right\} \tag{3.5}$$

where $\alpha (x_i) = - x_i c' (x_i) / c (x_i) = $ elasticity of unit cost reduction with respect to R & D; $\alpha (x_i) > 0$.

Now, although equation (3.5) should not be read as implying causality, we can see that, *ceteris parabus*, high R & D-to-sales ratios for the firm are associated with smaller market shares. Moreover, a greater effectiveness of R & D (a larger value for $\alpha (x)$) is associated with an R & D-to-sales ratio that is higher. We cannot however solve directly for x, and thus we cannot say whether small firms will generate greater or lesser reductions in unit costs.

The analysis above assumes that the firm can only invent a new process technology, but one does not need to be so limiting. Needham (1975) constructs a model similar to that discussed immediately above in which R & D generates new products. In Stoneman and Leech (1980) a model that allows the firm to develop both is constructed, with the direction of research as an endogenous variable. In this model also it is realized that the nature of the cost function assumed in Dasgupta and Stiglitz requires that, in each time period, a firm has to re-invent the technology it had in the previous period, for the static nature of that model prevents consideration of the accumulation of technical knowledge over time.

Consider again firm i, which can now undertake R & D expenditure that can be directed towards the production of cost-reducing (process) innovations or demand-stimulating (product) innovations. Allow that the stock of knowledge of each type can be represented by $\hat{K}^1$ and $\hat{K}^2$ (representing cost and demand technologies, respectively). Define gross additions to the stocks of knowledge in a time period as K^1 and K^2 and allow technological knowledge to decay at rates δ_1 and δ_2. Using the dot notation for a time derivative, we may then write

$$K^1 = \dot{\hat{K}}^1 + \delta_1 \hat{K}^1 \tag{3.6}$$

$$K^2 = \dot{\hat{K}}^2 + \delta_2 \hat{K}^2. \tag{3.7}$$

K^1 and K^2 are considered to be joint products from a black box production process given by

$$x = x (K^1, K^2), \quad \frac{\partial x}{\partial K^1} > 0, \frac{\partial x}{\partial K^2} > 0 \tag{3.8}$$

where x = R & D expenditure. Thus, the genration of new technologies of either type requires R & D expenditure. K^1 and K^2 are to be considered as choice variables for the firm. Equation (3.8) is an invention possibility function similar to that proposed by Nordhaus (1969), except that in this case we have distinguished between product and process technology advances. It is assumed that this function is continuously differentiable and is homogeneous of degree θ, in which case $1/\theta$ represents the degree of scale economies in the R & D process. We would consider that, for a given industry, technological opportunity would determine the actual form of (3.8). Thus, an industry with little opportunity for technological advance would require higher R & D for a given level of K^1 or K^2 than one with numerous opportunities. Consider these opportunities, as reflected in, for example, basic scientific knowledge, to be exogenous to the firm and the industry.

Allow that the firm has a total cost function

$$C_i = C_i(Q_i, \hat{K}_i^1), \quad \frac{\partial C_i}{\partial \hat{K}_i^1} < 0 \qquad (3.9)$$

where C_i = total production cost for firm i. The firm is assumed, following Nickell and Metcalf (1978), to face an inverse demand function that is additive in Q_i and Q_r and is separable between output and technology:

$$p_i = f(Q_i + Q_r) + g_i(\hat{K}_i^2, \hat{K}_r^2), \quad \frac{\partial g_i}{\partial K_i^2} > 0 \qquad (3.10)$$

where p_i = price of the output of firm i;
 $\hat{K}_r^2$ = product technology of rival firms.
Define (3.11) on the inverse demand function:

$$\frac{dp_i}{dQ_i} = (1 + \lambda_i) \frac{\partial f}{\partial Q_i}, \quad \lambda_i = \frac{dQ_r}{dQ_i} \geqslant 0. \qquad (3.11)$$

Then λ_i allows for expected retaliation (conjectural variation) on output. (If $\lambda = 0$, then we revert to Cournot conjectures.) It is assumed that the firm expects its rivals to react to its output changes immediately.

As to conjectural variation on R & D spending, it is considered that δ_2 is related to the response of rivals to the introduction of a new product technology and thus allows for conjectural variations on technology to match those on output. This rationale for δ_2 fits in with the reasoning of Nerlove and Arrow (1962). One would not, however, expect new process technology to be retaliated against, and thus δ_1 is not related to retaliation. It is allowed (with λ) that, if a firm changes its output in reaction to a new process technology, then it may expect other firms to change their output, and that seems to be sufficient. However, both δ_2 and δ_1 may be greater than zero because of changes in tastes or changes in basic knowledge.

The firm is assumed to maximize its present value by choosing Q_i, K_i^1 and K_i^2 subject to (3.6), (3.7), and (3.8). The present value is given by (3.12), with *r* as the discount rate:

$$V_i = \int_0^\infty e^{-rt} \{ p_{it} (Q_{it}, Q_{rt}, \hat{K}_{it}^2) Q_{it} - C_{it} (Q_{it}, \hat{K}_{it}^1) - x_{it} \} \, dt. \qquad (3.12)$$

From the first-order maximization conditions one can generate that, for firm *i* (dropping the *i* subscript) (3.13) holds:

$$\frac{x}{pQ} = -\frac{1}{\theta} \frac{\eta_{C1}}{\eta_{CQ}} \frac{1}{r + \delta_1} + \frac{\eta_{p2}}{\theta} \frac{1}{r + \delta_2} - \frac{(1+\lambda)\,\eta_{C1}}{\eta_D \cdot \eta_{CQ}} \frac{1}{\theta} \frac{1}{(r + \delta_1)} \qquad (3.13)$$

where η_{C1} and η_{p2} are the elasticities of costs and price to technologies 1 and 2, respectively, η_{CQ} is the elasticity of cost with respect to output, and η_D is the firm's own price elasticity of demand.

Two special cases can arise from (3.13). The first would be when $\eta_{C1} = 0$, $\eta_{p2} > 0$, and advances in technology do not reduce costs. In this case all advances would be of the product innovation type. In the second case $\eta_{p2} = 0$ and $\eta_{C1} < 0$, and all advances would be of the process innovation type. In this case we would revert to a Dasgupta and Stiglitz type model. In the first case we would revert to a Needham type model. From (3.13) we can confirm the Dasgupta and Stiglitz result that if $\eta_{C1} < 0$, then smaller market shares, *ceteris parabus*, will be associated with higher R & D-to-sales ratios. To show this one can write $1/\eta_D$ in current terminology as equal to $\epsilon (Q) (Q_i/Q)$ in the Dasgupta and Stiglitz terminology and the result follows.[1] However *if* $\eta_{C1} = 0$, then the R & D-to-sales ratio will equal $(\eta_{p2}/\theta) (1/r + \delta_2)$, which is not (directly) related to market shares.

If we define $z \equiv x_i/p_i Q_i$, the model allows us to draw out a number of implications, each of which represents an association with no direction of causality implied.

1   The greater is expected retaliation, the lower is the R & D-to-sales ratio. From (3.13) it is clear that $\partial z/\partial\lambda < 0$ and $\partial z/\partial\delta_2 < 0$. Thus, the more quickly other firms are expected to imitate new product technology, or the greater is their expected response to an output change, the smaller is the firms research effort as a proportion of sales.
2   The more productive is the research process, the greater is the R & D-to-sales ratio. From (3.13) we can see that $\partial z/\partial\theta < 0$, and $1/\theta$ is a measure of the productiveness of the R & D process.
3   The greater is the impact of a unit addition to knowledge, the higher is the R & D-to-sales ratio for $\partial z/\partial\eta_{p2} > 0$ and $\partial z/\partial (-\eta_{C1}) > 0$.
4   A higher R & D-to-sales ratio is associated with a higher degree of scale economies in production $(\partial z/\partial\eta_{CQ} > 0)$ and lower discount rate $(\partial z/\partial r < 0)$.

From (3.13) we can make some comments on how the market structure of the industry in which the firm operates may affect the R & D-to-sales ratio. One can argue, following Cowling (1972), that retaliation may increase with

---

[1] The argument does not allow, however, that λ and δ_2 may differ with firm size, or even that δ_1 may vary with firm size.

concentration, at least up to the duopoly level. If this means that both λ and δ_2 are higher, then z will be lower and the R & D-to-sales ratio may thus be negatively related to the level of concentration in the industry. Using similar arguments as to how λ, δ_2, θ, etc. may vary with firm size, one could expand further on the relationship between R & D and firm size. For example, λ and δ_2 may vary with firm size, because one might argue that small firms will not expect retaliation from large competitors but large firms will. However if, as Freeman (1965) argues, large firms undertake protective R & D or hold excess capacity (as argued by Spence, 1977), small firms may expect very large retaliation. One is getting into an area where results are very uncertain, and thus we would be unwise to draw any firmer conclusions.

This extension of the original Dasgupta and Stiglitz model thus introduces the possibility of retaliation or emulation, allows that technological knowledge can be retained, and caters for different types of technological advance. Still incorporated, however, is the hypothesis that there is a continuum of technological advances available and that the extent of the advance made is related to current R & D spending. The other strand in the literature is one that considers that money is spent on R & D to achieve a given predetermined technological advance, and the greater is the R & D expenditure, the earlier is the date at which the advance is made. This enables one to analyse both the expenditure on R & D and the date of innovation. The approach has some intuitive appeal, firms' R & D programmes often being directed towards some specific target rather than advance in general. The approach is perhaps best detailed in the work of Kamien and Schwartz (1980). The analysis considers one firm faced by rivals in the innovation process (these rivals may or may not be extant competitors in the product market). Prior to any innovation occurring, the firm is selling a product yielding a profit flow Π_0, which will change over time at a constant rate g in the absence of innovation. Thus at time t the profit flow is $e^{gt} \Pi_0$. If the firm introduces a new product at time T, it is assumed to receive a return $e^{gt} R_0$ if no imitation occurs. If imitation occurs at time v, the capitalized value at time T of the future profit stream will be $e^{gv} R_1 (v - T)$. If, however, the firm is an imitator, so a rival introduces the new product first, then during the period until its own new product is launched its profit flow falls to $e^{gt} \Pi_1$ for each t, and the capitalized value of profits after its own innovation date T is given as $e^{gt} R_2 (T-v)$ at time T.

The firm's problem is to determine T, and to do so it must hold expectations on rivals' innovation dates. These expectations are represented by $F(t)$, which is the firm's assessment at time $t = 0$ of the probability that rivals will have innovated by time t. We then define the hazard rate, $H(t)$ as

$$H(t) = \frac{F'(t)}{1 - F(t)} \tag{3.14}$$

where $F'(t)$ is the derivative of $F(t)$. Kamien and Schwartz write the hazard rate as the multiple of a hazard parameter h, and a hazard function $u(t)$, thus:

$$H(t) = hu(t), \tag{3.15}$$

and an increase in h results in a constant proportionate increase in the hazard rate. If we now define

$$v(t) \equiv \int_0^t u(s)\, ds, \tag{3.16}$$

one can show that the expected time of rival entry can be written as

$$E(t) = \int_0^\infty e^{-hv(s)}\, ds \tag{3.17}$$

and an increase in the hazard parameter will hasten the expected date of rival entry. Given that $1 - F(t)$ is the probability that innovation by rivals has not occurred by time t, and $F(t)$ is the probability that it has, then with the returns schedule detailed above we may write the discounted value of the firm's expected profit stream at time t from innovation of time T, $w(t)$ (with discount rate r) as

$$w(t) = \int_0^T e^{-(r-g)t} \left[\Pi_0 \{1 - F(t)\} + \Pi_1 F(t) \right] dt$$

$$+ \int_T^\infty e^{-(r-g)t} \left[R_0 \{1 - F(t)\} + R_1 (t - T) F'(t) \right] dt$$

$$+ \int_0^T e^{-(r-g)T} R_2 (T - t) F'(t)\, dt \tag{3.18}$$

where the first integral gives expected returns from its old product prior to introduction of its new ($\Pi_0 e^{gt}$ if other firms do not innovate, $\Pi_1 e^{gt}$ if they do), the second integral gives post-introduction returns ($R_0 e^{gt}$ prior to imitation, $R_1 e^{gt}$ after), and the third integral gives returns if another firm innovates first ($R_2 e^{gt}$). The costs of introduction by date t, i.e., R & D expenditures, are assumed to be $x(T)$, all incurred at time 0 ($x'(T) < 0$, $x''(T) > 0$), and the firm chooses T to maximize

$$V(T) = w(T) - x(T). \tag{3.19}$$

The assumption is that the development is undertaken under contract, and so the choice of T cannot be varied even if events should prove that not to be the best policy. The maximization of $V(T)$ yields T^*, the optimal innovation date. Given that T^* is finite (and thus the product will be introduced), it is shown that

$$\frac{\partial T^*}{\partial \Pi_0} > 0, \frac{\partial T^*}{\partial \Pi_1} > 0, \frac{\partial T^*}{\partial (\Pi_0 - \Pi_1)} < 0, \frac{\partial T^*}{\partial R_0} < 0, \frac{\partial T^*}{\partial R_1} < 0.$$

Thus, the greater are the innovational profits, the earlier will the innovation occur. However, the greater are the profits on the existing good, the later will innovation occur. As to the impact of rivalry, it is argued that, if $R_1 > R_2$ (if it is preferable to be an innovator rather than imitator), and $R_1' = R_2' = 0$ (i.e., if R is independent of $v - T$), then two possible cases arise: (1) $h = 0$ (i.e., no rivalry) produces earliest innovation; (2) earliest innovation occurs when $h = h^* > 0$, and if $h > h^*$ (greater rivalry), then innovation will be slower.

After considering further modifications to their model, Kamien and Schwartz summarize as follows.

1   High profits on an existing product slow down innovation if the innovation is a substitute for the existing product. A firm may thus have greater incentive to develop diversified products.
2   Rivalry will have a definite influence on the date of innovation.
3   The firm's introduction date may be premature or late relative to the social optimum (that time at which development costs just balance expected benefits).
4   As rivalry increases, innovations with expected modest returns will be introduced more rapidly. Innovations with major returns will be introduced more rapidly up to some level of rivalry, after which the innovation date starts to slip back.

The cost relationship underlying this model has been represented as $x = x(T)$ ($x'(T) < 0$, $x''(T) > 0$, for all $T > 0$). In the literature Mansfield *et al.* (1971) found some empirical support for a relationship of this form, with greater R & D expenditures being required for faster innovation. However, in both the Kamien and Schwartz model and in the two previous approaches discussed before, no allowance has been made for possible inefficiencies in this research process. It is quite possible that, if the efficiency of the research process of the firm varies with firm size and market structure, then this will be an avenue whereby these two factors could influence the firm's R & D spending and its rate of technological advance. However, this is just another reflection of the Fisher and Temin (1973) warning discussed in the previous chapter, whereby it is argued that R & D is the input and not the output. By similar arguments, one can also conclude that any welfare conclusions as to the effect of parameter changes should concentrate not on R & D expenditure but on the cost reductions or innovation date changes that result from the parameter changes.

3.2.3 Invention and innovation at the industry level

Thus far we have considered only individual firms, although we have allowed, through conjectural variation terms, that these firms realize their interdependence with other firms. In order to discuss the relationship of R & D to market structure, we really need to extend to the level of the industry. The conflict that arises at this level is that we are interested in how the rate of technological advance will be related to market structure — and, although certain structures may encourage firms to undertake research, at the same time a large number of firms doing research may imply excessive repetition and a low ratio of output to input from the research process at the industry level. High industry R & D spending may not imply fast advance. By seeking the market structure that is optimal we are seeking a balance between forces promoting research and forces generating excessive repetition.

The industry-level approach is to consider models of firm behaviour of the type discussed above and the resolution of the oligopoly game between the firms in an industry. The characteristics of the industry in this equilibrium yield for us relationships between technological advance and market structure.

We start by referring back to the Dasgupta and Stiglitz (1980a) model, with the firm's first-order profit maximization conditions given by (3.3) and (3.4). Dasgupta and Stiglitz then explore market equilibria that are symmetric (all firms are the same). Let the number of firms in the equilibrium be N^* (which is endogenous); then using N^*, Q^*, x^* to characterize the symmetric equilibria, we may write

$$p(Q^*) \left\{ 1 - \frac{\epsilon(Q^*)}{N^*} \right\} = c(x^*) \tag{3.20}$$

and

$$-c'(x^*)Q^* = N^* \tag{3.21}$$

where Q^* is total output in equilibrium and Q^*/N^* is output per firm.

If we assume that there is free entry and represent this by a zero-profit condition, then

$$\{p(Q^*) - c(x^*)\} Q^* = N^* x^* \tag{3.22}$$

will also hold. With (3.20), (3.21), and (3.22) we can determine N^*, Q^*, and x^*. From (3.20) and (3.22) we get

$$\frac{\epsilon(Q^*)}{N^*} = \frac{N^* x^*}{p(Q^*) Q^*}. \tag{3.23}$$

This equation states that the ratio of industry R & D to industry sales equals the inverse of the elasticity of demand divided by the number of firms. If one considers $1/N^*$ as a measure of concentration, then, *ceteris paribus*, the R & D-to-sales ratio is linearly related to concentration. One should especially note, however, that this is not a *causal* relationship. The number of firms and R & D are both endogenous variables determined simultaneously.[2]

To move further on, (3.21) and (3.22) yield

$$\frac{N^* x^*}{p(Q^*) Q^*} = \frac{\alpha(x^*)}{1 + \alpha(x^*)} \tag{3.24}$$

where $\alpha(x) = - xc'(x)/c(x) =$ elasticity of unit cost reduction with respect to R & D. Equation (3.24) implies that the R & D-to-sales ratio is the same in different industries if $\alpha(x^*)$ is the same.

From (3.23) and (3.24) we get

$$N^* = \epsilon(Q^*) \left\{ \frac{1 + \alpha(x^*)}{\alpha(x^*)} \right\}. \tag{3.25}$$

Thus, the greater is the elasticity of demand ($1/\epsilon$), the smaller is the equilibrium number of firms.

---

[2] This result is partly due to the instantaneous effect on cost of current R & D expenditures. In a more realistic world with lags, market structure might be determined more by past than by current R & D (see Chapter 16).

From (3.25) and (3.20) we can derive

$$\frac{p\,(Q^*)}{c\,(x^*)} = 1 + \alpha\,(x^*). \tag{3.26}$$

From (3.26) we may argue that the price cost margin (or degree of monopoly) is a positive function of $\alpha\,(x^*)$. From (3.24) the R & D-to-sales ratio is also a function of $\alpha\,(x^*)$. Thus we may argue that high R & D-to-sales ratios are associated with high price–cost margins. If we use the price–cost margin as an index of monopoly power, we then have a positive association between the R & D-to-sales ratio and monopoly power. Given from (3.25) that N^* is inversely related to $\alpha\,(x^*)$, we also have that R & D per firm will be higher, and thus unit cost reduction will be greater, the greater is monopoly power. Once again, however, we should stress that these relationships are not causal; the degree of monopoly is endogenous.

One should note that in this symmetric equilibrium all firms undertake the same amount of R & D and each independently discovers the same reduction in costs. There is no licensing by one firm to another. Moreover, the Cournot conjectures imply that each firm makes its decisions in the belief that its behaviour will not affect the behaviour of rivals.

The conclusion, however, is that greater monopoly power is associated with greater technological advance, although no direction of casuality is implied.

To proceed further we will assume that the demand and unit cost functions take the forms

$$p\,(Q) = \sigma Q^{-\epsilon} \tag{3.27}$$

and

$$c\,(x) = \beta x^{-\alpha} \quad (\alpha, \beta > 0) \tag{3.28}$$

With these specific functions one can show that in equilibrium the greater is the size of the market (σ), the greater is R & D expenditure per firm and thus the greater is unit cost reduction. However, the greater is β (the costlier is the R & D process), the smaller is equilibrium R & D per firm if demand is elastic, the greater if demand is inelastic.

However, the main reason for introducing these particular cost and demand functions is to enable us to make welfare comparisons, and thus to generate some results comparable to those of Arrow. With these specific functions, Dasgupta and Stiglitz show that the net social return to innovation is maximized if x is determined such that

$$x = x_s = (\alpha^\epsilon\,\sigma\,\beta^{\epsilon-1})^{1\,/\{\epsilon-\alpha\,(1-\epsilon)\}}$$

holds and

$$Q = Q_s = (\alpha\beta)^{-1}\,(\sigma\alpha^\epsilon\,\beta^{\epsilon-1})^{(1+\alpha)/\{\epsilon-\alpha\,(1-\epsilon)\}}.$$

Dasgupta and Stiglitz then show that, with these specific functions, $Q_s > Q^*$ and

$x_s > x^*$. However, if ϵ is large $N^*x^* > x_s$. Thus, compared with the socially optimal position, it will generally be the case that the rate of unit cost reduction will be too low ($x_s > x^*$), but if ϵ is large there will be excessive expenditures on R & D ($N^*x^* > x_s$) because of too much repetition.

In this framework the number of firms was endogenous because there was free entry. If we assume that there are entry barriers, then N is predetermined and fixed. We thus drop condition (3.22) but maintain (3.20) and (3.21), which, with the specific cost and demand functions, can be solved to yield

$$x^* = [\sigma (\alpha/N)^\epsilon \beta^{\epsilon-1} \{1 - (\epsilon/N)\}]^{1/\{\epsilon-\alpha (1-\epsilon)\}} \tag{3.29}$$

and

$$Q^* = (N/\alpha\beta) [\sigma (\alpha/N)^\epsilon \beta^{\epsilon-1} \{1 - (\epsilon/N)\}]^{(1+\alpha)/\{\epsilon-\alpha (1-\epsilon)\}}. \tag{3.30}$$

We assume $N > \epsilon$, $N \leqslant \epsilon (1 - \alpha)/\alpha$, $\epsilon > \alpha (1 - \epsilon)$. Then $\partial Q^*/\partial N > 0$, industry output increases with the number of firms, but the degree of monopoly will be falling. However, from (3.29) $\partial x^*/\partial N > 0$; thus, if the number of firms increases, each firm in equilibrium spends less on R & D and so the unit cost of production if higher. However, Nx^* will increase with N. The increased competition thus means higher industry R & D, but less R & D per firm — the increased R & D is just greater repetition. Moreover, $x^* < x_s$. Note, therefore, that more spending under competition does not mean faster advance. In fact, once again, less competition, a lower N, implies a faster rate of unit cost reduction. Both the Dasgupta and Stiglitz cases thus imply that market power is associated with faster technological advance, but in general that rate of advance will be less than is socially desirable.

One could perform a similar analysis at the level of the market for the Stoneman and Leech model discussed above, but it is more relevant at this stage to consider the third approach above, whereby the firm determines its date of innovation rather than the extent of innovation. To extend the Kamien and Schwartz model itself is rather too complicated, so we will exemplify the nature of the results that can be achieved by considering another model of Dasgupta and Stiglitz (1980b) (which itself is related to the work of Loury, 1979).

The basic assumption is that there is only one innovation to be made; this innovation will reduce unit costs from c_0 to c_1. The time it will take to make the innovation T is related to R & D spending by $T = T(x)$. If we have an industry with a monopolist supplier protected by entry barriers, then the innovation will yield an increase in profits of Π_m per period for $t > T$, where Π_m can be defined as $EFHG - ABDC$ in Fig. 3.1. Defining r as the discount rate, we can define the present value of the increment to the monopolist's profits as (3.31)

$$V_m = \frac{\Pi_m}{r}. \tag{3.31}$$

In a socially managed market, the gain in net social surplus per period after innovation will be Π_s where Π_s can be defined as $CJLG$ in Fig. 3.1, and the present value of this we may write as

$$V_s = \frac{\Pi_s}{r}.\tag{3.32}$$

In the socially managed market, x will be determined to maximize

$$V_s e^{-rT(x)} - x\tag{3.33}$$

whereas as in the monopoly market x will be determined to maximize

$$V_m e^{-rT(x)} - x\tag{3.34}$$

and given, from Fig. 3.1, that $\Pi_s > \Pi_m$ and thus $V_s > V_m$, the volume of R & D expenditures, and thus the speed of innovation undertaken by the monopolist, is less than is socially desirable.

If we have a competitive market we may think of a number of firms who can undertake R & D in order to invent the new technology. If one assumes, perhaps unrealistically, that a patent system will prevent any imitation, then the inventor of the new technology will get a gain in profits from the invention of $\Pi_c = EFGH$ for a large invention or $CJKG$ for a small one (ref. Fig. 3.1). We have shown that $\Pi_c > \Pi_m$, but $\Pi_c < \Pi_s$.

If we have a patent system, the patent, may be considered to have a life of $\bar{T}$ years and thus the present value of the profit flow may be defined as

$$V_c = \frac{\Pi_c}{r} (1 - e^{-r\bar{T}})\tag{3.35}$$

Under a patent system only one firm can obtain any return from his R & D expenditures: the researcher who first makes the invention. In a competitive market with free entry, it is argued that (assuming no uncertainty) any potential monopoly rents will be competed away in the R & D process. Thus x will be determined to make the net return to R & D equal to zero. In other words,

$$x_c = V_c e^{-rT(x_c)}.\tag{3.36}$$

Moreover, in the equilibrium only one firm will be undertaking R & D.

On the basis of these arguments, Dasgupta and Stiglitz (1980b) argue that more elastic demand curves are likely to be associated with excessive research in the competitive market while there will be inadequate research for 'big innovations'. The excessive R & D expenditure in the market economy does not stem [as in their previous model] from a duplication of research effort, but rather arises from the pressure of competition. Competition forces the single firm to innovate earlier than is socially desirable.

These results depend upon there being no uncertainty, and to this point we have not really considered this complication to any extent. We now wish to introduce uncertainty in the sense that, for a given expenditure on R & D, the date at which the innovation will be made is not known with certainty. In particular, we consider that, for a given R & D expenditure at time 0 of x, the probabilty that the invention is made at or before t is $1 - e^{-\lambda(x)t}$. In such a world Dasgupta and Stiglitz show that the entry-protected monopolist always delays innovation relative to the socially managed market and this parallels the certainty result. In the competitive market it is argued that

1   for small inventions the market always provides inadequate research;
2   for sufficiently long patent lives the market spends too much on research.

This work refers to a world in which only one innovation is considered. In a third paper Dasgupta and Stiglitz (1981) consider continuous flows of innovations. Rather than working through the whole paper we just state the point emphasized by the authors:

The model suggests that if the growth in demand for the product is high the forces of competition result in too frequent a set of innovations, none of which on its own is sufficiently large; and there are reasons for believing that a mono-polist, protected fully by entry barriers, undertakes innovation less frequently than is socially desirable, but that when it does, it undertakes unduly large innovations. (Dasgupta and Stiglitz, 1981)

These R & D models may usefully be considered in the light of Freeman's (1965) trilogy of R & D expenditures:

1   offensive R & D, which is R & D undertaken to attack, ultimately, a rivals' market, or open up new markets;
2   defensive R & D, which is R & D undertaken to retaliate against attack;
3   protective R & D, which is R & D undertaken to develop technology that will not generally be marketed, unless or until the firm's market is threatened by rivals.

The models above consider R & D in categories 1 and 2. Category 3 requires further analysis. Protective R & D may be rationalized on the grounds of the creation of entry barriers of the capacity type discussed by Spence (1977), and as independently analysed by Newbery (1978). The basic principle is that, by invention or innovation, entry to the market through invention or innovation is prevented, thereby allowing the inventor more profitably to exploit his current market position. Such R & D is the prerogative of the large dominant firm, and if carried out has undesirable welfare consequences. One may argue with such R & D that monopoly power can lead to excessive R & D spending.

We have now investigated a number of approaches to the determination of R & D spending. At the level of the firm our analysis suggests that correlations exist (and thus we are not necessarily implying causality) as follows.

1   R & D expenditure as a proportion of the firm's sales will be positively correlated with the effectiveness of the R & D in reducing costs, and nega-tively correlated with market share. The latter result may not follow if product innovation is being discussed.
2   The greater is the expected response of rivals to output change or technology change, the lower will be the firms R & D-to-sales ratio. This may provide

a link between R & D spending and market structure.
3   Early innovation will be associated with greater expected returns and lower current profitability.
4   The optimum degree of rivalry for early innovation with major expected returns involves a market structure intermediate between monopoly and competition. Competition will favour early introduction for innovations with modest expected returns.

These results arise from what Kamien and Schwartz call a decision-theoretic approach, whereby we concentrate on one firm. When we consider the interaction of firms at the industry level we can argue as follows.

1   Gross returns to invention are greater under competitive market structures than monopolistic market structures but both are less than in a socially managed market.
2   A comparison of incentives to innovation should involve net (of R & D), and not gross, returns.
3   In models comparing net returns, a monopolist protected by entry barriers will always under-invest in R & D relative to the socially managed market.
4   In an oligopoly with free entry, one can argue that market structure and R & D are jointly determined and one cannot talk of causality running from market structure to technological advance. In such a market, with R & D yielding contemporaneous advance, we have shown that in a symmetric equilibrium monopoly power and the R & D-to-sales ratio will be positively related, but the rate of unit cost reduction achieved by the R & D will be too low relative to the socially managed market.
5   In an oligopoly with free entry and no uncertainty, the date of innovation may be 'too early' (relative to the socially managed market) if the elasticity of demand is large, competition forcing firms to bring forward innovation dates. Under uncertainty small inventions bring forth inadequate research, long patent lives generate excessive research.

The models that we have constructed, especially those at the industry level, rely to a great extent on firms being able to protect their inventions. Mansfield's paper (Mansfield *et al.*, 1981) suggests that in most cases this is not actually the case, and in Chapter 2 we have considered this matter. Given also that market equilibrium may not be a frequent state under oligopoly, one may be somewhat more inclined to concentrate on the firm rather than on industry-level results. Even here, however, it seems that the models take inadequate consideration of the process of diffusion that follows innovation. The models tend to assume that innovations are accepted by the market immediately. In Part II of this book we consider the factors that determine the rate at which new technology is accepted and illustrate that in general take-up rates are slow. In particular, it is argued that the rate at which new technology replaces the old depends upon its 'superiority' to the old. If this is the case, then major advances may be taken up more quickly than minor advances which could influence the nature of the R & D process.

Moreover, the rate at which new technology is accepted will depend upon its price. In these R & D models this effect of pricing is ignored completely. It seems therefore that a logical advance on these models would be to link the R & D process to the diffusion process. As far as we know this has not been done.

One might also argue that these models are somewhat vague as to whether invention or innovation is being discussed. We have suggested above that, to the extent that R & D is mainly development, then innovation is the main topic of enquiry. Moreover, given the tendency of the models to allow that advances are marketed immediately, this may be the appropriate interpretation to put on them. However, to the extent that patents give protection, that protection, it can be argued, is more likely to apply to inventions than to innovations. It may be the case that the apparent emphasis on innovations is in contradiction to a belief in protection through patenting.

However, we do have a number of theoretical results detailed above, which we can now consider more fully in the light of empirical work.

3.3 The empirical evidence

The models presented above yield for us a number of predictions as to the determinants of R & D spending. To a great extent this body of theory is not reflected in empirical work, although there is a large body of empirical results. However, much empirical work is based on its own *ad hoc* theorizing, and to some extent the principles underlying this are not reflected in the theory above. It is therefore advantageous when discussing empirical results to refer also to the author's justifications for the relationships estimated. We present below an overview of the empirical literature (for a much more detailed survey see Kamien and Schwartz, 1980), but first we should note that, just as Cowling (1976) has pointed out with respect to research on price cost margins, and as is clear from the work above, the relationship between R & D and firm size and market structure will vary across industries and across firms. We have seen that inter-industry variations in elasticities of demand, research efficiency, conjectural variations, technological opportunity, etc., are all important influences on the R & D sales ratio. Only if such variation is accounted for can one really place much faith in empirical estimates.

What one finds is that much of the empirical literature in this area is based on limited *ad hoc* theorizing, without any attempt, for example, being made to determine theoretically the appropriate measure of concentration. The prime objective of the literature is to test a broadly defined 'Schumpeterian hypothesis', or the Galbraith (1952) version of the hypothesis. As an example of the theorizing, it is argued that the risky operation of financing R & D can be afforded only by large firms possessing quite high degrees of market power and having profit enough to do so. In addition to this, these large firms have the advantages of large secure markets and the possibility of achieving economies of scale in their R & D departments, factors that are conducive to high returns from R & D expenditures. Hence these firms are better suited to promoting new technology. As is obvious, such *ad hoc* reasoning does not lead to explicit testable hypotheses. However, it does represent a view of the world not reflected in the models above. We start our review by considering the role of technological opportunity.[3] Scherer (1967a) demonstrated the importance of accounting for technological differences in explaining inter-industry variation in R & D intensity and in relating R & D intensity to concentration levels. He used 'technology class' dummy variables, rationalized so as to account for inter-industry differences in 'technological opportunity'. Two of the four 'technology class' dummy variables were found significant determinants of inter-industry differences in R & D intensity — measured by the ratio of industry R & D employment to total industry employment — while the ones he used to distinguish between durable and consumer goods were found statistically insignificant. Generally, the conclusions he came to were unfavourable for Schumpeter's suggestions.

Comanor (1967) looked at the average number of research personnel in 1955 and 1966 corrected for firm size and performed an inter-industry analysis of the influence of a number of market factors, including, among others, product differentiation and barriers to entry, on the level of industrial R & D. There is some evidence that, in many industries, smaller firms undertake proportionately more research, and Comanor concludes that concentration levels were more likely to play a significant role where product differentiation, based on product design differences, was not a significant element of market behaviour. These results have been analysed further by Shrieves (1974) using intra-industry methods (after indicating shortcomings in Comanor's methodology), but he finally came to the same conclusion, that

it is not structure *per se* which influences the allocation of resources to innovation, but structure combined with product market and technological characteristics . . .

and that

the Schumpeterian and neo-Schumpeterian theories are, in themselves, quite deficient in explaining the relationships between firm size, market structure and innovation. (Shrieves, 1974)

Freeman sums up the results of his research with six conclusions:

1) R & D programmes are highly concentrated in all countries for which statistics are available.
2) These programmes are mainly performed in large firms with more than 5000 employees but the degree of concentration is significantly less by size of firm than by size of programme.
3) The vast majority of small firms (probably over 95%) do not perform any specialised R & D programmes.
4) Amongst those firms which do perform R & D there is a significant correlation between size of total employment and size of R & D programme in most industries.

[3] This survey of results is based on an exercise carried out by V. Georgoulis, a Warwick MA student (Georgoulis, 1976).

5) There is a much weaker correlation between . . . research intensity and size of firm and it is not significant in many industries . . .
6) Among the larger firms there is also evidence that in some industries research intensity diminishes with size above a certain level. (Freeman, 1974)

An index of diversification is often included in cross-section studies to capture inter-firm differences in expected returns from R & D. Its performance is not however particularly startling. Grabowski (1968) found a significant positive effect for two out of three industries, but Comanor's (1967) results were much less encouraging. Scherer (1965) found no influence.

Barriers to entry have often been included in the models, usually expressed as dummy variables. These barriers may be an important determinant of R & D through their effect on expectations or for reasons discussed in the theory above. Comanor (1967) found that industries with moderate barriers have the highest levels of R & D compared with industries with either high or low entry barriers.

When examining R & D intensity, an index of research productivity, usually defined over some earlier period, is sometimes included among the variables (Grabowski, 1968). This measure is usually represented by a proxy variable, like the number of patents granted to the firm, new-product sales, and the number of significant inventions made by firms. Grabowski (1968) also tried, but without success, the first difference of sales, searching for some accelerator mechanism in R & D. Past scale of R & D may also have a positive influence on current R & D, because of the 'tendency of research to feed upon itself' and also because of the claim that continuity and stability of research are desirable (Hamberg, 1966, p. 127). Moreover, through past R & D projects new knowledge is acquired which may form the basis for later projects.

Most studies do develop some theory before setting up their estimating equations, but these are not usually derived explicitly. One exception is Howe and McFetridge (1976), who have treated R & D decisions as asset acquisition decisions, taken in a profit-maximizing firm, so that 'investment in R & D can be assumed to proceed to the point at which the marginal rate of return to R & D is equal to the marginal cost of funds'. But then, even these authors have not derived explicitly the functional form of the relationship determining R & D.

A further criticism of much work has been raised following arguments on directions of causality. For instance, technical change could lead to concentration or to barriers to entry through the reduction of cost for existing firms. This is a view expressed above in the theoretical section and by Kennedy and Thirlwall (1972), but it has not been extensively examined.

One can argue that there is a serious problem of interpretation of a coefficient in a regression of R & D on concentration because of the presence of simultaneity problems; and of course the application of ordinary least squares would not give accurate estimates of the coefficients. The problem of simultaneity might not be serious, for it is likely that there are lags in the influence of technical change on market structure, while in the opposite direction such lags may be less probable. Howe and McFetridge (1976), comparing the results obtained from studies using single equations and simultaneous models, concluded that, 'having estimated a more general model which treats the R & D decision as one of a number of inter-related financial decisions, we find the single-equation approach adequate to the analysis of the determinants of R & D expenditures.'

Buxton (1975) also considers a simultaneous approach, formulating a theory of the process of technical change to provide evidence on the determinants of technical change in the UK. This he tested with data from eleven manufacturing industries from 1956 to 1965 using the three aspects of his theory incorporated in an implicitly defined three-equation model: one equation expressing the influence of market structure on R & D, another the relationship of technical change with R & D and market structure together, and the last one expressing how technical change affects changes in industrial structure.

His results support the Schumpeterian side of the coin on concentration, but on the inverse causality he concludes that the growth of concentration is reduced by technical change. The same results do not hold for diversification, whose growth is increased by technical change which, in turn, is enhanced by low levels of diversification but is unaffected by the height of barriers to entry. These results, even if indicative of how a particular mechanism works, are subject to serious criticism because of the absence of lags, especially in the second and third equations of the model, and the use of ordinary least squares as a method of estimation.

To summarize the empirical work, let us turn to Kamien and Schwartz (1975). After a survey of the literature they consider first whether R & D activity increases more than proportionately with firm size. They conclude that:

The bulk of empirical findings do not support it, with the notable exception of the chemical industry. Relative R & D activity, measured either by input or output intensity, appears to increase with firm size up to a point, then level off or decline beyond it.

On the impact of concentration they conclude that:

Little support has been found for the hypothesis that R & D activity increases with monopoly power. Instead recent evidence suggests that rivalry in R & D may be nonlinearly related to industry concentration. A new empirically inspired hypothesis has emerged to the effect that a market structure intermediate between monopoly and perfect competition would promote the highest rate of activity. Some theoretical support for this has been advanced. (Kamien and Schwartz, 1975)

These conclusions seem to be a fair reflection of the empirical literature. We should note, however, that this empirical literature considers a number of aspects of the R & D process that could usefully be incorporated in the theoretical models. For example, liquidity constraints, learning economies, diversification, lags, and barriers to entry arising from R & D all have merited attention and empirical investigation. Perhaps this indicates a need for further theoretical investigation.

3.4 Conclusions

We have provided conclusions to sections 3.2 and 3.3 which do not really need repetition. A one-line conclusion would state that the evidence to date leaves the Schumpeterian hypothesis unproven (whichever version of that hypothesis we take). Perhaps more useful here is to consider the link between the two aspects of the literature on invention and innovation — the input and output approaches. To what extent does the literature on R & D give us any insight into the appropriate analysis of output-orientated measures? First, we can argue that in the models at the industry level we see that high industry R & D does not necessarily mean high rates of technological advance (high rates of output) if there is excessive repetition; following Dasgupta and Stiglitz, we could argue that in certain cases competition encourages higher R & D but also inefficiency. Second, we can argue that the analysis of patenting activity is somewhat un-sophisticated relative to these R & D models, but that the R & D models, because of their inclusion of functions relating R & D to a measure of the output from R & D (the research process cost function), could be used to analyse patenting behaviour. It would seem that this would be a profitable line for future research. Finally, these R & D models predict that the efficiency of R & D in terms of input per unit of output at the industry level will be related to market structure. One way to test this may be to look at patenting activity.

References

Arrow, K. (1962), 'Economic Welfare and the Allocation of Resources for Inventions', in R.R. Nelson (ed.), *The Rate and Direction of Inventive Activity*, Princeton University Press.

Buxton, A. (1975), 'The Process of Technical Change in UK Manufacturing', *Applied Economics*, 7, 53–71.

Comanor, W.S. (1967), 'Market Structure, Product Differentiation, and Industrial Research', *Quarterly Journal of Economics*, 81, 639–57.

Cowling, K. (1972), 'Optimality in Firms' Advertising Polices: An Empirical Analysis', in K. Cowling (ed.), *Market Structure and Corporate Behaviour*, Gray Mills, London.

Cowling, K. (1976), 'On the Theoretical Specification of Industrial Structure–Performance Relationships', *European Economic Review*, VIII, 1–14.

Dasgupta, P. and Stiglitz, J. (1980a), 'Industrial Structure and the Nature of Innovative Activity', *Economic Journal*, 90, 266–93.

Dasgupta, P. and Stiglitz, J. (1980b), 'Uncertainty, Industrial Structure and the Speed of R & D', *Bell Journal of Economics*, 11, 1–28.

Dasgupta, P. and Stiglitz, J. (1981), 'Entry, Innovation, Exit: Towards a Theory of Oligopolistic Industrial Structure', *European Economic Review*, 15, 137–58.

Demsetz, H. (1969), 'Information and Efficiency: Another Viewpoint', *Journal of Law and Economics*, 12, 1–22.

Fisher, F.M. and Temin, P. (1973), 'Returns to Scale in Research and Development: What does the Schumpeterian Hypothesis Imply?' *Journal of Political Economy*, 81, 56–70.

Freeman, C. (1965), 'Research and Development in Electronic Capital Goods',

National Institute Economic Review, 34, 40–91.

Freeman, C. (1974), *The Economics of Industrial Innovation*, Penguin, Harmondsworth.

Galbraith, J.K. (1952), *American Capitalism*, Houghton Mifflin, Boston.

Georgoulis, V. (1976), 'R & D, Technological Progress and Market Structure', MA dissertation, Warwick University, Coventry.

Grabowski, H. (1968), 'The Determinants of Industrial Research and Development', *Journal of Political Economy*, 76, 292–306.

Hamberg, D. (1966), *R & D: Essays on the Economics of Research and Development*, Random House, New York.

Howe, J. and McFetridge, D. (1976), 'The Determinants of R & D Expenditures', *Canadian Journal of Economics*, IX, 57–71.

Kamien, M. and Schwartz, N. (1975), 'Market Structure and Innovative Activity: A Survey', *Journal of Economic Literature*, 13, 1–37.

Kamien, M. and Schwartz, N. (1980), *Market Structure and Innovation*, Cambridge University Press.

Kennedy, C. and Thirlwall, A. (1972), 'Surveys in Applied Economics: Technical Progress', *Economic Journal*, 82, 11–72.

Loury, G. (1979), 'Market Structure and Innovation', *Quarterly Journal of Economics*, XCIII, 395–410.

Mansfield, E., *et al.* (1971), *Research and Innovation in the Modern Corporation*, W.W. Norton, New York.

Mansfield, E., *et al.* (1981), 'Imitation Costs and Patents: An Empirical Study', *Economic Journal*, 91, 907–18.

Needham, D. (1975), 'Market Structure and Firms' R & D Behaviour', *Journal of Industrial Economics*, 23, 241–55.

Nerlove, M. and Arrow, K. (1962), 'Optimal Advertising Policy Under Dynamic Conditions', *Economica*, 29, 129–42.

Newbery, D. (1978), 'Sleeping Patents and Entry-deterring Inventions', Churchill College, Cambridge, mimeo.

Nickell, S. and Metcalf, D. (1978), 'Monopolistic Industries and Monopoly Profits or, Are Kellogg's Cornflakes Overpriced', *Economic Journal*, 88, 254–68.

Nordhaus, W.D. (1969), *Invention Growth & Welfare*, MIT Press, Cambridge, Mass.

Scherer, F.M. (1965), 'Firm Size, Market Structure, Opportunity and the Output of Patented Inventions', *American Economic Review*, 55, 1097–125.

Scherer, F.M. (1967a), 'Market Structure and the Employment of Scientists and Engineers', *American Economic Review*, 57, 524–31.

Scherer, F.M. (1967b), 'Research and Development Resource Allocation Under Rivalry', *Quarterly Journal of Economics*, 81, 359–94.

Shrieves, R. (1974), 'Firm Size, Market Structure and Innovation: Further Evidence', University of Tennessee, Working Paper no. 19.

Spence, A.M. (1977), 'Entry, Capacity, Investment and Oligopolistic Pricing', *Bell Journal of Economics*, 8, 534–44.

Stoneman, P. and Leech, D. (1980), 'Product Innovation, Process Innovation and the R & D/Market Structure Relationship', paper presented at the EARIE Conference, Basle, September.

Usher, D. (1964), 'The Welfare Economics of Invention', *Economica*, 31, 279–87.

Chapter 4

Induced Bias and Learning

4.1 Introduction

In the two chapters above we have discussed the factors influencing research and development and patenting activity, and thus, by implication, the direction of inventive and innovative activity in terms of which industries will experience the greatest level of inventive and innovative activity. In this chapter we will consider a different approach to the question of direction. We ask, can one model the direction of inventive and innovative activity, in the sense of its labour- or capital-saving bias, and if so what directions would one expect technological advance to follow in a capitalist system? In section 4.2 we will discuss the induced bias hypothesis and then in section 4.3 we detail one application of the hypothesis, investigating rates of labour- and capital-saving advances in British and American industry in the nineteenth century. In section 4.4 we consider learning by doing. We should note that this literature is essentially macroeconomic, whereas our previous two chapters are micro-orientated.

4.2 The induced bias hypothesis

Kennedy (1964), von Weizsacker (1966), and Drandakis and Phelps (1966) have introduced into the literature on technological change an hypothesis that the capital- or labour-saving bias in technological advance is endogenous to the economic system. The original proposition however is due to Hicks:

A change in the relative prices of the factors of production is itself a spur to invention, and to invention of a particular kind — directed to economising the use of a factor which has become relatively expensive. (Hicks, 1932, pp. 124–5)

Salter refutes this argument on the grounds that

the entrepreneur is interested in reducing costs in total, not particular costs such as labour costs or capital costs. When labour costs rise any advance that reduces total costs is welcome, and whether this is achieved by saving labour or capital is irrelevant. (Salter, 1966, pp. 43–4)

The Kennedy–Weizsacker approach, however, enables one to derive theoretically a version of the Hicks induced bias hypothesis. We will first detail the model and then discuss its implications. The model is essentially considered to be a macro-model, and there are two inputs; capital (K) and labour (L). We write the production function as

$$Y = F\{A(t)K, B(t)L\}$$

where Y is output, and, using the dot convention for a time derivative, $\dot{A}(t)/A(t)$

$\equiv a_t$ is the rate of capital augmentation through technical change and $\dot{B}(t)/B(t) = b_r$ is the rate of labour augmentation. It is now assumed that there exists an invention possibility frontier (IPF) that gives a relationship between attainable rates of factor augmentation. We will discuss below exactly what this frontier is supposed to represent, but for now we write it as (4.1) and draw it as in Fig. 4.1:

$$b = \Psi(a)$$

$$\Psi'(a) < 0 \quad \Psi''(a) < 0 \tag{4.1}$$

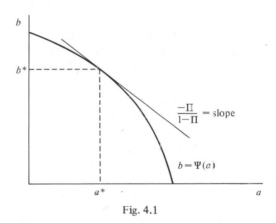

Fig. 4.1

The entrepreneur is assumed to maximize the current rate of unit cost reduction. If we let Π and $1 - \Pi$ be the shares of capital and labour in total cost (and also total output), then the rate of unit cost reduction (C) can be written as

$$C = \Pi a + (1 - \Pi)b. \tag{4.2}$$

Maximizing (4.2) subject to (4.1) requires that a is chosen so that

$$\frac{\partial C}{\partial a} = \Pi + (1 - \Pi)\frac{db}{da} = \Pi + (1 - \Pi)\Psi'(a) = 0. \tag{4.3}$$

Second-order conditions for a maximum are satisfied given $\Psi''(a) < 0$. From (4.3), C is maximized where

$$\Psi'(a) = \frac{-\Pi}{1 - \Pi}, \tag{4.4}$$

i.e., where the slope of the invention possibility frontier is equal to the ratio of the shares of capital and labour in national income. This is shown in Fig. 4.1 at a^*, b^*. One may immediately observe that, the greater is Π (the higher is the capital share), the lower will be b^* and the higher will be a^*. Thus the chosen ratio of labour to capital augmentation is dependent on factor shares, and the nature of the solution is that there is a tendency to have more of a bias towards saving a particular factor the greater is its share. This statement should be distinguished from the original one stating that higher prices for a factor lead towards savings of that factor — it is shares that matter (although if the elasticity of substitution is less than unity factor shares and factor prices move in the same direction).

Thus far this theory could apply to either a firm or the economy as a whole. The next stage of the argument, however, must refer to the whole economy. In this second stage it is considered that factor shares will react to the different rates of labour and capital augmentation which will then feed back on the choice of $a*$ and $b*$. Thus, one can show that, in a world with a constant propensity to save, if the elasticity of substitution of the production function σ is less than one, then the changes in shares will generate a time path for a_t^* and b_t^* such that eventually a_t^* tends to zero. In such a position technical progress is Harrod-neutral (for proofs see Burmeister and Dobell, 1970, pp. 83–9). This argument is picked up again in Chapter 14.

The question that we must now consider is, exactly what process is the choice procedure modelling? As the basic assumption is one of disembodied technological change and a world in which new technology instantaneously affects production relations, there is no diffusion question being discussed. We thus have the choice of whether we are to take this as a model of invention or of innovation. Given that no resource costs are involved in introducing the chosen technology, this implies that the technologies represented on the IPF are available for immediate use. This implies that the IPF is best considered as representing a set of developed *innovation* possibilities, or a book of blueprints, from which users choose those that maximize their objective functions.

Taking this interpretation then raises the issue of where the IPF comes from. If we think of it as representing developed innovation possibilities, then these innovations represent the end of a process that has involved invention, development, and R & D. The story being told therefore must be that the invention and development process yields a book of blueprints that can be summarized by the IPF, and from the IPF the innovations to be used are chosen. This is somewhat heroic. Over the last two chapters we have spent a considerable amount of time looking at such development processes, and there is nothing there to suggest that they will yield an IPF of the above form. In fact, the hypothesis used in earlier chapters that invention direction is motivated by expected profitability would suggest (1) that firms would develop new technologies with the greatest chance of use, and thus the book of blueprints would be full around $a*b*$ but relatively empty elsewhere; and (2) that the type of new technologies that are developed will depend upon the cost of development. In fact, Binswanger (1974) has developed a microeconomic approach to induced innovation whereby R & D expenditure is needed to advance technologies and suggests that the relative cost of labour-saving and capital-saving advances will influence the direction of technological change. His analysis throws considerable doubt on the validity of the macro-approach.

Moreover, if we consider that new technologies have to be developed, then the continued existence over time of the IPF implies that, as soon as an a^*b^* position is selected, another new technology is developed that will replace that one removed from the book of blueprints. This is again rather heroic. Moreover, given that the IPF is not shifting over time, this implies that *only* replacements for chosen a^*b^* positions are being developed. This might imply that the model could be used to consider selection at a moment in time, but to use it to model a continuing process of selection over time is misleading.

Overall, therefore, this approach to induced bias is probably best considered as a process of selection by new technology users from a set of blueprints that represent fully developed technologies. We may thus think of invention and R & D as yielding innovations, some of which will be a success and some a failure. The successful ones will be those selected by potential users. This analysis is best represented as one aspect of that particular selection process. The main problem, however, is that the supply of innovations coming forward for selection will depend upon expectations of success and cost of development, and this suggests that the IPF may not have the form assumed above.

However, even if we assume that the IPF can be defined as the model requires, there is still one further objection to the analysis. The model assumes maximization of the rate of cost reduction at given factor prices. But in general, prices will be changing over time and it seems inappropriate for entrepreneurs to ignore this. Minimization of the present discounted value of expected future costs may be more appropriate. The price expectations that enter this minimization will be crucial, however. One might think of one alternative to maximizing myopically by assuming rational expectations. If the system is stable it will tend to an equilibrium in which $a^* = 0$, $b^* = \Psi(0)$. Associated with this position is a certain Π, $\hat{\Pi}$. Under rational expectations one might argue that entrepreneurs knowing that the system will tend to $\Pi = \hat{\Pi}$ will maximize assuming $\Pi = \hat{\Pi}$; then one would have $a^* = 0$, $b^* = \Psi(0)$ for all t. Despite these objections, however, the basic idea underlying the approach, namely that the selection of new technology by users is an economic process, is an important one. We will turn now to consider one particular application of the theory.

4.3 American and British technology in the nineteenth century

In his book on this subject, Habakkuk (1962) initiated probably the most extensive argument for which the theory of induced innovation has been used. In the midst of a discussion on how relative factor prices will affect the choice of techniques and speeds of diffusion, he states:

Labour scarcity . . . gave Americans an incentive . . . to attempt to invent methods specifically to save labour. (Habakkuk, 1962)

The scarcity of labour, it is further argued, is evidenced by higher industrial wages in the USA than in England, this difference being due to the abundant supply of land in the USA (relative to the UK), making average wages high in agriculture:

In order to attract labour ... industry had to assure the workers in industry a real wage comparable to average earnings in agriculture. English industry, by contrast, could acquire labour from agriculture at a wage equal to the very low product of the marginal agricultural labourer. (Habbakuk, 1962)

Basically, therefore, the proposition being considered here is that the relative abundance of land in the USA led to attempts to invent more labour-saving methods for industry there than in the UK.

This proposition has merited much consideration in the literature, the whole being well summarized and advanced by David (1975). We will first ask whether, if we compare two economies that differ only in their land endowment, and in the less well endowed one industrial labour is in greater supply, technological change will be biased towards labour-saving in the industrial sector of the more favourably endowed. We then ask whether a greater land endowment will imply labour scarcity in the industrial sector. This enables us to derive what conditions are required to satisfy the Habbakuk hypothesis. We then ask whether putting the analysis in this form really tells us anything about factor-saving bias in the USA and the UK.

Consider the bias question first. If we follow David and assume that land is t used by industry (only capital and labour are relevant inputs), then, assuming at the elasticity of substitution of the production function is less than unity, a relative labour scarcity will imply a high labour share. To show this, assume that industrial sector technology is the same for both economies and fixed and is represented by a CES production function. Then we may write

$$\frac{\dot{\Pi}_L}{\Pi_L} - \frac{\dot{\Pi}_K}{\Pi_K} = \frac{1-\sigma}{\sigma}\left(\frac{\dot{K}}{K} - \frac{\dot{L}}{L}\right) \tag{4.5}$$

where $\Pi_{L,K}$ = shares of labour and capital;

$\quad\ \sigma$ = elasticity of substitution.

Given Π_L = $1 - \Pi_K$, we may rewrite this as

$$\frac{\dot{\Pi}_L}{\Pi_L\,(1 - \Pi_L)} = \frac{1-\sigma}{\sigma}\left(\frac{\dot{K}}{K} - \frac{\dot{L}}{L}\right). \tag{4.6}$$

If we compare a scare labour to abundant labour economy, we may say that in the scarce relative to the abundant $\dot{K}/K - \dot{L}/L$ is positive, and thus, if $\sigma < 1$, then $\dot{\Pi}_L > 0$; i.e., the scarce labour economy will have a higher labour share.

Following the induced bias hypothesis above, we have shown that the higher labour share will generate for a given IPF a greater labour saving bias in technological advance. Theoretically, therefore, if both countries face the same IPF and have the same production function, and if land is not used in industry, then the country with the higher capital–labour ratio in industry would tend to have the higher labour share in industry, given $\sigma < 1$, and thus greater labour-saving bias in technological advance.

Let us turn now to the problem of labour scarcity. Temin (1971) discussed whether an abundant supply of land will lead to an increase in the capital-labour ratio in industry, generating a higher marginal product of labour in industry and a higher real wage. He considers a model with an agricultural and an industrial sector with factors paid their marginal product, and a given stock of capital and labour supply. He shows that the proposition on the effect on industrial wages of a higher land stock is generally valid only if capital is an input specific to manufacturing and land an input specific to agriculture. In this case a greater land stock will mean a higher marginal product for any given labour input in agriculture. This higher marginal product means a higher real wage in agriculture. This attracts labour from industry, leaving the given capital stock spread over fewer workers, yielding a high capital–labour ratio and thus a higher wage–rental ratio in industry. In such circumstances also the marginal product of capital would be lower in US than UK industry.

The position one reaches, therefore, is that the Habakkuk hypothesis, in terms of the bias of technological advance (which does not necessarily mean invention), can be maintained in a neoclassical view of the world under the assumption of common technologies, common labour forces, common capital stocks, and common IPF, as long as there are sector-specific inputs and an elasticity of substitution in industry less than unity.

However, the sector specificity of inputs is not really acceptable either on its own grounds or because of its implication that the US marginal product of capital would be lower than that in the UK. The evidence suggests that the return to capital was greater in the UK than the USA. Jones (1971) has shown that such a ranking of returns is consistent with land abundance in the USA if land is also an input to industry, for the abundant land would raise the marginal product of capital in industry. But if land is an input to industry the Temin conditions are not satisfied. Basically, therefore, the Habbakuk hypothesis of the effect of land abundance on factor shares cannot be derived as a general result in a neoclassical framework.

Even if we accept this, but argue that in US industry the wage share was higher, which David argues is consistent with the evidence, can the induced bias part of the hypothesis detailed still be maintained? The problem here is the empirical evidence.

David first refers to the empirical work of Asher (1972), from which some empirical estimates of rates of capital- and labour-saving in the textile industries are derived. Assuming a CES production function, we can derive

$$\frac{\dot{K}}{K} - \frac{\dot{L}}{L} = \frac{\sigma}{1 - \sigma} \frac{\dot{\Pi}_L}{\Pi_L} - \frac{\dot{\Pi}_K}{\Pi_K} + (a_t - b_t) \qquad (4.7)$$

where σ = elasticity of the production function;
Π_L, Π_K = factor shares;
a_t, b_t = rates of factor augmentation.

Asher's results indicate that, defining $\beta \equiv a_t - b_t, \beta_{GB} > \beta_{US} > 0$.

If $\sigma < 1$, then as we have argued, a higher real wage would be associated with a higher labour share in national income Π_L. Using the IPF analysis, this implies a higher a_t and lower b_t. Given that all the empirical evidence indicates that $\sigma < 1$, this argument would imply that, if the USA and UK had the same IPF, then $\beta_{US} > \beta_{GB}$, which contradicts the evidence. The other problem with finding $\beta_{GB} > \beta_{US}$, if the IPF is common to both countries, is that this would essentially state that the USA did not have a bias towards labour-saving. $\beta_{GB} > \beta_{US}$ is consistent with a relative labour-saving bias in the USA only if the IPF for the UK lay outside that of the USA. In fact, if the induced bias hypothesis of Kennedy–Weiszacker is to be used and Asher's empirical evidence is correct, then only different IPFs will make the two consistent. If one reaches this point then one must explain why the IPFs are different, and as we have already detailed above, the origins of the IPF are not detailed in the theory.

Overall, therefore, the main test of an empirical observation for which the induced bias hypothesis has been used has indicated that it is left unproven.

This leaves us in a somewhat undesirable position. If US technological advance was more labour-saving, we still have not explained why. All we have shown is that the neoclassical approach does not really provide an answer. David proposes an explanation that is summarized on the following lines. Think of the UK and USA having basically similar arrays of technologies from which to choose in time period zero. However, the capital-intensive technologies required greater resource or raw material inputs per unit of output than did the labour-intensive. The USA was abundant in raw materials and resources which were therefore comparatively cheap. This encouraged the use of capital-intensive techniques in the USA relative to the UK.

To this static story we now add a dynamic element. It is argued, following Atkinson and Stiglitz (1969), that a major part of the technological change process comes from learning on existing technologies; i.e., technological change will mean changes 'close' to rather than 'distant' from current technologies. This implies that the USA would generate changes in capital-intensive technologies, the UK in less capital-intensive technologies. The bias in technological change is the result of the starting point because change is localized. David argues that such a story is consistent with the experience patterns of the UK and USA.

The bias in technological advance therefore in this view of the world is basically the result of learning by doing. Before finishing this chapter, we will consider the question of learning by doing, for it will also be used in other places in the book (e.g., Chapter 9).

4.4 Learning by doing

Learning by doing has been introduced into the main line of economic theory by Arrow (1962). He proposed an 'endogenous theory of the changes in knowledge which underlie intertemporal and international shifts in production functions'. He argued (1) that learning is the product of experience; (2) that learning by repetition of a problem is subject to diminishing returns. Thus, to have continuous learning one needs continually to meet new problems. Arrow refers to the work of Wright on airframes, Verdoorn on national outputs, and Lundberg on the Horndahl iron works, all suggesting that increasing productivity is allied with experience. Arrow's hypothesis is that 'technical change in general can be ascribed to experience, that it is the very activity of production which gives rise to problems for which favourable responses are selected over time'.

To formalize the model Arrow considers cumulative output as an index of experience but rejects it on the grounds that, if the rate of output is constant, then the stimulus to learning would seem to be constant. He thus takes cumulative gross investment as an index of experience, because each new machine produced and put into use is capable of changing the environment in which production takes place, implying that learning can occur because of continually new stimuli.

Assuming that technological change is embodied in new capital goods, letting G equal cumulative gross investment, and giving capital goods a serial number G, appropriate to cumulative gross investment at their date of construction, we define
$\lambda(G)$ = labour used in production with a capital good number G;
$\gamma(G)$ = output capacity of capital good serial number G;
Y = total output;
L = labour force employed.

Let capital goods have a fixed lifetime T, and define G' as the serial number of the machine constructed T periods before. Then

$$Y = \int_{G'}^{G} \gamma(G)dG \tag{4.8}$$

and

$$L = \int_{G'}^{G} \lambda(G)dG. \tag{4.9}$$

Allowing that $\lambda(G)$ takes the special form

$$\lambda(G) = \delta G^{-n}, \tag{4.10}$$

and $\gamma(G) = \epsilon$, Arrow shows that one can write that

$$Y = \epsilon G \left\{ 1 - \left(1 - \frac{L}{ZG^{1-n}} \right)^{1/1-n} \right\} \quad \text{if } n \neq 1 \tag{4.11}$$

where $Z = \delta/(1-n)$ and

$$Y = \epsilon G (1 - e^{-L/b}) \quad \text{if } n = 1 \tag{4.12}$$

Expressions (4.11) and (4.12) correspond to the basic production functions of neoclassical analysis.

The learning effect in this model is essentially represented by (4.10). However, this is not the only form that learning functions have been allowed to take. Spence (1981), for example, allows that costs per unit of output are related to cumulative output

$$E_t = \int_0^t Y_v \, dv$$

by

$$c(E_t) = m_0 + c_0 e^{-\lambda E_t}. \tag{4.13}$$

This is just one possible alternative.

Holding Arrow in mind we now make two diversions.

1 In the field of management consultancy, the concept of the experience curve seems to carry much weight. The results of the Boston Consulting Group, as detailed in for example HMSO (1978), show reductions in cost per unit of output against cumulative output for a number of products over long time-profiles. This work should not be confused with the true concepts of learning by doing, for experience curve results agglomerate effects of scale, new products, factor input price changes, etc., with true learning effects to generate the empirical results (see Hart, 1983).

2 Arrow refers to the work of Verdoorn. In the literature on economic growth Verdoorn's Law has assumed a major role in a side debate. Kaldor (1966) suggested that, allied with other factors, the operation of Verdoorn's Law in the manufacturing sector was a contributing factor to the UK's poor growth performance after the Second World War. Verdoorn's Law for these purposes is considered as an association between the rate of growth of labour productivity and the rate of growth of output, based if one wishes on the same principles of learning as Arrow discusses. The outcome of the debate (see McCombie, 1981, for references and the flavour of the debate) is that the Kaldor hypothesis does not really carry much empirical support.

We now return to Arrow, who argues that his production assumptions 'are designed to play the role assigned by Kaldor to his "technical progress function" which relates the rate of growth of output per worker to the rate of growth of capital per worker', but in Arrow the relations between rates of growth are derived from more fundamental relations between the magnitudes involved. Moreover, Arrow stresses gross rather than net investment. (In Kaldor's latest formulation (Kaldor and Mirlees, 1962), gross investment is in fact stressed.) The technical progress function can be represented by

$$\frac{d(Y/L)}{dt} \cdot \frac{1}{Y/L} = F \left\{ \frac{d(K/L)}{dt} \cdot \frac{1}{K/L} \right\} \tag{4.14}$$

with $F(O) > 0$, $F' < 0$, $F(\infty) < \infty$. Black (1962) shows that if (4.14) is linear, then the underlying technology is the same as a Cobb–Douglas production function, but if it is nonlinear then it is not possible in general to derive any specific underlying aggregate production function.

The technical progress function, although not in the mainstream of the literature any more, does provide a nice link for closing this chapter. We have gone from Kennedy-Weizsacker through David to Arrow to Kaldor. We can now go back to Kennedy, who argues

surprisingly enough, in view of the role that it has played in the above analysis, our innovation possibility function is really a disguised form of Kaldor's famous technical progress function. Kaldor relates the proportional change in output per man year to the proportional change in capital per man. But it is, of course, possible to derive from these two variables the proportional change in output per unit of capital, and this means that, if the technical progress function is known, the innovation possibility function can be derived from it. (Kennedy, 1964)

Our discussion has now gone full circle.

4.5 Conclusions

We started this chapter with a discussion of the induced bias hypothesis, and via an application of this theory were led into technological change as a learning phenomenon. We have isolated a number of views illustrating the endogeneity of the process of technological advance in the course of this discussion, and if the chapter has no other outcome this result makes the trip worthwhile.

We have considered that it is possible to consider the labour- or capital-saving bias of new technology as determined endogenously, although the theory and the empirical support are not completely acceptable. We have introduced specific forms of learning functions and discussed their relevance. At the same time the concept of the technical progress function was discussed. The induced bias hypothesis will reappear in Chapter 14, when income distribution is discussed.

We have now come to the close of a three-chapter investigation of the process of generating new technology. The underlying principle of these chapters is the view that the rate and direction of advances in technology will, at least in part, be subject to rational economic behaviour, and it is therefore a suitable topic for economic analysis.

Although we have shown that a proportion of inventive activity occurs outside the market economy, we have also argued that patenting activity will be influenced to some degree by considerations of profitability. This indicates that those industries with prevalent technological opportunities and with the greater expected returns will experience the faster changes in technology. We have spent considerable space looking also at the relationships of R & D and patenting activity to market structure and firm size, attempting to see if there is any strong support for a proposition that bigness and fewness encourage technological advance. The proposition remains unproven. In this chapter we have attempted to consider whether the direction of advance in technology, i.e., the labour–capital savings bias of new technology, can be considered to be endogenous to the economic system and if so in what direction technology will develop in a capitalist system.

The outcome of all this is to argue that there are strong theoretical reasons to believe that from economic analysis we can make predictions as to the determinants of the rate of technological advance and the nature of technological advance in different industries, as well as saying something about the bias of technological advance at the macro-level. However, the empirical support for many of these theories is not really as encouraging as one might hope.

References

Arrow, K. (1962), 'The Economic Implications of Learning by Doing', *Review of Economic Studies*, XXIX, 80, 155–73.

Asher, E. (1972), 'Industrial Efficiency and Biased Technical Change in American and British Manufacturing Industry: The Case of Textiles in the Nineteenth Century', *Journal of Economic History*, 32, 431–42.

Atkinson, A. and Stiglitz, J. (1969), 'A New View of Technological Change', *Economic Journal*, 78, 573–8.

Binswanger, H.P. (1974), 'A Microeconomic Approach to Induced Innovation', *Economic Journal*, 84, 940–58.

Black, J. (1962), 'The Technical Progress Function and the Aggregate Production Function', *Economica*, 30, 166–70.

Burmeister, E. and Dobell, J. (1970), *Mathematical Theories of Economic Growth*, Macmillan, New York.

David, P. (1975), *Technical Choice, Innovation and Economic Growth*, Cambridge University Press.

Drandakis, E. and Phelps, E. (1966), 'A Model of Induced Invention, Growth and Distribution', *Economic Journal*, 76, 823–39.

Habakkuk, H. (1962), *American and British Technology in the Nineteenth Century*, Cambridge University Press.

Hart, P. (1983), 'On Experience Curves', *International Journal of Industrial Organisation*, 1, forthcoming.

Hicks, J.R. (1932), *The Theory of Wages*, Macmillan, London.

HMSO (1978), *A Review of Monopolies and Mergers Policy*, Cmnd 7198, HMSO, London.

Jones, R.W. (1971), 'A Three Factor Model in Theory, Trade and History', in J. Bhagwati, *et al.* (eds), *Trade, Balance of Payments and Growth*, North Holland, Amsterdam.

Kaldor, N. (1966), *Causes of the Slow Rate of Growth in the United Kingdom: An Inaugural Lecture*, Cambridge University Press.

Kaldor, N. and Mirlees, J. (1962), 'A New Model of Economic Growth', *Review of Economic Studies*, 29, 174–92.

Kennedy, C. (1964), 'Induced Bias in Innovation and the Theory of Distribution', *Economic Journal*, 74, 541–7.

McCombie, J. (1981), 'What Still Remains of Kaldor's Law?' *Economic Journal*, 91, 206–16.

Salter, W. (1966), *Productivity and Technical Change* (2nd ed.), Cambridge University Press.

Spence, A. (1981), 'The Learning Curve and Competition', *Bell Journal of Economics*, 12, 49–70.

Temin, P. (1971), 'Labor Scarcity in America', *Journal of Interdisciplinary History*, 1, 251–64.

von Weizacker, C. (1966), 'Tentative Notes on a Two-Sector Model with Induced Technical Progress', *Review of Economic Studies*, XXXIII, 245–51.

Part II
The Spread of New Technology

Chapter 5

Diffusion of Innovations:
An Introduction

An economy is not affected in any material way by new technology until the use or ownership of that technology is widespread. This Part of the book is concerned with the spread of new technology, a topic usually labelled 'technological diffusion'. We shall be making substantial reference to this section when we come to investigate the impact of new technology on the economy in Part III.

The underlying basis of the study of diffusion is to rationalize why, if a new technology is superior, it is not taken up immediately by all potential users. The technology discussed here can be either of the product or process innovation type; the general principles of analysis applied are very similar, the major difference being that the analysis of the diffusion of new products labels the economic decision-makers as consumers, whereas in the analysis of new processes the decision-makers are defined as firms. The theory of diffusion starts at the point where the 'user' has already innovated. This is worthy of some further clarification below. The theory of diffusion, moreover, has been concerned primarily with the demand for new products or processes rather than with their supply. We shall argue below that one cannot empirically investigate the spread of new technology without considering supply, but it is quite reasonable theoretically to discuss demand in its own right. The supply of products embodying new technology is discussed in Chapter 9.

To begin the presentation of the diffusion concept, let us consider the case of a new product that is a consumer durable (although durability is by no means necessary — the argument is simplified by this assumption). Let the stock of that consumer durable that will be owned by households in aggregate when the diffusion is complete be defined as S^*, and let S_t be the stock that households in aggregate own at time t. The diffusion problem concerns how S_t tends to S^* over time. In the case where diffusion is immediate, $S_t = S^*$ for all t. In any other case, S_t may differ from S_t^* for any t.

The first problem that arises is what exactly S^* represents. There are basically two alternative representations in the literature, which I will label the equilibrium and the satiation approaches. The equilibrium approach is best explained by considering a population of size P, made up of households $i, (i = 1, \ldots, P)$ and defining S_{1i}^* as the stock of the new good that household i would own if the new good were not new at all but were like any other good. (S_{1i}^* may therefore be compared to the demand for the good that would be derived from the standard utility-maximizing, perfect-information models of consumer behaviour found in a standard microeconomic textbook.) We then define

$$S^* = \sum_{i=1}^{P} S_{1i}^*.$$

The standard models predict that S_{1i}^* will be a function of at least relative prices (p) and income (Y), and thus we may write in a general form that $S^* = S^* (p, Y, P)$. Given that relative prices and income may change over time, we may argue that S_{1i}^* is time-dependent. Because of this we may also argue that this concept of S^* is also time-dependent and ought to carry a time subscript.

The second or satiation approach is to define S_{2i}^* as the stock of the new good that household i will own when it is saturated with the good; i.e.,

$$S_{2i}^* = \lim_{t \to \infty} S_{1it}^*.$$

Then we consider

$$S^* = \sum_{i=1}^{P} S_{2i}^*.$$

Basically, S_{2i}^* is not time-dependent, but because population may be changing over time this second concept of S^* may still be considered time-dependent. We will tend therefore always to write S^* with a time subscript as S_t^*.

As should be realized, the definition of S_t^* employed in any particular circumstance will be important for any results derived. With the first concept of S_t^*, S_t *can* equal S_t^* in every period and in effect no diffusion is taking place. Any change in S_t will be due to changes in prices, income, or population. However, the same circumstances would be viewed in the second scenario as a continuing process of diffusion, and any changes in prices or incomes would reflect in the speed of adjustment of S_t to S_t^* (the diffusion speed proper).

We can extend the above to consider an alternative measure used in the study of diffusion processes. We may define under either the equilibrium or satiation approach that, if N_t^* equals the number of owners of the new technology when the diffusion is complete and N_t the number of owners in time t, then $S_t^* = N_t^*$ (S_t^*/N_t^*) and $S_t = N_t (S_t/N_t)$. If one assumes that $S_t^*/N_t^* = S_t/N_t$, or that S_t^*/N_t^* and S_t/N_t are functionally related independently of N_t^* and N_t, then one can look at the relationship of S_t to S_t^* by investigating the relationship of N_t to N_t^*, i.e., the number of households owning rather than the level of ownership. This condition essentially requires that the diffusion proceeds by the extension of ownership rather than the deepening of individual households' level of ownership. However, as we really expect both to be occurring during a diffusion process, we cannot fully analyse the approach of S_t to S_t^* by analysing the approach of N_t to N_t^*. This is a point we shall be making again below in the context of the diffusion of process innovations. The study of N_t and N_t^* can be further extended to looking at N_t/P_t and N_t^*/P_t, i.e., the proportion of households owning. A common procedure in studies using the satiation definition of S_t^* is to consider

$N_t^*/P_t = 1$; i.e., all households will eventually own.

These introductory remarks have been made in the context of the spread of ownership of a new consumer durable, for that simplifies the explanation of the concepts. However, it is fair to state that, although not ignoring the consumer side, the majority of the work on diffusion is concerned with the spread of new processes across firms. An exact equivalent of the consumer study would be the economy-wide study of the spread of use of a new process embodied in a new capital good. Thus one would define S_t^* as the post-diffusion level of the stock of the new capital good and S_t as the current stock, and the analysis would concern the relationship of S_t to S_t^* as time proceeds. Both equilibrium and satiation concepts of S_t^* are used in such analyses.

However in the study of the diffusion of new processes much lower levels of aggregation than this have been studied. The analysis splits quite neatly into three parts:

1 intra-firm diffusion;
2 inter-firm diffusion;
3 economy wide-diffusion.

Consider a firm i in industry, j, which has already used a new technology for the first time (has already innovated), which produces a level of output using the new technology in time t, Y_{ijt} with a total output X_{ijt}, and which will, when the diffusion is complete, produce a proportion of its output $(Y_{ij}/X_{ij})^*$, which may be time-dependent, on new technology. The theory of intra-firm diffusion concerns how Y_{ijt}/X_{ijt}, the proportion of a firm's output produced on new technology, changes over time and approaches $(Y_{ij}/X_{ij})^*$.

One may note here that the proportions are expressed as flows rather than as the stocks used in the consumer example. That is not important. One could just as easily define S_{ijt} to represent the stock of the new capital good owned by firm i in time t, and S_{ijt}^* as the post-diffusion stock, and then consider how S_{ijt} approaches S_{ijt}^* without materially changing the problem. The post-diffusion concepts can again be interpreted as satiation or equilibrium concepts, although now equilibrium refers to the theory of the firm section of the standard microeconomic text.

Given Y_{ijt} and X_{ijt}, we may sum over i to obtain Y_{jt} and X_{jt} as the industry output produced on the new technology and total industry output. It is an obvious step then to consider that the theory of inter-firm (or intra-industry) diffusion will concern an analysis of how, once innovation has occurred in the industry, Y_{jt}/X_{jt} approaches $(Y_j/X_j)^*$, the post-diffusion ratio of Y_{jt}/X_{jt}. Although this is a reasonable supposition, it is not the main variable analysed in inter-firm studies. Define N_t as the number of firms in the industry using a new technology at at least some level k (i.e., for whom $Y_{ijt}/X_{ijt} \geqslant k$) at time t; then define N_t^* as the post-diffusion level of N_t. The theory of inter-firm diffusion is often concerned with how N_t approaches N_t^* or, in proportion terms, how N_t as a proportion of the total number of firms approaches N_t^* expressed as a similar proportion. A reasonable question to ask is how a study of the number of users relates to the variable that is really of interest, the proportion of industry output produced on the new technology. We may write that

$$\frac{Y_{jt}}{X_{jt}} \equiv \frac{Y_{jt}}{N_t} \cdot \frac{N_t}{X_{jt}}$$

and

$$\frac{Y_{jt}^*}{X_{jt}} = \frac{Y_{jt}^*}{N_t^*} \cdot \frac{N_t^*}{X_{jt}}.$$

If we know of the relationship of Y_{jt}/N_t to Y_{jt}^*/N_t^*, then a study of N_t to N_t^* will tell us how the proportion of an industry's output produced on the new technology will vary over time. Knowledge of Y_{jt}/N_t, the average output on new technology of users of new technology, and Y_{jt}^*/N_t^*, the post-diffusion level of this variable, is of course provided from the intra-firm diffusion studies. It is therefore necessary to analyse both inter-firm and intra-firm diffusion if one is to obtain insight into the general problem of how new technology is adopted by an industry. This is exactly the point made above in the consumer case.

The next level up from the industry is the meso-economic level, or economy-wide diffusion. We might label this 'inter-industry diffusion'. If we aggregate Y_{jt} over j we get an economy-wide measure of the level of use. However, although such aggregation *within* an industry may be acceptable, given a certain homogeneity of the product, it will probably not be acceptable *across* industries. The tendency therefore is to consider stocks of the new capital good at the economy-wide level rather than output produced on the new capital good, just as we have discussed in the consumer case above. Of course, stocks could be used in the intra-firm and inter-firm studies as well.

We thus have diffusion studies at the three levels when it comes to processes: intra-firm, inter-firm (or intra-industry), and economy-wide (or inter-industry). We could of course define a similar trilogy at the household level (intra-household, intra-group — e.g., socioeconomic group — and economy-wide). The final stage of the diffusion analysis will of course be inter-country studies, on which we will also have something to say later.

Given this breakdown of approaches to the analysis of the spread of a new technology, what questions are asked of the analysis? These can be detailed as follows.

1   What determines the post-diffusion level of use or ownership of new technology by either the firm, the industry (i.e., the number of users), the economy, or the household?
2   Why are some firms, households, or industries early users and some late?
3   What time-path will the firm's, household's, industry's, or economy's level of use follow?
4   Why will it follow that time-path?

5   What characteristics of the firm, household, industry, or technology will be
the key factors in influencing that time-path?

We attempt in the following chapters to indicate the answers to these questions
that have been proposed. For now, however, we confine ourselves to consideration of question 3, the nature of the time-path followed, for much of the
theoretical literature especially is concerned with this point. From numerous
empirical studies it has been found that, in a large majority of the studies of the
diffusion of new technology, the diffusion path is sigmoid (i.e., S-shaped). It is
worth investigating this in greater detail at this early stage. Consider the problem
of inter-firm diffusion where we consider N_t as the number of firms in a fixed
population P using new technology in time t. It is argued that, if one plots
N_t/P against time, the profile will follow an S-shaped curve. The upper limit of
the curve will be N_t^*/P (which itself has a maximum of 1). Initially consider
that the concept of N^* in use here is the satiation concept, and write proportions
of the population as n_t and n^*. We may then plot the diffusion curve in Fig. 5.1.

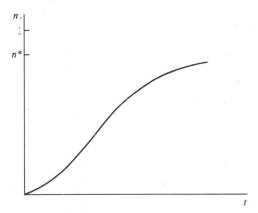

Fig. 5.1 A sigmoid diffusion curve

This general sigmoid shape may also apply when the equilibrium concept of
n_t^* is being used, in which case one would consider that, for any given n_t^*, the
economy would follow a diffusion curve such as that shown in Fig. 5.1, but at
the same time n_t^* is changing over time. The accumulation of observed combinations of n_t and n_t^* will trace out the whole curve describing the spread of
ownership.   ·

These sigmoid curves can take a number of mathematical forms. The most common characterization of the curve is by the logistic growth curve. To define this
strictly, n^* must be taken as a constant, and thus we will consider the satiation
view of n^*. The logistic curve is then described by

$$n_t = \frac{n^*}{1 + \exp\{-G(t)\}} \tag{5.1}$$

where $G(t)$ is some (usually linear) function of time. This curve can also be characterized by the differential equation

$$\frac{dn_t}{dt} = g(t) n_t \left(1 - \frac{n_t}{n^*}\right).$$ (5.2)

In (5.2),

$$G(t) = \int_0^t g(v)dv.$$

The diffusion speed itself is $g(t)$. The higher is the diffusion speed, the steeper is the curve in Fig. 5.1. This logistic diffusion curve has the property that the point of inflexion occurs at $\frac{1}{2}n^*$ in the case where $g(t) = $ constant. Although the case where n^* is not a constant does not lead to a growth curve that is strictly of the logistic form, a time-dependent variant of n^* can be introduced (and has been introduced) into (5.1) above. One great advantage of the logistic curve is that it can be written in a form as $\log\{n_t/(n^* - n_t)\} = G(t)$, the right-hand side of which is linear if $g(t)$ is a constant. We ought to mention here that Chow (1967) has introduced into the literature as a logistic curve

$$\frac{dN}{dt} = g(t)N_t (N^* - N_t)$$ (5.2a)

which we will call the Chow logistic, being somewhat different from the standard form. In proportion terms we get

$$\frac{dn}{dt} = g(t)n_t (N^* - N_t)$$ (5.2b)

which differs from (5.2) in that

$$1 - \frac{n_t}{n^*} = \frac{N^* - N_t}{N^*}.$$

The sigmoid curve need not be logistic, of course; another common form that has been used is the Gompertz curve. This has a differential equation

$$\frac{dn_t}{dt} = h(t)n_t (\log n^* - \log n_t)$$ (5.3)

where $h(t)$ can be called the speed of diffusion; and in the case where $h(t) = $ a constant, the curve has an inflexion point at $n_t = 0.37n^*$.

The logistic and Gompertz curves, however, are just two of a whole class of curves that may be labelled sigmoid. We can approach others by noting that the logistic curve is no more than the cumulative density curve derived from a sech-squared (or logistic) frequency distribution. If we consider a variable, the distribution of which over the population has a bell shape, then the cumulative density will have an S-shape. We may then consider that, for each bell-shaped distribution, there exists an S-shaped curve, the form of which can be used in a diffusion study. This observation will become particularly important when we come to the probit models below, but for the moment we wish to draw attention to the most frequently used sigmoid curve derived from this source — the lognormal curve. If we consider the cumulative density of a bell-shaped distribution it is clear that the inflexion point of the cumulative density must occur at the mode of the distribution. As in the lognormal, $\Lambda(x)$, the mode occurs at $x = e^{\mu-\sigma^2}$; where μ is the mean and σ^2 the variance of the distribution, the point of inflexion on the sigmoid curve will vary with μ and σ^2. Thus the cumulative lognormal will in fact produce a whole family of sigmoid curves, each with different inflexion points. A useful linear approximation to the lognormal curve is

$$\frac{dn_t}{dt} = \frac{f}{t} n_t \left(1 - \frac{n_t}{n^*}\right) \tag{5.4}$$

where f may be called the speed of diffusion. Although this detail has been provided for the lognormal distribution, any bell-shaped distribution will do. In fact, the cumulative normal is often used.

Given these different sigmoid curves, it is worth asking whether there is any one general form that, by the appropriate selection of parameters, can represent any sigmoid curve. The answer is no. The next obvious question concerns whether any individual curve such as the lognormal, because it can take different amounts of skewness, can subsume the logistic or Gompertz. Again the general answer is no (see Friedman, 1974). Finally, we can ask whether, given a set of data, it is always possible to say that one particular sigmoid curve fits that data better than any other. Here, within limitations, we may say yes, but it is not a question that is really of much interest in its own right. For we are interested not in summarizing, but rather in explaining, diffusion data, and the explanation itself should tell us what specific mathematical form should be taken by the sigmoid curve. Our statistical tests should then tell us as to the validity of our hypothesis by indicating whether the sigmoid form suggested does or does not fit the data.

We have proceeded some distance now into the theory of diffusion without actually approaching the primary question of why, if a new technology is superior, it is not used by all immediately. The task of unravelling the answer in detail is confined to the next three chapters, but as a basis of these chapters we will present one view of the answer here. This answer is that the spread of technology can be likened to an epidemic. Diseases are assumed to spread by contact between individuals, and by implication the use of new technology will be spread as individuals make contact with one another. If we allow N_t to be the number in a constant population P infected at time t and N^* the number in the population not immune to the disease, then at time t there are $P - N_t$ individuals not affected, of whom $N^* - N_t$ are susceptible to the disease. If we allow that the disease is spread by contact, and if $g(t)$ represents infectiousness and the frequency of contact, then the chance of a non-infected individual catching the disease in time t will depend upon $g(t)$ and the chance of meeting a currently affected member of the population (call this z_t). Then the expectation of the change in N_t will be the multiple of the chance of a non-infected individual catching the disease and the number of non-infected individuals; i.e.,

$$\frac{dN}{dt} = g(t)z_t \, (N_t^* - N_t).$$ (5.5)

Given $n_t = N_t/P$, we may write (5.5) as

$$\frac{dn}{dt} = g(t)\frac{z_t}{P}(N_t^* - N_t).$$ (5.6)

If z_t, the chance of meeting a currently affected individual, is n_t, the proportion of the population currently infected, then we get

$$\frac{dn}{dt} = g(t)n_t \, (n^* - n_t).$$ (5.7)

But if the chance of meeting a currently affected individual is n_t/n^*, the proportion of the non-immune population currently affected, then we get

$$\frac{dn}{dt} = g(t)n_t \left(1 - \frac{n_t}{n^*}\right)$$ (5.8)

which is the standard logistic discussed above. For z_t to equal n_t/n^* really requires that one meets only that part of the population that is not immune, which seems unlikely unless $n^* = 1$. Thus, although the theory of epidemics can generate logistic curves, they require special characteristics to do so. Despite this objection, it is worth enquiring further into these epidemic models, for they have been used extensively in the diffusion literature.

The question is, to what extent is an epidemic approach a reasonable explanation of economic phenomenon? Davies (1979, pp. 10–11) suggests that it may be a reasonable model for fashion goods or a piece of gossip, but casts doubts on its relevance in general. To support this explanation of diffusion we need to ask what it is that is being spread like a disease by contact between individuals. The most logical argument is that, by meeting users, a non-user learns about the existence of the technological advance. It is information that is being spread in these models.[1] The user must then make a decision on whether to use the new technology. The factors determining this decision will enter into $g(t)$. As we shall see below, Mansfield's seminal work on diffusion uses this epidemic type of approach.

We proceed then to the study of diffusion in greater detail. We will take the logical sequence, starting with intra-firm diffusion and moving on to inter-firm diffusion, and will then consider economy-wide or inter-industry diffusion.

[1] Lekvall and Wahlbin (1973) distinguish between internal and external information sources to predict diffusion curves that are modified logistics.

Although these labels are orientated towards process innovations, we will also consider the consumer side in the appropriate places.

References

Chow, G.C. (1967), 'Technological Change and the Demand for Computers', *American Economic Review*, 57, 1117–30.

Davies, S. (1979), *The Diffusion of Process Innovations*, Cambridge University Press.

Friedman, A. (1974), 'Choosing Among S-Shaped Curves For Diffusion Processes', CIEBR Working Paper no. 45, University of Warwick, Coventry.

Lekvall, P. and Wahlbin, C. (1973), 'A Study of Some Assumptions Underlying Innovation Diffusion Functions', *Swedish Journal of Economics*, 75, 362–77.

Chapter 6

Intra-firm Diffusion

6.1 Introduction

We start this analysis by considering a firm that has already used a new technology for the first time, and proceed to analyse the time-path of use of the new technology from that point until the diffusion is complete. The new technology concerned could be a new input, a new form of work organization, or a new management style, but we present the analysis for the case of a new technology that is embodied in a new type of capital good. Two basic theoretical approaches are considered, the Mansfield model and a Bayesian learning model. Both consider that information and uncertainty are the keys to explaining why it is rational for firms not immediately to switch to the new technology. Empirical work is presented in the later part of the chapter.

6.2 Theory

6.2.1 The Mansfield model

The Mansfield (1968) model, using our own notation, defines S_i^* (independent of time) as the post-diffusion stock of the new capital good for firm i. S_i^* is taken as given and fixed and should thus be treated as a satiation concept. Let S_{it} be the stock of the new capital good in time t used by firm i. Define w_{it} as

$$w_{it} \equiv \frac{S_{it+1} - S_{it}}{S_i^* - S_{it}};$$

(6.1)

i.e., w_{it} is the addition to the stock of new technology in time t as a proportion of the gross additions still be be made. Mansfield then 'supposes that' w_{it} is a positive function of the expected profitability of a change in technology (Π_i, assumed constant over time), and a negative function of the risk involved in making the change. The latter is defined as U_{it} varying over time. He also supposes that w_{it} will be a positive function of a firm's liquidity (C_i) and a function (with ambiguous sign) of the size of the firm (I_i), both assumed constant over time. One may note at this stage that Mansfield does not present any real theoretical or *a priori* justification for the inclusion of these variables, or for the assumptions of constancy over time, except that 'there would [seem to] exist an important economic analogue to the classic psychological laws relating reaction time to the intensity of the stimulus' (Mansfield, 1968, p. 190). We thus have

$$w_{it} = f(\Pi_i, U_{it}, I_i, C_i, \dots).$$

(6.2)

The next step is to argue that the risk involved in changing technology will depend upon two factors: (1) the date when the new technology was first used by the firm, L_i (measured from an origin given by the date of first use by any firm), on the grounds that 'the longer the firm waited before beginning to use the new technology the more knowledge it had derived from other firm's experience and thus the less uncertainty it has with regard to profitability when it began to use the new technology'; and (2) it is also argued that U_{it} will also depend on how close the firm is to complete diffusion, i.e., on S_{it}/S_i^*. The argument is that L_i will yield a value for U_{it} at the date of first use, $\hat{U}_i$. We may then say that $U_{it} = \hat{U}_i (U_{it}/\hat{U}_i)$. Mansfield then *assumes* that $U_{it}/\hat{U}_i$ is inversely related to S_{it}/S_i^*. We thus have

$$U_{it} = J\left(L_i, \frac{S_{it}}{S_i^*}, \dots\right) \quad J_1 < 0, J_2 < 0. \tag{6.3}$$

Substituting into (6.1) we obtain

$$w_{it} = \frac{S_{it+1} - S_{it}}{S_i^* - S_{it}} = H\left(\Pi_i, L_i, I_i, C_i, \frac{S_{it}}{S_i^*}, \dots\right). \tag{6.4}$$

Writing (6.4) as a differential equation, we have

$$\frac{dS_{it}}{dt} = H(\ )(S_i^* - S_{it}). \tag{6.5}$$

If one could write $H(\ )$ as solely a linear multiple of S_{it}/S_i^*, say $g\, S_{it}/S_i^*$, then (6.5) would correspond exactly to the logistic curve discussed in the previous chapter. Mansfield assumes that

$$H(\ ) = \phi_i \frac{S_{it}}{S_i^*} \tag{6.6}$$

where $\phi_i = c_1 + c_2\Pi_i + c_3 L_i + c_4 I_i + c_5 C_i + \epsilon_i$. To do this he takes a Taylor series expansion of $H(\ )$, dropping third- and higher-order terms for all variables and also the quadratic term in S_{it}/S_i^*, and then assumes that

$$\lim_{t\to -\infty} S_i(t) = 0.$$

The effect however is to assume that H is of the form above, i.e., linear in S_{it}/S_i^*. We may now write

$$\frac{dS_{it}}{dt} = \phi_i S_{it}\left(1 - \frac{S_{it}}{S_i^*}\right) \tag{6.7}$$

which is the equation of the standard logistic curve with the speed of diffusion given by the linear combination ϕ_i, and has the solution

$$S_{it} = \frac{S_i^*}{1 + \exp\left(-\phi_i t - \eta\right)} \tag{6.8}$$

where

$$\eta = \log\left(\frac{S_{i0}}{S_{i0} - S_i^*}\right).$$

One might note that ϕ_i is defined as including an error term ϵ_i. We are not told how or why this appears, or why the error should appear here rather than else-where. The prediction of this model then is as follows.

1   The diffusion curve will be sigmoid, and the sigmoid curve will in fact be the logistic curve.
2   The level of use of the new technology in time t by a firm will be positively related to Π_i, L_i, C_i, and S_i^* with the sign of the effect of I_i uncertain.
3   The speed of diffusion is a linear function of Π_i, L_i, I_i, and C_i, and will vary across firms and technologies as these variables change across firms and technologies.

These results are neat and informative, and as we shall see below carry a certain amount of empirical verification. However, there is a problem with this model. As Davies (1979) states, 'To all intents this model is merely an application of the simple epidemic model,' the 'uninfected' part of the firm being the more likely to catch the disease (adopt), the more of the firm has been infected (has adopted); the infectiousness of the innovation is determined by its financial characteristics (profitability and cost). This epidemic model, as we argued in the previous chapter, is most logically justified, although Mansfield never presents it as such, when we consider that information is being transmitted epidemically and firms then decide on whether to use the new technology. This seems rather odd in an intra-firm study. Unless there are no formal internal information routes within the firm, there is no need to rely on an epidemic information-spreading mechanism.

One would tend to think that, if a firm had information, the whole firm would have access to that information so that the information-spreading part of the epidemic model is irrelevant (although this might not be so in a multi-plant or multi-divisional firm). This suggests that we must emphasize the technique choice aspects of Mansfield's model. Here we have a problem, for Mansfield 'supposes' the basis of his technique choice rule. The decision on use depends on risk and profitability, but why or how is not made clear. The driving force of Mansfield's approach is that risk will reduce with usage. By relating risk to S_{it}/S_i^*, Mansfield manages to derive his logistic curve. If risk is so related it is all right, but there is very little justification for it. Moreover, Mansfield's risk is risk as to the uncertainty attached to the profitability of the new technology. This uncertainty reduces over time, but the firm's estimate of expected profit-ability Π_i, does not change over time; i.e, the firm holds a constant estimate of Π_i, the uncertainty regarding which reduces over time. It must therefore be only learning that its estimate of Π_i is the right one. This seems a very strange story. In fact, as we have stated, the justification for assuming Π_i, C_i, and I_i are constant over time is not given to us. We might also note that S_i^* is taken as a satiation stock, and thus in this view of the diffusion process any factors that might influence the 'equilibrium stock' show themselves as influences on the diffusion speed. We can argue further on S^*. As we shall detail below in the empirical section, S^* is assumed to be such that eventually all the firm's output is produced on new technology. For this there is, again, no *a priori* reason.

What we wish to do at this point is to consider an alternative approach, which departs from the epidemic model but at the same time maintains the wholly reasonable basis of Mansfield's model that in some way the choice of technique is related to risk and profitability. We wish to do this, however, by making the technique choice decision specific. We are also in sympathy with Mansfield in believing that, as the firm uses a technology, it will learn more about it and thus might consider riskiness to be lower. However, we want to justify the nature of this learning mechanism. We also consider it desirable that, as the firm learns, it may well adjust its estimate of the expected profitability of a new technology; otherwise there does not seem any advantage in learning. We consider then an alternative approach.

6.2.2 A Bayesian learning model

In both Stoneman (1981) and in Lindner, Fischer, and Pardy (1979), a model of diffusion is constructed in which firms learn in a Bayesian way from their experience. We will concentrate on the version presented in Stoneman (1981). The starting point is to construct a model of the choice of technique in which the desired level of usage of new technology can be determined endogenously. Over time the firm is assumed to learn about the characteristics of the new technology and this affects the desired level of use. It is then allowed that there are costs of adjustment involved in changing the level of use of the new technology. The interaction of learning, the choice of technique procedure and adjustment costs generate a time-path for usage of the new technology that is the diffusion path.

To be more explicit, we consider that the firm faces a choice between using an old technology and a new technology. Define α_t as the proportion of the firm's (fixed) output produced on the new technology in time t. The time path of α_t is the diffusion path. To determine the desired level of α_t, defined as α_t^*, the firm is assumed to choose according to a means–variance approach to technique choice. Consider that at time t the two technologies, new (n) and old (o) have returns that are anticipated by the entrepreneur to be normally distributed as follows: .

$$\text{New: } N(\mu_{nt}, \sigma^2_{nt}) \qquad (6.9)$$

$$\text{Old: } N(\mu_{ot}, \sigma^2_{ot}) \qquad (6.10)$$

The entrepreneur considering the appropriate proportions in which to use the two technologies will face the additive distribution of returns, $N(\mu_t, \sigma^2_t)$, with mean and variance given by (6.11) and (6.12) below, where $\alpha_t : 1 - \alpha_t$ is the ratio of the proportion of (fixed) output produced on the new and old technologies, respectively, and σ_{not} is the covariance of the returns to the new and old technologies:

$$\mu_t = \alpha_t \mu_{nt} + (1 - \alpha_t) \mu_{ot} \tag{6.11}$$

$$\sigma_t^2 = \alpha_t^2 \sigma_{nt}^2 + (1 - \alpha_t)^2 \sigma_{ot}^2 + 2\alpha_t (1 - \alpha_t) \sigma_{not}. \tag{6.12}$$

The entrepreneur chooses α_t by maximizing a utility function (6.13) in which C is defined as the disutility of the adjustment costs that arise when α_t is changed:

$$U = H(\mu, \sigma^2) - C. \tag{6.13}$$

The firm is assumed to obtain positive utility from increased returns but dislikes risk; i.e., $\partial H/\partial \mu > 0$, $\partial H/\partial \sigma^2 < 0$. For simplicity it is assumed that $H(\mu, \sigma^2)$ can be written as (6.14), which is a convenient form with some pedigree (see Chipman, 1973):

$$H(\mu, \sigma^2) = a\mu - \frac{1}{2} b \sigma^2 \quad b > 0, a > 0. \tag{6.14}$$

Without loss of generality one can set $a \equiv 1$. Maximizing (6.13) subject to (6.11) and (6.12) determines the desired level of usage of the new technology. With α_t^* defined as that level of α_t that would be desired if there were no adjustment costs, α_t^* is given by

$$\alpha_t^* = \frac{\mu_{nt} - \mu_{ot} + b (\sigma_{ot}^2 - \sigma_{not})}{b (\sigma_{nt}^2 + \sigma_{ot}^2 - 2\sigma_{not})}. \tag{6.15}$$

When there are non-zero adjustment costs, maximization of (6.13) with respect to α_t yields

$$\frac{\partial C}{\partial \alpha_t} = \{\mu_{nt} - \mu_{ot} + b (\sigma_{ot}^2 - \sigma_{not})\} \left(\frac{\alpha_t^* - \alpha_t}{\alpha_t^*}\right). \tag{6.16}$$

In the investment theory literature (see Nickell, 1978) a growing amount of work considers the existence of adjustment costs and their effect on the level of investment. The choice of technique decision is very much an investment decision. It is specified that the disutility of adjustment is related to the change in α_t according to

$$C_t = \frac{\theta}{2} \frac{(\alpha_t - \alpha_{t-1})^2}{\alpha_{t-1}}. \tag{6.17}$$

This form has two desirable properties.

1   For a given level of past use (α_{t-1}), disutility increases at an increasing rate as the current level of use (α_t) increases. This property is consistent with the suppositions of the literature as discussed by Nickell.

2   For a given increase in use, the greater is the existing level of use (α_{t-1}),

the lower is the disutility of adjustment. This is logically appealing on the grounds that, the greater is α_{t-1}, the more experience the firm has and thus the less costly is any given adjustment likely to be.

Assuming that the firm acts in a myopic manner, such that given α_{t-1} it chooses α_t to maximize (6.13), one can derive (6.18) from (6.16) and (6.17):

$$\frac{d\alpha_t}{dt} \frac{1}{\alpha_t} = \frac{1}{\theta} \left\{ \mu_{nt} - \mu_{ot} + b\,(\sigma_{ot}^2 - \sigma_{not}) \left(\frac{\alpha_t^* - \alpha_t}{\alpha_t^*} \right) \right\} . \qquad (6.18)$$

The theory of technique choice and the existence of adjustment costs thus yield (6.18). The final step is to consider a learning mechanism. We have stated that $\mu_{ot}, \mu_{nt}, \sigma_{ot}^2, \sigma_{nt}^2$ all refer to anticipated returns. The simplifying assumption is made that the returns to the old technology are known, fixed, held with certainty, and have the distribution $N(\mu_o, \sigma_o^2)$. One may justify the assumption of certainty on the grounds that this technology has been used for some time and entrepreneurs know all about it. However, one could argue that the appearance of the new technology may lead to changes in the return to the old technology. For example, Harley (1973) argues that a new technology can further stimulate the development of the old technology and thus that the return to the old technology (μ_{ot}) may increase over time. By the same token, but in the reverse direction, one can argue that the appearance of a new technology may lead to a greater demand for primary inputs and a greater supply of outputs. The resultant price changes would obviously lead to reductions in the returns to the old technology. The analysis may thus be criticized on the grounds that these effects are ignored. However, at a number of stages it is clear that this framework is capable of development in many directions, and the varying of this assumption is one of them. Defining

$$\sigma_{not} \equiv \rho \sigma_{nt} \sigma_{ot} \qquad (6.19)$$

where ρ is the correlation between the returns to new and old technologies, it is further assumed that ρ remains constant over time. This implies that

$$\frac{d\sigma_{not}}{dt} = \rho \sigma_o \frac{d\sigma_{nt}}{dt} . \qquad (6.20)$$

With the new technology it is allowed that the entrepreneur learns over time and changes his anticipations. The behaviour assumed is that in time period zero the entrepreneur uses new and old technology in proportions $\hat{\alpha} : (1 - \hat{\alpha})$; as time proceeds, he monitors the performance of this initial batch of new technology and he adjusts his anticipations of the returns to the new technology in a Bayesian manner. These changes in anticipations will lead to changes in α_t^* and α_t.

This Bayesian learning process is formalized in the appendix to this chapter. The outcome of the learning process is that σ_{nt}^2, the anticipated variance of the returns to the new technology, will fall over time, but that μ_{nt}, the anticipated mean returns, may rise or fall depending on whether the initial estimate of this mean was greater or less than the true mean return. As σ_{nt}^2 and μ_{nt} change over time, then α_t^* will also change (see (6.15)). Substituting into (6.18) for μ_{nt}, σ_{nt}^2, σ_{not}, and α_t^* yields a differential equation for α_t in terms of

1   the true mean and variance of the returns to the new technology;
2   the mean and variance of returns to the old technology;
3   the firm's initial estimate of μ_{nt} and σ_{nt};
4   the firm's risk coefficient b; and
5   the correlation between the returns to the new and old technologies.

The value to which α_t will tend as $t \to \infty$, $\bar{\alpha}$, is given by

$$\bar{\alpha} = \frac{\bar{\mu}_n - \mu_o + b\,(\sigma_o^2 - \rho\bar{\sigma}_n\sigma_o)}{b(\bar{\sigma}_n^2 + \sigma_o^2 - 2\rho\bar{\sigma}_n\sigma_o)} \tag{6.21}$$

where $\bar{\mu}_n$ and $\bar{\sigma}_n^2$ are the true mean profitability of the new technology and the variance of its returns, respectively. In Stoneman (1981) it is shown under what conditions the path of α_t, as it tends towards $\bar{\alpha}$, will be sigmoid. We will not go into details here. However, what is worth noting is that the model is capable of explaining why some technologies do not diffuse. In this model, at the beginning of the diffusion process, the firm has an initial estimate of the mean profitability of the new technology, although this estimate is not held with certainty. As time proceeds and it learns, the firm becomes more certain of its estimate of the mean (thus uncertainty reduces stimulating use) but it may also revise downward its estimate of mean profitability. This can mean that its preferred level of use α_t^* may decline over time, and once α_t^* falls below α_t the diffusion is halted and perhaps reversed.

If, however, α_t^* remains above α_t for all $t < \infty$, then the diffusion proceeds. It is shown in Stoneman (1981) that the diffusion is faster the greater is the true profitability of the new technology, and that the diffusion path will depend on entrepreneurs' attitudes to risk, the initial uncertainty attached to the new technology, adjustment costs, and covariances of returns.

To explain what is happening in the model we argue as follows. At a moment in time the firm has an anticipation of what returns and risks are involved in the use of the two technologies. Being risk-averse, but gaining positive utility from profits, the firm chooses that mix of the two technologies which will maximize its utility. Because of costs of adjustment it moves to this level of usage at a rate that will not be infinitely fast. While using the new technology, the firm gains experience that will lead it to update its conceptions of the returns available from and risk involved in the use of the new technology, which in turn will lead it to modify its desired level of use, to which new level it will again adjust at a less than infinitely fast speed. As time proceeds the firm will move towards estimates of the return and risk of the new technology that are 'true', and will then establish its post-diffusion level of use. Note that learning is learning from experience, not infection, and the non-instantaneous diffusion is rational in that at all times the level of use maximizes the utility of the decision-maker, given his anticipations of returns and risks and the costs of adjustment.

We have here then a theory of intra-firm diffusion grounded solidly within economic theory rather than relying upon supposition or psychological laws. The means–variance theory of the choice of technique has the advantage over the Mansfield approach in that it is firmly based on rational maximizing behaviour. The Bayesian theory of learning replaces Mansfield's theory of learning (expression (6.3)) and is the closest one can get to an 'economic' theory of learning. Finally, allowing that there are adjustment costs in the economy that affect the level of usage the firm wishes to achieve in any time period, one is reflecting current approaches in the general investment theory literature. In some ways the predictions of the model (e.g., on profitability) are similar to those of the Mansfield model, but it appears to have a much richer menu of factors that can influence diffusion and is much more closely linked to those parts of economics to which economists are so attached, such as choice theory. Its main failing, however, is that it is somewhat intractable empirically.

6.3 Empirics

We now turn from these theoretical models to some empirical work. The first is Mansfield's (1968) own work on the spread of diesel locomotives in US railroads. As noted above, Mansfield's theory predicts that the diffusion of the new technology will follow a logistic curve with a speed of diffusion,

$$\phi_i = c_1 + c_2 \Pi_i + c_3 L_i + c_4 I_i + c_5 C_i + \epsilon_i.$$

The sample covered thirty railroads' dieselization experiences from 1925 to 1960. S_i^* is determined as the number of diesel locomotives required when all steam locomotives are replaced by diesels, and S_{it} is the number of diesel locomotives the firm owns in time t. One might note here that S_i^* is determined only because it is assumed that firms will replace all steam locomotives by diesels. Mansfield then has data on S_{it}/S_i^*, I_i, firm size measured as freight ton miles travelled in 1949, L_i, date of first use, measured from a 1941 base as the year the firm's diesel stocks reached 10 per cent of its total locomotive stock (which does not really relate closely to the theoretical definition), and C_i, liquidity, measured as the ratio of the firm's current assets to liabilities in the two years prior to and including the year when it 'began' to use diesel locomotives (i.e., 10 per cent of the stock was dieselized). The definition of date of first use (10 per cent dieselization) seems rather arbitrary. The estimate of profitability, Π_i, was obtained as an *ex post* measure of realized profitability for twenty-two firms by obtaining data on the reciprocal of the payout period from the firms themselves. From this sample of twenty-two a regression was run on Π_i against R_i, the number of steam locomotives replaced by one diesel. The coefficients were then used to predict Π_i for the remaining eight firms.

To give some idea of the time scales being discussed in these diffusion models, we present in Table 6.1 Mansfield's data on the length of the time interval between 10 per cent usage and 90 per cent usage. As can be seen, it differs considerably between firms (one of the facts to be explained), but the average time period is approximately nine years.

Table 6.1   *Dieselization in US railroads:*
time interval between 10 and 90 per cent usage

Time interval (years)	Number of firms
14 or more	3
11–13	7
8–10	11
5–7	3
3–4	6

Source: Mansfield (1968), p. 178.

As will be recalled, Mansfield's model requires that Π_i, I_i, and C_i remain constant over time for each firm. Mansfield admits that Π_i did not remain constant, and there is some evidence that firm sizes, although reasonably stable, did change (Mansfield's index of railroad traffic was 100 in 1929, 144 in 1946, and 136 in 1956). We do not really have any data on changes in firm's liquidity.

With the data available Mansfield first transforms the logistic expression into its linear form:

$$\log\left(\frac{S_{it}/S_i^*}{1 - S_{it}/S_i^*}\right) = \eta_i + \phi_i t. \tag{6.22}$$

For each firm he calculates the correlation between the transformed dependent variable and time, getting an average R^2 of 0.90. This, he suggests, means that the logistic function is a fair representation of the data. The sample sizes over which these correlations are measured is not really clear. It seems that the sample sizes vary from three to nineteen observations.

Second, Mansfield fits the equation above, using weighted least squares (the weights being those suggested by Berkson — see below) to generate an estimate of ϕ_i, $\hat{\phi}_i$ for each firm. The $\hat{\phi}_i$ is then regressed cross-sectionally for the thirty firms, using least squares to generate the estimate in (6.23), with standard errors in parentheses:

$$\hat{\phi}_i = -0.163 + 0.900\Pi_i + 0.048L_i - 0.0028I_i + 0.115C_i \quad R = 0.83.$$
$$\qquad\quad (0.492) \quad (0.008) \quad (0.0023) \quad (0.040) \tag{6.23}$$

In this case it is clear that, for the estimation of $\hat{\phi}_i$, the sample sizes do vary from three to nineteen with an average size of thirteen. As can be seen, the coefficients on Π_i, L_i, C_i are significant and of the expected sign. The sign of the coefficient on I_i was always considered ambiguous. In addition to these results Mansfield considers that ϕ_i may also be related to

1   the age distribution of the firm's steam locomotives when it first began to dieselize: the variable was a statistically insignificant determinant of $\dot{\phi}_i$;
2   the absolute size of investment required by firm i to fully dieselize, with an expected negative coefficient: again, the coefficient was statistically insignificant;
3   the average length of haul of the railroad, on the grounds that diesels were more profitable on long hauls: again, the coeffient was statistically insignificant;
4   the overall profitability of the firm: again, no significance was found.

All these variables were added individually to (6.23) above for estimation purposes.

Because this is the first example of estimating a logistic diffusion curve that we have come upon, it is worth at this stage investigating the appropriate econometric estimation procedures. Kmenta (1971) has a preliminary discussion of the estimation of a logistic. Consider

$$S_t = \frac{S^*}{1 + \exp(\eta + \phi t)}, \tag{6.24}$$

which we have specifically written as error-free. Kmenta then specifies two possible ways in which an error term, ϵ_t, may enter (6.24). These are:

$$S_t = \frac{S^*}{1 + \exp(\eta + \phi t)} + \epsilon_t \tag{6.25}$$

and

$$S_t = \frac{S^*}{1 + \exp(\eta + \phi t + \epsilon_t)}. \tag{6.26}$$

The importance of this distinction is that (6.26) may be transformed into

$$\log\left(\frac{S^*}{S_t} - 1\right) = \eta + \phi t + \epsilon_t \tag{6.27}$$

which, if ϵ_t satisfies the standard requirements for OLS regression, is an equation to which OLS regression can be applied. However, if we take (6.25), then the above transformation is not feasible and Kmenta suggests the use of maximum likelihood estimators.

In Mansfield's study ϕ carries the error term, ϵ_t; thus, if we define $\bar{\phi} = \phi - \epsilon_t$, we can write his version of the logistic as

$$\log\left(\frac{S^*}{S_t} - 1\right) = \eta + \bar{\phi} t + t\epsilon. \tag{6.28}$$

Although a simple time trend regression yields an efficient estimate for (6.28), this property is an asymptotic one, and thus to estimate (6.28) one would need to use weighted least squares regression with weights given by $1/t$; i.e., one applies OLS to

$$\frac{\log \{(S^*/S_t) - 1\}}{t} = \frac{\eta}{t} + \bar{\phi} + \epsilon_t \qquad (6.29)$$

and the intercept is the estimate of $\bar{\phi}$.

These differences regarding error terms are not just statistical niceties, however. The first form of the logistic (6.25) states that S_t follows a logistic time trend with white noise around that trend owing to random factors. The second form of the logistic (6.26) says that S_t follows a logistic curve without any random variation, but the intercept or slope of the logistic suffers from white noise owing to random factors. Given the differences in interpretation, it is clear that the researcher's choice of form ought to be justified *a priori* (even if in most cases the justification is on grounds of econometric tractability).

Berkson (1953) recommends that to estimate the logistic curve one should undertake a regression of

$$\text{logit } p_t = \log \left(\frac{S^*}{S_t} - 1\right) = \eta + \phi t \qquad (6.30)$$

using weighted least squares. The weights are to be $n_t\, p_t\, q_t$, where p_t is the observed proportion of the exposed population n_t affected by the 'disease' in time t and $q_t = 1 - p_t$.

This is the weighting scheme used by Mansfield. However, Berkson states that his estimation method applies to a statistical model in which 'it is assumed that the observation on p_t at t can be considered a random variable binomially distributed around the "true" p_t at t with the variance $\{p_t\, (1 - p_t)\}/n_t$'. I do not see that this should necessarily apply to the diffusion logistic in general or to Mansfield's model in particular.

This method of Berkson's is detailed for a logistic in which S^* is known (and may therefore be called a two-parameter logistic). If S^* is not known, then we have a three-parameter logistic. Oliver (1964) considers the appropriate methods for estimating such a logistic and comes to the conclusion that only an iterative nonlinear least-squares method is efficient. However, this result is dependent on the error term being homoscedastic. If it is not, then an appropriate transformation may be necessary. This is important when we are discussing the three-parameter logistic. Say, for example, S^* is being interpreted as an equilibrium concept and

$$S_t^* = aY_t + \epsilon_t$$

where ϵ_t is $N(0, \sigma^2)$ and there are no other errors in the regression. Then we have

$$S_t = \frac{aY_t + \epsilon_t}{1 + \exp(-\eta - \phi t)} \qquad (6.31)$$

which may be written as (6.25) but with an error term $\epsilon_t/\{1 + \exp(-\eta - \phi t)\}$, which does not satisfy the classical regression requirements. Again, the point is an obvious one — it is important to be clear about the error structure of the model.

The next question to consider is the two-stage estimation procedure used by Mansfield and others. The procedure is first to estimate ϕ_i, yielding $\hat{\phi}_i$, and to then regress $\hat{\phi}_i$ on the characteristics of firm i. Mansfield uses OLS techniques for this second-stage estimation, 'assuming that the errors in these estimates of $\hat{\phi}_i$ are uncorrelated with Π_i, L_i, I_i and C_i'. As Dixon (1980) points out, the dependent variables in these explanatory regressions are themselves estimates with different standard errors. He suggests that, by the use of an appropriate weighting mechanism (see Saxonhouse, 1977), this problem can be overcome. But weighted least squares regression does seem to the order of the day.

The points that derive from this discussion are as follows.

1   The specification of the stochastic structure of the model is crucial to the choice of estimating procedure.
2   As a general statement one should use weighted least squares to estimate the Mansfield model (probably with different weights than those used by Mansfield).
3   To estimate a three-parameter logistic where the error term has zero mean and is homoscedastic, one should use nonlinear least squares.
4   The two-stage procedure dictates the use of weighted least squares regression for the estimation of the coefficients in the second stage.

These comments cast considerable doubt upon the emphasis to be placed on Mansfield's results. To be fair, however, Mansfield's results were generated in a period when econometrics had not made many of its recent advances. We shall also see, when we come to consider Dixon's (1980) work on hybrid corn, that the use of more sophisticated techniques does not necessarily lead to much change in coefficient estimates.

Before closing this chapter we ought to mention Romeo's (1975) work on the intra-firm diffusion of numerically controlled (NC) machine tools. Romeo uses the basic Mansfield model to argue that

$$S_{it} = \frac{S_i^*}{1 + \exp(-\eta_i - \phi_i t)} \tag{6.32}$$

will hold, but then wishes to redefine the model in terms of a displacement process, so that the dependent variable becomes D_{it}, the percentage of firm i's new machine tool purchases that were NC machines at time t, and S_i^* becomes h_i, the maximum possible percentage of new machine tool purchases that were NC, yielding

$$D_{it} = \frac{h_i}{1 + \exp(-a_i - z_i t)} \tag{6.33}$$

where z_i is now the estimate of the diffusion speed. This transformation is achieved by defining

$$\delta_{it} = \frac{D_{it+1} - D_{it}}{h_i - D_{it}}$$

and arguing that δ_{it} plays the role of w_{it} in Mansfield's model. The model is thus just as suspect as Mansfield's original. To estimate the above curve, h_i is *assumed* equal to unity for all i, which is arbitrary and worrying. In fact, Romeo states that his results are insensitive to changes in h_i (he tries other values as well), which is of equal concern. With the assumption on h_i, the logistic is estimated 'by least squares'. From above we can judge the validity of this. In the second stage the estimates of z_i, $\hat{z}_i$ are related to firm characteristics. For some reason the explanatory factors are multiplicatively related in Romeo's model. This multiplicative relationship is not consistent with Mansfield's derivation (based on a Taylor's series expansion), which yields a linear form. However, using ordinary least squares (unweighted) Romeo generates the following results (*t*-statistics in parenthesis):

$$\log \hat{z}_i = -0.136 \log I_i + 0.024 \ \log \Pi_i + 0.0343 \log L_i - 1.371$$
$$\phantom{\log \hat{z}_i = } (-2.958) \qquad (0.266) \qquad (2.352) \qquad (-3.483)$$
$$\bar{R}^2 = 0.416$$

where I_i = firm size;
 Π_i = profitability of the innovation to the firm;
 L_i = the year the firm first began using numerical control (from 1950).
Other variables were included, without any significant result. One may note from these results that firm size and date of first use carry positive significant coefficients. Profitability has a positive insignificant coefficient.

These pieces of work by Mansfield and Romeo are two of a very small field of full intra-firm studies, and from the above it is clear that neither is totally satisfactory. Other work does exist but it is not in general very formal. Thus, for example, in Nabseth and Ray (1974) there is not only a fascinating study of NC machine tool diffusion, the detail of which shows up the validity of some of Romeo's assumptions, but also a study of intra-firm diffusion of special presses in paper-making. This study is carried out at both the plant and firm level. Nabseth and Ray confirm the role of profitability and also illustrate the sigmoid diffusion curve. They also study the basic oxygen process in steel-making and the continuous casting of steel at the intra-firm level. If anything general is to come from these studies, it is that the different production programmes, product mixes, and institutional characters of firms are key factors in the diffusion process.

6.4 Conclusions

We have considered two basic approaches to intra-firm diffusion. The first relies upon an epidemic approach, the second on learning by experience. Both approaches, however, give reductions in uncertainty a major role to play in the determination of the diffusion path. We have then presented and discussed some of the empirical work in this area and discussed both the results and the shortcomings. We now move to consider inter-firm diffusion.

Appendix: The Bayesian learning model

Consider that the new technology has true returns normally distributed, $N(\bar{\mu}_n, \bar{\sigma}_n^2)$, and constant over time, but although $\bar{\sigma}_n^2$ is known, $\bar{\mu}_n$ is not known with certainty. The entrepreneur holds a prior distribution on the mean returns to the new technology at time t that is normal $N(\mu_{nt}, Z_t)$. In time t the entrepreneur experiences a return Y_t; then, by Bayes theorem (see Lindley, 1965, p. 2, Theorem 1), the posterior density of the mean is $N(\mu_{nt+1}, Z_{t+1})$ where

$$\mu_{nt+1} = \frac{Y_t Z_t + \bar{\sigma}_n^2 \mu_{nt}}{Z_t + \bar{\sigma}_n^2} \tag{A6.1}$$

and

$$Z_{t+1} = \frac{\bar{\sigma}_n^2 Z_t}{Z_t + \bar{\sigma}_n^2}. \tag{A6.2}$$

Define that at time t, the uncertainty associated with the new technology, σ_{nt}^2, is equal to the inherent variance in its returns ($\bar{\sigma}_n^2$) plus the variance of the distribution of the mean (Z_t) (Johnson and Kotz, 1970, p. 87), yielding

$$\sigma_{nt}^2 = \bar{\sigma}_n^2 + Z_t. \tag{A6.3}$$

(The conclusions however would not be affected were we to define $\sigma_{nt}^2 \equiv Z_t$.) Solve (A6.2) to yield (A6.4) and substitute into (A6.3) to yield (A6.5):

$$Z_t = \frac{\hat{Z} \bar{\sigma}_n^2}{\bar{\sigma}_n^2 + t\hat{Z}} \tag{A6.4}$$

$$\sigma_{nt}^2 = \bar{\sigma}_n^2 + \frac{\hat{Z} \bar{\sigma}_n^2}{\bar{\sigma}_n^2 + t\hat{Z}} \tag{A6.5}$$

where $\hat{Z} = Z_t$ for $t = 0$. From (A6.1) one can obtain an expression for μ_{nt}. Equation (A6.1) contains a term in Y_t, the return actually realized by the new technology in time t. If we define $E(\mu_{nt})$ as the expected value of the entrepreneur's anticipated mean return, we may write (A6.6) with $E(Y_t) = \bar{\mu}_n$:

$$E(\mu_{nt}) = \frac{\bar{\mu}_n + (\bar{\sigma}_n^2/Z_{t-1})\{E(\mu_{nt-1})\}}{1 + (\bar{\sigma}_n^2/Z_{t-1})}. \tag{A6.6}$$

Solve (A6.6) as follows:

$$\mu_{nt} = \frac{t\hat{Z} \bar{\mu}_n + \hat{\mu}_n \bar{\sigma}_n^2}{t\hat{Z} + \bar{\sigma}_n^2} \tag{A6.7}$$

where μ_{nt} is now redefined as the expected anticipated mean return in time t.

From (A6.5) it is obvious that σ_{nt}^2 falls over time and, as $t \to \infty$, $\sigma_{nt}^2 \to \bar{\sigma}_n^2$. From (A6.7) we see that, as $t \to \infty$, $\mu_{nt} \to \bar{\mu}_n$, and that if $\bar{\mu}_n > \hat{\mu}_n$, μ_{nt} will increase over time, but if $\bar{\mu}_n < \hat{\mu}_n$, μ_{nt} will fall over time ($\hat{\mu}_n = \mu_{nt}$ for $t = 0$).

Redefine α_t^* to represent the expected value of desired use in time t when adjustment costs are zero. Then from (6.15) in the text, (A6.5), and (A6.7) one can obtain an expression for the time-path of α_t^*. Let α_t^* for $t = 0$ be defined as $\hat{\alpha}^*$ and for $t = \infty$ be defined as $\bar{\alpha}$, and one may then write $\bar{\alpha}$ as

$$\bar{\alpha} = \frac{\bar{\mu}_n - \mu_o + b\,(\sigma_o^2 - \rho\,\bar{\sigma}_n\,\sigma_o)}{b\,(\bar{\sigma}_n^2 + \sigma_o^2 - 2\rho\,\bar{\sigma}_n\,\sigma_o)}. \tag{A6.8}$$

Redefine α_t as the expected level of usage in time t and then the three parts of the model when put together imply (from (6.15), (6.19), and (6.20) in the text, and (A6.5) and (A6.7)) that α_t will be given by (A6.9), where α_t^* is given by (A6.10) and Π_t and Ω_t by (A6.11) and (A.612):

$$\frac{d\alpha_t}{dt}\frac{1}{\alpha_t} = \frac{\Omega_t}{\theta}\left(\frac{\alpha_t^* - \alpha_t}{\alpha_t^*}\right) \tag{A6.9}$$

$$\alpha_t^* = \frac{\Omega_t}{\Pi_t} \tag{A6.10}$$

$$\Omega_t = \frac{t\,\hat{Z}\,\bar{\mu}_n + \hat{\mu}_n\,\bar{\sigma}_n^2}{t\hat{Z} + \bar{\sigma}_n^2} - \mu_o + b\left\{\sigma_o^2 - \rho\,\sigma_o\left(\bar{\sigma}_n^2 + \frac{\hat{Z}\,\bar{\sigma}_n^2}{\bar{\sigma}_n^2 + t\hat{Z}}\right)^{1/2}\right\}$$

$$\Pi_t = b\left\{\bar{\sigma}_n^2 + \frac{\hat{Z}\,\bar{\sigma}_n^2}{\bar{\sigma}_n^2 + t\hat{Z}} + \sigma_o^2 - 2\rho\,\sigma_o\left(\bar{\sigma}_n^2 + \frac{\hat{Z}\,\bar{\sigma}_n^2}{\bar{\sigma}_n^2 + t\hat{Z}}\right)^{1/2}\right\}. \tag{A6.11}$$

$$\tag{A6.12}$$

Equation (A6.9) is the differential equation referred to in the text.

References

Berkson, J. (1953), 'A Statistically Precise and Relatively Simple Method of Estimating the Bio Assay with Quantal Response, Based on the Logistic Function', *Journal of the American Statistical Association*, 48, 565–99.

Chipman, J.S. (1973), 'The Ordering of Portfolios in Terms of Mean and Variance', *Review of Economic Studies*, 40, 167–90.

Davies, S. (1979), *The Diffusion of Process Innovations*, Cambridge University Press.

Dixon, R. (1980), 'Hybrid Corn Revisited', *Econometrica*, 48, 1451–62.

Harley, C.K. (1973), 'On the Persistence of Old Techniques: the Case of North American Wooden Shipbuilding', *Journal of Economic History*, 33, 372–98.

Johnson, N. and Kotz, S. (1970), *Continuous Univariate Distributions*, vol. 1, Houghton-Mifflin, Boston.

Kmenta, J. (1971), *Elements of Econometrics*, Macmillan, New York.

Lindley, D.V. (1965), *Introduction to Probability and Statistics*, Part 2, Cambridge University Press.

Lindner, R., Fischer, A. and Pardey, P. (1979), 'The Time to Adoption', *Economic Letters*, 2, 187–90.

Mansfield, E. (1968), *Industrial Research and Technological Innovation*, W.W. Norton, New York.

Nabseth, L. and Ray, G.F. (1974), *The Diffusion of New Industrial Processes: An International Study*, Cambridge University Press, London.

Nickell, S. (1978), *The Investment Decisions of Firms*, Cambridge University Press and Nisbet, Welwyn Garden City.

Oliver, F.R. (1964), 'Methods of Estimating the Logistic Growth Function', *Applied Statistics*, 13, 57–66.

Romeo, A.A. (1975), 'Interindustry and Interfirm Differences in the Rate of

Diffusion of an Innovation', *Review of Economics and Statistics*, 57, 311–19.

Saxonhouse, G.R. (1977), 'Regressions for Samples Having Different Character-istics', *Review of Economics and Statistics*, 59, 234–7.

Stoneman, P. (1981), 'Intra Firm Diffusion, Bayesian Learning and Profitability', *Economic Journal*, 91, 375–88.

Chapter 7

Intra-sectoral Diffusion

7.1 Introduction

The material in this chapter carries a number of labels. It is perhaps most commonly known as 'inter-firm diffusion', or 'intra-industry diffusion'; but as we can also apply much of the argument to product as well as process innovation, and because industries may not always define the appropriate boundaries, we prefer the general title of 'intra-sectoral diffusion'. Another common label is that we shall be looking at the rate of imitation. The starting point is that some economic agent in the sector under consideration has already adopted a new technology and we are interested in how it spreads across all potential users in the sector.

We will begin by looking at Griliches's (1957) work on the spread of hybrid corn in the USA. Griliches fits the basic logistic equation,

$$P_{it} = \frac{P_i^*}{1 + \exp(-\eta_i - \phi_i t)}, \tag{7.1}$$

to the spread of the use of hybrid corn in different US states, where P_{it} is the proportion of total corn acreage of a state i that is planted with hybrid seed at time t, and

$$P_i^* = \lim_{t \to \infty} P_{it},$$

i.e., the post-diffusion proportion of acreage planted. Griliches proceeds to find estimates of P_i^*, η_i and ϕ_i and relates them to economic variables.

In his original paper P_i^* was estimated by considering the familiar transformation

$$\log\left(\frac{P_{it}}{P_i^* - P_{it}}\right) = \eta_i + \phi_i t \tag{7.2}$$

and choosing P_i^* by visual inspection as that which led to the highest degree of linear association of equation (7.2) for each state. This procedure may be suspect. Given that we are not really provided with a justification of why the diffusion curve is logistic, it may be that any deviation of the curve from the logistic is being hidden by the choice of P_i^*.

Given P_i^*, (7.2) was estimated for thirty-one states using OLS procedures and generating R^2 in all cases above 0.89. Using the estimates $\hat{\eta}_i$, $\hat{\phi}_i$, and $\hat{P}_i^*$, Griliches then proceeds to investigate the role of profit-related variables in influencing $\hat{P}_i^*$, $\hat{\eta}_i$, and $\hat{\phi}_i$. (It should be noted that $\hat{\eta}_i$ is essentially a measure of date of first use.) Griliches contends that (1) 'the fraction of acreage ultimately planted by hybrid seed (P_i^*) depends upon expectations of profits to be realised from the change': (2) 'a large fraction of the variability between areas in the rate of acceptance of hybrid corn (ϕ_i) can be explained with help of these two profitability variables (pre-hybrid yield per acre and average corn acres per farm)'; and (3) 'area differences in date of first planting (η_i) can be explained in terms of differences in date of availability of hybrid corn seed. Innovators among the seed producers first entered those areas where the expected profits from the commerical production of hybrid corn seed were largest.'

Dixon (1980) has taken a number of steps to update Griliches's results.

1   He argues that P_i^* eventually reached 100 per cent in every state and thus that Griliches's estimates of ϕ_i are no longer correct.
2   In order to estimate ϕ_i one needs to use more sophisticated estimation procedures. Dixon uses weighted least squares and nonlinear least squares with the results illustrated in Table 7.1 (b_0 is Griliches's original estimate; b_1 and b_2 the estimates derived from the other estimating procedures). The difference that the weighting scheme makes to the estimates is slight (e.g., Spearman's rank correlation coefficient between b_0 and $b_1 = 0.76$, and $r_{b_0 b_1}^2 = 0.73$). The nonlinear least squares estimate also incorporated a first-order autoregressive scheme.
3   The logistic curve did not explain the pattern of diffusion very well and so a Gompertz curve was fitted, and the Gompertz curve was preferred in twenty-seven of the thirty-one states, suggesting that the logistic is an inappropriate summary device.
4   Using appropriate weighting schemes, second-stage regressions on profitability indicators were undertaken to explain the diffusion parameters. The non-linear estimates and Gompertz estimates appear to perform best in terms of significance, suggesting a positive and significant influence of profitability on diffusion speed.

In his reply, Griliches (1980) does not refute Dixon's findings; however he queries the relevance of his work, suggesting that the appropriate diffusion curve is that in which P_i^* should be an equilibrium and not a satiation concept. Essentially, it is argued that continued improvements in the new hybrid seeds led to changes in P_i^* over time, and giving P_i^* a time subscript one should estimate

$$P_{it} = \frac{P_{it}^*}{1 + \exp(-\eta_i - \phi_i t)}. \tag{7.3}$$

This early work of Griliches is important, for it suggests a number of questions that have to be answered, the most basic being why the diffusion curve is sigmoid. However, it also raises other questions. Griliches looks at the diffusion of one innovation within a sector (state) by undertaking a cross-section analysis of data on different sectors (states). We might ask ourselves whether different innovations will have different diffusion curves. We might also ask what characteristics of the sectors other than profitability are important; for example, is the degree of concentration of the product market in the sector important in the diffusion process? We also wish to know, within a sector, which economic units diffuse faster than others, and Griliches's work tells us little, in fact nothing, about these points.

Table 7.1   *Hybrid corn: estimates of the rate of acceptance*

State	Original logit[a] b_0	Updated and weighted logit b_1	(s.e.)	Nonlinear least squares b_2	(s.e.)	r.s.s.[d]
NY[b]	0.36	0.29	(0.014)	0.40	(0.043)	0.014
NJ	0.54	0.35	(0.045)	0.80	(0.166)	0.051
Pa	0.48	0.32	(0.024)	0.46	(0.052)	0.026
Ohio	0.69	0.74	(0.049)	0.75	(0.046)	0.006
Ind.	0.91	0.87	(0.058)	0.84	(0.052)	0.003
Ill.	0.71	0.77	(0.050)	0.80	(0.076)	0.005
Mich.	0.68	0.30	(0.032)	0.73	(0.085)	0.018
Wisc.	0.69	0.44	(0.037)	0.75	(0.048)	0.006
Minn.	0.79	0.48	(0.048)	0.84	(0.037)	0.002
Iowa	1.02	0.96	(0.026)	0.95[c]	(0.016)	0.001
Mo.	0.57	0.53	(0.035)	0.72	(0.078)	0.008
ND[b]	0.43	0.14	(0.014)	0.42	(0.073)	0.035
SD	0.42	0.19	(0.016)	0.50	(0.062)	0.022
Neb.	0.62	0.53	(0.023)	0.60	(0.035)	0.002
Kan.	0.45	0.35	(0.026)	0.56	(0.060)	0.040
Del.	0.47	0.46	(0.027)	0.51	(0.043)	0.014
Md.	0.55	0.44	(0.026)	0.47	(0.035)	0.014
Va.	0.50	0.28	(0.028)	0.70	(0.057)	0.011
W Va.[b]	0.39	0.23	(0.015)	0.37	(0.043)	0.014
NC[b]	0.35	0.28	(0.021)	0.30	(0.041)	0.023
SC[b]	0.43	0.24	(0.018)	0.24	(0.025)	0.020
Ga.[b]	0.50	0.36	(0.017)	0.43	(0.062)	0.016
Fla.[b]	0.38	0.34	(0.018)	0.38	(0.029)	0.026
Ky.	0.59	0.31	(0.027)	0.64	(0.036)	0.059
Tenn.[b]	0.34	0.24	(0.013)	0.24	(0.025)	0.014
Ala.[b]	0.51	0.36	(0.019)	0.47	(0.040)	0.006
Miss.[b]	0.36	0.22	(0.014)	0.38	(0.062)	0.025
Ark.[b]	0.41	0.25	(0.022)	0.37	(0.052)	0.016
La.[b]	0.45	0.22	(0.014)	0.29	(0.045)	0.015
Okla.[b]	0.56	0.26	(0.033)	0.59	(0.079)	0.025
Tex.[b]	0.55	0.24	(0.023)	0.47	(0.090)	0.031

[a] Data source Griliches (1957).
[b] Data collection discontinued prior to attainment of the 95 per cent level.
[c] After adjustment for heteroscedasticity.
[d] Residual sum of squares.

Source: Dixon (1980), p. 1454.

To get at them we are going to go back to the question of why diffusion curves are sigmoid; then, from the theories proposed to explain this, we can suggest answers to the above questions.

Before moving on, however, it is useful to give some idea of the time periods being discussed. In Table 7.2 we present data on the number of years taken for various innovations to reach a certain percentage of use in different countries.

Table 7.2 *Speeds of diffusion across innovations and countries*

Innovation	Percentage of output produced with new technology	Years from own date of innovation		
		UK	West Germany	Sweden
Special presses (Papermaking)	10	3	2	2
Tunnel kilns (Brickmaking)	10	n/a	2	8
Basic oxygen process (Steel)	20	5	3	9
Gibberalic acid (Brewing)	50	4	n/a	3
Continuous casting (Steel)	1	6	9	3
Shuttleless looms (Textiles)	2	6	6	9
Automatic transfer lines (Vehicles)	30	10	1	2

Source: Nabseth and Ray (1974), p. 17.

As can be seen, there are significant differences between countries and innovations but the overall impression is that complete diffusion will be a lengthy process.

7.2 Theory

The theory in this area splits neatly into three camps. The first, labelled by David (1969) — the psychological approach — posits that diffusion takes time because actors respond to stimuli only with a lag and these lags differ across the population. The second approach, which we will label the probit approach, posits that 'across the population of potential adopters the magnitude of at least one economic variate influencing the outcome of individual adoption decisions is not a constant but, instead, can be described by a more or less continuous frequency density function' (David, 1969). The third approach is a game theoretic approach. We consider the three approaches separately.

7.2.1 The Mansfield approach

We start by considering Mansfield's (1968) 'psychological approach'. This defines diffusion on the number of users of the new technology, rather than on the intensity of use, as is common in this area. This was discussed in Chapter 5.

Mansfield's contribution in this area follows almost exactly his theory of intra-firm diffusion, but the variables are appropriately re-specified. Mansfield presents his model in both deterministic and stochastic forms. We will discuss the deterministic form. Consider that, in an industry, N^* firms will use an innovation eventually (the satiation concept), but at time t only N_t firms are users. Define ζ_t as

$$\zeta_t = \frac{N_{t+1} - N_t}{N^* - N_t}. \tag{7.4}$$

Then ζ_t equals the proportion of non-users who become users in the time interval t to $t + 1$. Mansfield then hypothesizes that

$$\zeta_t = f\left(\frac{N_{t+1}}{N^*}, \Pi, K, \ldots\right) \tag{7.5}$$

where Π is the profitability of installing the innovation (assumed constant over time) and K is the size of the investment required to install it (also assumed constant over time). The justification for this relationship is exactly the same as in the intra-firm model. Following the same methods as the intra-firm study, Mansfield derives

$$N_t = \frac{N^*}{1 + \exp\left(-\eta - \phi t\right)} \tag{7.6}$$

where $\phi = b_1 + b_2 \Pi + b_3 K + \epsilon$ (ϵ is the error term); i.e., the diffusion curve is logistic. This basic model of Mansfield's is open to all the objections raised against its intra-firm version. However, within an industry, an information-spreading mechanism based on personal contact does have some appeal, more than at the intra-firm level. In fact, Nabseth and Ray (1974, p. 300) find that there are considerable lags in the information process.

Mansfield tests his model by investigating a cross-section data sample covering three innovations ($j = 1, 2, 3$) in four industries ($i = 1, 2, 3, 4$). He fits (7.7) to the data using the same weighting schemes as in his intra-firm study:

$$\log\left(\frac{N_{ijt}}{N^*_{ij} - N_{ijt}}\right) = \eta_{ij} + \phi_{ij}t. \tag{7.7}$$

In a second stage, Mansfield regresses the estimates of ϕ_{ij}, $\hat{\phi}_{ij}$ on Π_{ij} and K_{ij} using ordinary least squares (we have already cast doubts on this method). Mansfield's results suggest that Π_{ij} and K_{ij} have respectively positive and negative but significant effects on ϕ_{ij}. He also investigates the effect on ϕ_{ij} of the durability of previous equipment, the rate of growth of industry sales, a time trend (owing to better information channels), and the stage of the business cycle, with mixed results. It seems however that this latter variable is merely a proxy for arguing that Π_{ij} and K_{ij} may vary over time.

Mansfield's hypothesis then is that:

1   diffusion will follow a logistic curve;
2   technologies yielding higher expected profits with lower absolute capital requirements will diffuse fastest;
3   industries that can gain greater profit from an innovation will diffuse it faster; and
4   the durability of an industry's capital stock, its rate of growth of sales, and its stage in the business cycle will affect the diffusion speed.

Thus diffusion will vary both across industries and technologies.

Romeo (1977) has applied the Mansfield model to the spread of numerically controlled machine tools. Apart from representing ϕ_i by a multiplicative function (which is inconsistent with Mansfield's model), his work is very much a rerun of Mansfield's own. He finds that differences in diffusion speeds are explainable in part by certain industry characteristics, the two most interesting being the two variables measuring degree of competitiveness – the number of firms (N) and the variance of the log of firm size (σ_x^2). He finds a positive coefficient on the former and a negative coefficient on the latter. This result suggests that competitive pressures lead to higher rates of diffusion. One may illustrate this by arguing that, if firm size is lognormally distributed and we measure the degree of concentration by the Herfindahl index H, then concentration is related to σ_x^2 and N:

$$H = \frac{\exp \sigma_x^2}{N}. \tag{7.8}$$

Thus, if diffusion is positively related to N and negatively related to σ_x^2, when H increases the diffusion speed falls. This result will be discussed again later in the context of Davies's work. The work of Romeo is subject to the usual criticisms applied to the Mansfield-type analysis. It should also be pointed out that in certain of his industries he has very little data from which to estimate ϕ_i. For example, in large steam turbines he estimates a two-parameter logistic from *two* observations (not surprisingly generating an $\bar{R}^2 = 1.000$, but most surprisingly not estimating an infinite t-statistic). Three of his other industries have four, six, and seven observations respectively. No allowance for these small samples is made in the estimation procedures at the second stage.

The Mansfield model is 'psychological' in the sense that the speed of response is assumed to be related to the size of the stimulus. It is also epidemic in the sense of learning by infection. The firm if assumed to adopt the new technology if profitability is high enough or uncertainty sufficiently low. Uncertainty is reduced in proportion to the number of firms already using the new technology. The rational for less than instantaneous diffusion is thus that uncertainty prevents this, and only as use extends will uncertainty be reduced. In the previous chapter we subjected such an approach to considerable criticism.

Given that in the Mansfield model one is looking at diffusion by considering the proportion of firms using the new technology, it is not surprising that he should also then proceed to look at differences between firms within a sector that will provide information on why some firms are early adopters and some late adopters. In fact, if one can explain the date of adoption by individual firms, then by aggregation one should have the inter-firm of intra-sectoral diffusion curve. Mansfield (1968) proposed that expression (7.9) will hold if

d_{ij} = number of years the jth firm waits before beginning to use innovation i;

X_{ij} = size of the jth firm adopting the ith innovation;

Π_{ij} = measure of profitability of the jth firm's investment in innovation i; and

ϵ_{ij} = random error term.

$$d_{ij} = \theta_i \Pi_{ij}^{a_{i2}} X_{ij}^{a_3} \epsilon_{ij}. \tag{7.9}$$

In (7.9) θ_i is a scale factor differing across innovations and $a_{i2} < 0, a_3 < 0$. Later further terms are added. Estimates are significant and generally of the expected sign.

This approach is getting so close to the probit approach that we will not take it any further here. However, it is worth noting that his study of d_{ij} leads obviously through an aggregation procedure to a sector-wide diffusion curve even though its underlying principles do not seem to be at all consistent with Mansfield's earlier epidemic model. There seems to be no connection at all with information-gathering or uncertainty. It is in fact an approach that is very divorced from that model.

The psychological approach of which Mansfield's former approach is an example is not usually displayed as formally as in Mansfield's work. It has often been the reserve of sociologists and psychologists and as such is peripheral to our main line of argument. The literature is surveyed in Rogers (1962) but is neatly summarized by Fig. 7.1 from Norris and Vaizey (1973). They argue that:

1 innovators are venturesome and eager to try out new ideas; they have cosmopolitan relationships and sufficient financial resources to bear any losses that may occur from time to time;
2 early adopters are an integrated part of their social community and tend to act as opinion leaders;
3 the early majority tend to be deliberate in their actions;
4 the late majority are always initially sceptical about an innovation;
5 the laggards are highly localized and traditional.

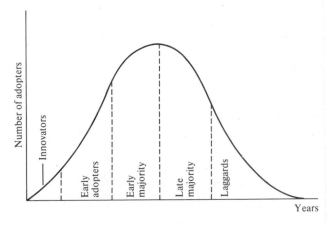

Fig. 7.1

Unfortunately, this really tells us little more than that early adopters adopt early and late adopters adopt late.

7.2.2 *The probit approach*

We will turn now then to the probit approach to the diffusion of innovations. This approach concentrates on the characteristics of the individuals in a sector and as such not only is suitable for generating a diffusion curve but also will give indications of which firms (individuals) will be early adopters and which late.

The principle of these probit models is stated succinctly by P.A. David:

Whenever or wherever some stimulus variate takes on a value exceeding a critical level, the subject of the stimulation responds by instantly determining to adopt the innovation in question. The reason such decisions are not arrived at simultaneously by the entire population of potential adopters lies in the fact that at any given point of time either the 'stimulus variate' or the 'critical level' required to elict an adoption is described by a distribution of values, and not a unique value appropriate to all members of the population. Hence, at any point in time following the advent of an innovation, the critical response level has been surpassed only in the cases of some among the whole population of potential adopters. Through some exogenous or endogenous process, however, the relative position of stimulus variate and critical response level are altered as time passes, bringing a growing proportion of the population across the 'threshold' into the group of actual users of the innovation. (David, 1969)

Let X be the stimulus variate with a relative density function $f(X)$ and $\bar{X}_t$ represent the critical value of the stimulus variate in time t. Then, in a non-stochastic version of the model, the proportion of the population who are adopters by time t is given as

$$\int_{\bar{X}_t}^{\infty} f(X_i)\, dX_i.$$

It should be clear that only once one has specified $f(X)$ and $\bar{X}_t$ does this model really have operational significance.

David suggests that prior to his contribution there were essentially four classes of models that fitted within this framework:

1   the entrepreneurial inertia model: in this X is defined as the return necessary to persuade an entrepreneur to change technology;
2   the information cost model: in this case the return necessary to induce change is the same for each entrepreneur, but it is considered that entrepreneurs face different search costs;
3   fixed capital replacement model: here, entrepreneurs differ in terms of the gain to be made from new technology, for they own capital equipments of different productive capacities;
4   the vintage model approach, which is really a version of class 3; we will be considering this in the next chapter.

We will not pursue all these models in the present chapter. We will instead discuss three variants of the probit approach. The first will be David's firm size model and Davies's (1979) independently produced variant, the second will be a learning model, and the third will be an application of the approach to the diffusion of a product innovation based on income differences.

David's model defines firm size as the critical variable, and proceeds to discuss how firm size is distributed within an industry and how the threshold value of firm size is determined. It is argued that firm size is lognormally distributed. To justify the choice of firm size, and to determine the critical value of firm size, David argues as follows.

Consider a capital-embodied process innovation that involves fixed costs above and variable costs below those of the replaced technique. Define C as the purchase cost of the new equipment and R as its imputed rental rate (defined as $r(1 - e^{-rd})^{-1}$ where r is the interest rate and d the expected service life of the equipment). Define $c = CR$. For simplicity let C be zero for the replaced technique. Let the new technique save labour input relative to the old technique such that, for each unit of output produced, the labour saving is L_s. Let w be the wage rate. As should be obvious, there will be some level of scale (output) at which the labour savings will compensate for the increased capital cost. This defines the critical value of firm size $\bar{X}$.

We may define $\bar{X}$ as that value of X where

$$\bar{X}wL_s = CR; \tag{7.10}$$

i.e.,[1]

$$\bar{X} = \frac{1}{L_s}\frac{CR}{w} \equiv \frac{1}{\omega}\frac{1}{L_s}. \tag{7.11}$$

To generate a diffusion path we need either $\bar{X}$ or the distribution of X_i to change over time. For the simplest case we will assume that $f(X_i)$ remains constant over time. David then argues that $\bar{X}$ will change over time as wages rise relative to capital costs.

He argues that

$$\frac{\partial \bar{X}}{dt} = -\bar{X}_t\frac{d\omega}{dt}\frac{1}{\omega} \tag{7.12}$$

and that $(d\omega/dt)(1/\omega) = \lambda$, a constant independent of time; i.e., the relative factor prices ω follow an exponential time trend.

Thus, defining D_t as the proportion of the population using the new technology in time t, we have

$$D_t \equiv \Pr\ (X_i \geqslant \bar{X}_t) \ = \int_{\bar{X}_t}^{\infty} f(X_i)dX_i; \tag{7.13}$$

thus

---

[1] Defining $\omega = w/CR$, i.e., relative prices of inputs.

$$\frac{dD_t}{d\bar{X}_t} = -f(\bar{X}_t) \tag{7.14}$$

and

$$\frac{dD}{dt} = \frac{dD}{d\bar{X}_t} \frac{d\bar{X}_t}{dt} = f(\bar{X}_t) \lambda \bar{X}_t. \tag{7.15}$$

David then goes on to show that D, as thus defined, given $f(X_i)$ is lognormal and λ is constant, will trace out the standard cumulative normal curve when plotted against a positive linear transformation of the time variable, which curve is of course sigmoid.

This diffusion curve is defined on the proportion of firms using the new technology. David attempts to define a similar diffusion curve for the percentage of industry output produced on the new technology. To do so, however, he assumes that there is no intra-firm diffusion problem, and thus we will not explore the model further.

Davies's (1979) model is similar but is some ways richer. He argues that because of uncertainty firms make decisions in a behavioural manner; specifically, a firm i will use a new technology if the expected pay-off period from its use, ER_{it}, $\leqslant \bar{R}_{it}$, some critical pay-off period. The expected pay-off period is then a function of firm size X_{it} and other firm characteristics (Y_{ijt}) and time.

Specifically, it is argued that

$$ER_{it} = \theta_{1t} X_{it}^{\beta_1} \epsilon_{1it}, \qquad\qquad \beta_1 \gtrless 0 \tag{7.16}$$

where

$$\epsilon_{1it} = \prod_{j=1}^{r} Y_{ijt}^{y(j)} > 0$$

and

$$\theta_{1t} > 0$$

$$\frac{d\theta_{1t}}{dt} \frac{1}{\theta_{1t}} < 0 \text{ for all } t.$$

$\bar{R}_{it}$ is related to firm size and other firm characteristics (Z_{ijt}) by

$$\bar{R}_{it} = \theta_{2t} X_{it}^{\beta_2} \epsilon_{2it} \qquad\qquad \beta_2 \gtrless 0 \tag{7.17}$$

where

$$\epsilon_{2it} = \prod_{j=i}^{u} Z_{ijt}^{\gamma(j)} > 0$$

and

$$\theta_{2t} > 0$$

$$\frac{d\theta_{2t}}{dt} \frac{1}{\theta_{2t}} > 0 \text{ for all } t.$$

In other words, both the expected pay-off period and the critical period vary with firm size.

Following from the above we may state that $ER_{it} \leqslant \bar{R}_{it}$ if

$$\frac{\theta_{1t} X_{it}^{\beta_1} \epsilon_{1it}}{\theta_{2t} X_{it}^{\beta_2} \epsilon_{2it}} \leqslant 1. \tag{7.18}$$

Define $\theta_t = \theta_{1t}/\theta_{2t}$, and $\epsilon_{it} = \epsilon_{1it}/\epsilon_{2it}$ and $\beta = \beta_1 - \beta_2$; then (7.18) may be rewritten as

$$\theta_t X_{it}^{\beta} \epsilon_{it} \leqslant 1. \tag{7.19}$$

We may thus state that a firm will be using the new technology in a deterministic model if

$$X_{it}^{\beta} \leqslant (\theta_t \epsilon_{it})^{-1}. \tag{7.20}$$

We then define critical firm size as

$$\bar{X}_{it} = (\theta_t \epsilon_{it})^{-1/\beta}. \tag{7.21}$$

Davies now argues that actual firm sizes are lognormally distributed. We thus have a distribution of firm sizes, and a definition of the critical threshold of firm size, so once we specify how critical firm size varies over time we have a theory of diffusion. For the moment, assume ϵ_{it} is constant and the same for all firms. Also, for the sake of the argument assume $\beta > 0$. Then the critical value of firm size will change as θ_t changes. Davies considers two types of innovations:

1 Group A: for which $\theta_t = \alpha t^{\psi}$ $\alpha > 0$
2 Group B: for which $\theta_t = \alpha e^{\psi t}$ $0 < \psi < 1$.

He then shows that the proportion of firms using the new technology will follow a cumulative lognormal time path for Group A innovations and a cumulative normal time path for Group B innovations.

Group A innovations are relatively cheap and simple innovations; they experience major improvements in their early years but thereafter there are fewer improvements. Group B innovations are expensive and technically complex, and experience improvement for many years after their first commercial introduction.

Through ϵ_{it} differences between firms can be allowed to influence the diffusion pattern. The assumption that $\beta < 0$ is shown not really to affect the argument. Davies also allows the model to be made more complex by allowing the firm size distribution to vary and shift over time.

The David and Davies models, because of their basic structure, both predict that the probability of a firm adopting an innovation in time t will be a linear function of the log of firm size. Davies tests this hypothesis and finds strong evidence to support it on a data sample of twenty-two innovations. He also finds that typically $\beta > 0$, which is consistent with the David model.

To test the model in more detail Davies proceeds to the estimation of a cumulative lognormal curve for Group A innovations and a cumulative normal curve for Group B innovations. Consider Group A innovations. Define D_t as the percentage of the population that has adopted the new technology in time t. Let firm sizes be lognormally distributed as in (7.22) and ϵ_{it} be lognormally distributed as in (7.23):

$$X \sim \Lambda(\mu_X, \sigma_X^2) \qquad (7.22)$$

$$\log \epsilon_{it} \sim N(0, \sigma^2). \qquad (7.23)$$

Then

$$D_t = N(\log t \mid \mu_D, \sigma_D^2) \qquad (7.24)$$

where

$$\mu_D = -(\log \alpha + \beta\mu_X)/\psi$$

and

$$\sigma_D^2 = (\sigma^2 + \beta^2 \sigma_X^2)/\psi^2$$

With the transform,

$$D_t = N(Z_t \mid 0,1), \qquad (7.25)$$

we may write

$$Z_t = \frac{\log t - \mu_D}{\sigma_D}. \qquad (7.26)$$

We estimate the coefficients in (7.26) from the following regression:

$$Z_t = a_1 + b_1 \log t, \qquad (7.27)$$

in which $a_1 = -\mu_D/\sigma_D$ and $b_1 = 1/\sigma_D$. In the case of Group B innovations, one undertakes a regression of

$$Z_t = a_2 + b_2 t. \qquad (7.28)$$

Z_t is in fact the normal equivalent deviate (normit) of D_t read off from the standard normal tables. D_t is measured as the proportion of the population using the innovation in time t; b_1 and b_2 are our estimates of the diffusion speed.

Fitting the appropriate curve to the appropriate innovations yields satisfactory results; fitting the 'wrong' cumulative curve gives less satisfactory results. In Davies's testing cyclical factors are also introduced, but 'there is only limited evidence that cyclical factors have much influence on the shape of the typical diffusion curve' (Davies, 1979, p. 108).

In Davies's second stage b_1 and b_2 are used as measures of the diffusion speed and Davies tests hypotheses on their determinants. Without going into detail on his estimation procedures (which as far as I can tell suffer only from ignoring that $\hat{b}_1$ and $\hat{b}_2$ (the estimates of b_1 and b_2) have a sampling variance), Davies finds that the diffusion speed is signficantly negatively affected by (1) the number of firms in the industry, (2) the typical pay-off period associated with adoption and (3) the variance of the log of firm size, and is significantly positively affected by (4) the labour intensity of the industry and (5) the rate of growth of the industry. Davies's results on the sign of the coefficients on N and σ_X^2 should be compared with those of Romeo (1977) discussed above.

Davies's results are that the diffusion speed is negatively related to both σ_X^2 and N, and therefore there is no way of saying whether diffusion speed is positively or negatively related to concentration until the appropriate weights are specified. Romeo's results have a positive coefficient on σ_X^2.

This contrast between Davies's and Romeo's results is not really surprising, for there does not seem to be any good *a priori* reasoning to suggest that higher levels of concentration should encourage faster diffusion. Scherer (1980) presents a simple theoretical apparatus, which he suggests indicates that competition is conducive to fast diffusion. In Fig. 7.2 we reproduce the Scherer diagram.

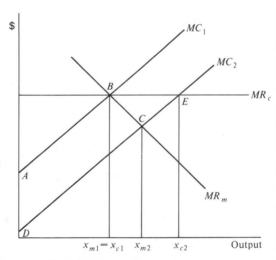

Fig. 7.2 Diffusion and market structure

MC_1 and MC_2 are short-run marginal cost functions using the old and new technologies respectively. They exclude investment costs and are the same for the monopolist and competitive firms. MR_c represents the marginal revenue curve for the competitive firms and MR_m the marginal revenue curve for the monopolist. Output is determined where $MC = MR$, and the diagram is drawn such that pre-innovation output is the same, whatever the market structure, at $x_{m1} = x_{c1}$. The post-innovation output for a monopolist is determined where $MR_m = MC_2$, i.e., at x_{m2}, yielding a profit increment of *ABCD* per period. The competitive post-innovation output is determined at x_{c2}, with a profit increase of *ABED* per unit of time, which exceeds the increment enjoyed by the monopolist by an amount *BCE* per period, *ceteris paribus*. Thus, Scherer argues, 'we should expect competitive producers to adopt new cost-reducing processes more rapidly than firms with market power, other things being equal.'

The problem with this analysis is that it is wrong; errors exist on three counts.

1   It is clear from the presentation that Scherer is comparing a monopolist with a competitive *industry*, not a competitive firm. Given this, the competitive industry cannot support a price equal to MR_c after innovation. As the competitive industry increases its output price must fall, and thus the profit increment to be achieved cannot be *BCE*.

2   In Scherer's analysis the monopoly firm is considered to be switching all its production to the new technology immediately and the competitive industry similarly to be' making a complete switch-over. All the analysis above shows this to be wrong. What we really wish to know is how the returns to an extension to use will vary with market structure for different levels of usage.

3   In this analysis the *ceteris paribus* assumption has been taken to an extreme. The pre-innovation output is taken as independent of market structure ($x_{c1} = x_{m1}$). This must imply that the competitive and monopolistic industries face different demand curves. This is not acceptable.

In total, therefore, the question of the influence of market structure is not settled at either the theoretical or the empirical level. The other factor usually considered in the industrial organization literature as an influence on technological change is the influence of firm size. In Mansfield's empirical results firm size does not carry a significant coefficient. In the David and Davies models, however, firm size is a crucial (by assumption) factor. The whole basis of their models is that large firms will use first and then diffusion will proceed down the firm size distribution. The fact that the Davies model works empirically lends support to the firm size hypothesis. Oster (1982), however, finds some evidence that in the US steel industry large firms were somewhat slower diffusers.

Returning to the David and Davies model, it is clear that a key factor in both models is that the returns to an innovation change over time. At each moment in time firms hold the stock of new technology that is appropriate to their estimates of the returns to that new technology (although these returns vary across firms and innovations). The returns then change over time. In a sense the equilibrium concept of the post-diffusion stock is held in each time period, but the equilibrium stock changes over time. The approach is thus very different from Mansfield's, where the *satiation* stock is *approached* over time, and returns are constant over time. In the David/Davies approach the reason why all firms do not use the new technology instantaneously is that it is not profitable for all firms to do so. Only as factor prices change will the proportion of the firms for whom the innovation is profitable change, and only then will the number of users change.

The model can be subjected to a number of criticisms. First, the diffusion process is driven by exogenous changes in factor prices and no real attempt is made to endogenize such price movements. This problem is tackled in Chapter 9. Second, learning is given a very minor role, if any role, in these models. Basically they assume complete information (although Davies does allow some internal learning to lead to changes in $\bar{R}$). This problem is considered below. Third, diffusion is defined in these models by the time-path of the proportion of the population of firms who have adopted in time t. It is by no means clear that the limiting value of this variable is unity, except by the appropriate definition of the population as those firms that finally adopt. If the limiting value is not unity, then one may be introducing some bias into the estimates of the diffusion speed by assuming, as is done in these models, that it is. Consider for example that we measure D_t, the proportion of firms using the new technology in time t, by N_t/V_t where V_t is the whole population and N_t is the current number of users; but in fact, as $t \to \infty$, N_t tends to $N_t^* < V_t$. Surely, then, we ought to measure diffusion against

$$\frac{N_t}{N_t^*} = \frac{N_t}{V_t} \frac{V_t}{N_t^*},$$

and not N_t/V_t.

Finally, before leaving these models of David and Davies we must say something about expectations. Davies has incorporated in his model expectations of returns from the use of new technology, but he is not really explicit as to how these are formed. Rosenberg (1976) puts great stress on the role that expectations of future technological advances might play in the diffusion process. We can illustrate his argument in the context of the David model. In the David model the critical value of firm size is determined by equation (7.10). One way to derive this is to consider the technology as described by David but to allow the date of adoption to be determined by a profit-maximizing firm. Define the date of adoption as t^*; then such a firm will choose t^* so as to maximize (assuming zero depreciation)

$$V_i = Ce^{-rt} + \int_t^\infty wL^s X_i e^{-r\tau}\, d\tau,$$

yielding as a first-order maximization condition

$$wL^s X_i = -\dot{C}^e + rC$$

where $\dot{C}^e$ is the expected change in the price of the capital good. We may now see that the rate at which $\bar{X}$ will fall, and thus the diffusion proceed, will depend on changes in C and on changes in expected capital gains and losses. If we argue, as is appropriate, that $\dot{C}^e$ reflects expectations of future technical advances, then we can agree that these will impinge on the diffusion process. Thus, if the price of a machine is expected to fall it may be worth postponing adoption until it does fall. David assumes $\dot{C}^e = 0$. However, to model the formation of expectations we really need to consider the industry supplying the new capital goods. In

Chapter 9 we do consider models where the supply side is incorporated, but even there we do not find it possible to proceed far while making $\dot{C}^e$ endogenous. This is a major task of future research.

What is more practical at this stage is to re-incorporate learning into the diffusion process. In Stoneman (1980), learning is again brought to the fore, and in this approach there is one further advance, which is that inter-firm and intra-firm diffusion are brought together.

In this approach the critical value of the characteristic determining the use of new technology is determined by the rational maximizing behaviour of individual entrepreneurs, rather than being generated in an *ad hoc* manner. The model involves uncertainty, as is appropriate in a world with technological change, and Bayesian learning, and it is capable of generating the often observed sigmoid diffusion curves. The approach is based on the intra-firm model detailed in Chapter 6. In Jensen (1979) a similar view of the diffusion process is developed, although in that paper risk and uncertainty do not affect technique choice.

Following Chapter 6, we consider an individual entrepreneur who at time t has two technologies available to him, the new and the old. These have perceived returns in time t normally distributed as $N(\mu_{nt}, \sigma_{nt}^2)$ and $N(\mu_{ot}, \sigma_{ot}^2)$ respectively, with the covariance of returns $\rho\sigma_{nt}\,\sigma_{ot}$ (we assume $\rho > -1$). At time t, the new and old technologies are used in the proportions $\alpha_t : 1 - \alpha_t$, α_t being chosen to maximize the quadratic utility function (7.29), in which μ_t and σ_t^2 are the parameters of the joint distribution of returns when the technologies are used in the proportions $\alpha_t : 1 - \alpha_t$:

$$U = \mu_t - \tfrac{1}{2}b\sigma_t^2. \tag{7.29}$$

From standard portfolio theory, we define the optimal value of α_t (α_t^*) as follows:

$$\alpha_t^* = \frac{(1/b)\,(\mu_{nt} - \mu_{ot}) + \sigma_{ot}^2 - \rho\sigma_{nt}\,\sigma_{ot}}{\sigma_{nt}^2 + \sigma_{ot}^2 - 2\rho\sigma_{nt}\,\sigma_{ot}}. \tag{7.30}$$

There is also the usual restriction that $0 \leqslant \alpha_t^* \leqslant 1$.

The means and variances of the returns are perceived returns about which the entrepreneur learns. We will assume that from past experience the entrepreneur knows with certainty that $\mu_{ot} = \mu_o$ and $\sigma_{ot}^2 = \sigma_o^2$. We further assume for simplicity that $\sigma_o^2 = 0$. However, the new technology, being new, has uncertain characteristics. We argue that the entrepreneur learns about the characteristics of the new technology in a Bayesian manner. Explicitly, the true returns to the new technology are assumed to be $N(\bar{\mu}_n, \bar{\sigma}_n^2)$. For simplicity, $\bar{\sigma}_n^2$ is assumed to be known but $\bar{\mu}_n$ is not known with certainty. At time t the entrepreneur holds a prior distribution on the difference in mean profitability to be gained from using the new rather than the old technology, that is, $N(\mu_{nt} - \mu_o, z_t)$. The entrepreneur observes in time t the change in returns that he actually achieved using the new technology, or would have achieved had he used the new technology, and uses this to update his prior.

Defining $\bar{z}$ as z_t for $t = 0$ and $\bar{\mu}$ for μ_{nt}, $t = 0$, then standard Bayesian results

(Lindley, 1965, p. 2) suggests

$$z_t = \frac{\tilde{z}\bar{\sigma}_n^2}{\bar{\sigma}_n^2 + t\tilde{z}}.$$ (7.31)

From Chapter 6, redefining $\mu_{nt} - \mu_o$ as the expected value of $\mu_{nt} - \mu_o$, one may also write

$$\mu_{nt} - \mu_o = \frac{t\tilde{z}\,(\bar{\mu}_n - \mu_o) + (\bar{\mu} - \mu_o)\bar{\sigma}_n^2}{t\tilde{z} + \bar{\sigma}_n^2}.$$ (7.32)

Realizing that $\sigma_{nt}^2 = \bar{\sigma}_n^2 + z_t$, and redefining α_t^* as the expected value of α_t^*, one may now write (7.30) as

$$\alpha_t^* = \frac{(1/b)\,\{t\tilde{z}\,(\bar{\mu}_n - \mu_o) + \bar{\sigma}_n^2\,(\bar{\mu} - \mu_o)\}}{\tilde{z}\bar{\sigma}_n^2 + \bar{\sigma}_n^2\,(\bar{\sigma}_n^2 + t\tilde{z})}.$$ (7.33)

Define a firm as using the new technology if the proportion of its output produced on the new technology is greater than some arbitrary value k (often taken as 5 per cent). Allow that in each time period, $\alpha_t = \alpha_t^*$, and that for the firm, $\bar{\mu}_n - \mu_o - bk\,\bar{\sigma}_n^2 > 0$ (i.e., that eventually it will wish to be a user); then from (7.33) a firm will be expected to use the new technology if

$$t \geqslant \frac{k\,\tilde{z}\bar{\sigma}_n^2 + k\,(\bar{\sigma}_n^2)^2 - (1/b)\,\bar{\sigma}_n^2\,(\bar{\mu} - \mu_o)}{\{(1/b)\,(\bar{\mu}_n - \mu_o) - k\bar{\sigma}_n^2\}\,\tilde{z}}.$$ (7.34)

Defining the right-hand side of (7.34) as J, we see that $\partial J/\partial\bar{\mu} < 0$, $\partial J/\partial\mu_o > 0$, $\partial J/\partial\bar{\mu}_n < 0$. Thus, the greater is the initial estimate of the mean profitability of the new technology, the greater its true mean return, and the lower the mean return to the old technology, the earlier will the firm be expected to use the technology. If $\bar{\mu} - \mu_o > 0$ and $\bar{\mu}_n - \mu_o > 0$, we may note that $\partial J/\partial(1/b) < 0$, and thus low risk aversion is associated with early use. Also, $\partial J/\partial\bar{\sigma}_n^2 > 0$; thus, a high variance of returns means late use. If $k > (\bar{\mu} - \mu_o)/b\bar{\sigma}_n^2$, then $\partial J/\partial\tilde{z} > 0$. We may thus argue that if $\bar{\mu}$ is high a low $\tilde{z}$ is associated with early use.

All the above is for one entrepreneur. Consider now that there are j entrepreneurs in the industry and that they differ in the returns that they can achieve and in their attitude to risk. Define for each entrepreneur J_i, $i = 1, \ldots j$, from (7.34). Let the J_i be distributed across all firms with a density function $g(J)$ and a distribution function $G(J)$. (J_i, for a given firm, may be positive or negative but will remain constant over time.) Define D_t as the proportion of firms in the population at time t that are using the new technology. The theory of imitation or inter-firm diffusion concerns how D_t changes over time. The probability that a firm chosen at random will use the new technology at time t is $\Pr\{t \geqslant J\} = G(t)$. Thus $D_t = G(t)$. The characteristic of diffusion paths that has been observed and noted above all is that they are sigmoid. To generate a sigmoid diffusion path, all one requires is that $d^2 D_t/dt^2$ changes sign from positive to negative as t increases. This can be achieved as long as $g(J)$ is bell-shaped. One would observe the point of inflexion as long as it occurs for a value of $J > 0$.

To obtain results on the determination of the diffusion speed we will carry out two exercises. In the first, b is considered to be large for all firms so that (7.34) reduces to (7.35), and in the second b is considered to be small for all firms so that (7.34) reduces to (7.36):

$$t - 1 \geqslant \bar{\sigma}_n^2/\bar{z} \qquad (7.35)$$

$$t \geqslant \frac{\bar{\sigma}_n^2 (\bar{\mu} - \mu_o)}{\bar{z} (\bar{\mu}_n - \mu_o)}. \qquad (7.36)$$

Let b in (7.29) be large for all firms to that (7.35) represents the condition for defining use. Assume that $\bar{\sigma}_n^2$ and $\bar{z}$ are lognormally distributed as $\Lambda(\mu_1, \sigma_1^2)$ and $\Lambda(\mu_z, \sigma_z^2)$ respectively. Then $\sigma_n^2/\bar{z}$ is $\Lambda(\mu_1 - \mu_z, \sigma_1^2 + \sigma_z^2)$. The pattern of diffusion will then be represented by a cumulative lognormal curve with an inflexion point at time $\hat{t}$ where $\hat{t} = \exp(\mu_1 - \mu_z - \sigma_1^2 - \sigma_z^2) + 1$ (the mode of distribution of $\bar{\sigma}_n^2/\bar{z}$, plus 1) and half the firms will be using the new technology when $t + 1$ equals the median value of $\bar{\sigma}_n^2/\bar{z}$ ($\exp(\mu_1 - \mu_z)$). Define this value of t as $\hat{t}$. From this we may see that, the greater is the initial uncertainty about the new technology, the earlier does the inflexion point occur and the faster is the take-up for $\partial \hat{t}/\partial \mu_z < 0$ and $\partial \hat{t}/\partial \mu_z < 0$. However, $\partial \hat{t}/\partial \mu_1$ and $\partial \hat{t}/\partial \mu_1$ are both positive. The more diverse are $\bar{\sigma}_n^2$ and $\bar{z}$, the earlier is the inflexion point, for $\partial \hat{t}/\partial \sigma_1^2 < 0$, and $\partial \hat{t}/\partial \sigma_z^2 < 0$.

In the second case, where b is considered small, allow that all firms face the same $\bar{\sigma}_n^2$ and $\bar{z}$ but $(\bar{\mu} - \mu_o)$ and $(\bar{\mu}_n - \mu_o)$ are lognormally distributed as $\Lambda(\mu_2, \sigma_2^2)$ and $\Lambda(\mu_3, \sigma_3^2)$ respectively. Then, following similar arguments to the above,

$$\hat{t} = \frac{\bar{\sigma}_n^2}{\bar{z}} \exp(\mu_2 - \mu_3 - \sigma_2^2 - \sigma_3^2)$$

and

$$\hat{t} = \frac{\bar{\sigma}_n^2}{\bar{z}} \exp(\mu_2 - \mu_3),$$

from which it is clear that the inflexion point is earlier the lower is $\bar{\sigma}_n^2$, the higher is $\bar{z}$, the smaller is μ_2 (the mean initial estimate of the advantage of the new technology), the greater is μ_3 (the mean true advantage of the new technology), and the greater are σ_2^2 and σ_3^2 (the variances of the initial estimate and true mean advantage of the new technology). Similarly, $\partial \hat{t}/\partial \bar{\sigma}_n^2 > 0$, $\partial \hat{t}/\partial \bar{z} < 0$, $\partial \hat{t}/\partial \mu_2 > 0$ and $\partial \hat{t}/\partial \mu_3 < 0$.

In this approach a firm chooses the proportions $(\alpha : 1 - \alpha)$ in which to use new and old technologies so as to maximize a utility function defined on the mean and variance of the returns to the joint use of new and old technology. As it learns from experience, it updates the perceived parameters of the distribution of returns to the new technology and thus changes the mix in which it wishes to use the two technologies. This will, under certain conditions, lead to α rising above some arbitrarily defined proportion k after which the firm is defined as a user. Because firms differ in the initial conceptions of the returns to the new technology and have different attitudes to risk, the date at which α will become greater than k will differ across firms, and because of this one obtains a diffusion curve. One may note that no exogenous changes in the economic environment are occurring; the diffusion results purely from learning, although it is learning by experience and not by infection. It is rational for the economy to not jump immediately to the post-diffusion level of use, because in the early years of the life of a new technology the firms will not correctly perceive the true returns and risk associated with the new technology.

The probit approach has been used not only for looking at process innovations, but also to look at the diffusion of product innovations. For example Bonus (1973) sets the problem up as follows. A household i will own a new good if its income y_{it} is greater than some critical level $\bar{y}_{it}$. Both household incomes and threshold levels are argued to be lognormally distributed; therefore,

$$y_{it} = \Lambda(\mu_t, \sigma_t^2)$$

$$\bar{y}_{it} = \Lambda(\bar{\mu}_t, \bar{\sigma}_t^2).$$

Then the probability of ownership is given by $\Pr\{\bar{y}_{it} \leqq y_{it}\}$. By allowing the distribution of income, or the distribution of thresholds, to shift over time, one generates a diffusion curve. The principle is of course the same as in the other probit diffusion models. For an earlier and more complex and complete model of the diffusion of household durables one may turn to Pyatt (1963).

Bain (1964) surveys the literature on the diffusion of product innovations showing that it relies heavily on epidemic models. He himself, however, argues that the learning process will lead to a skewed diffusion curve, which he represents, on empirical grounds, by the cumulative log normal. He argues that if D_t is the proportion of households owning a television set (the product innovation being studied), D^* is the saturation or equilibrium proportion and z_t the normal equivalent deviate of D_t/D^*, one may write the diffusion curve

$$\frac{dD_t}{dt} = \frac{\alpha}{(2\Pi^2)^{\frac{1}{2}}} \exp(-\mu - \sigma z - \tfrac{1}{2}z^2). \tag{7.37}$$

He considers that the parameters α, μ, and σ will be functions of social and economic variables that are likely to change over time, so that the actual growth curve of ownership will consist of a series of segments of cumulative lognormal curves, rather than the envelope of short-run diffusion curves being itself the lognormal curve. We will not however detail the empirical results that Bain derives.

7.2.3 The game theoretic approach

Reinganum (1981) considers a world where firms are assumed identical and information on technology is perfect. This is a world where our previous theories of inter-firm diffusion would not generate a diffusion curve. However, she considers that firms maximize their present value and undertake strategic behaviour, and shows that the equilibrium of the oligopoly game will in fact generate different adoption dates for firms and thus a diffusion curve.

A capital-embodied process innovation is considered and it is allowed

1   that the profit to be gained from introducing the new technology declines as the number of firms that have already adopted increases;
2   that the cost of adopting the new technology declines the later it is adopted.

It is then shown that a Nash equilibrium of adoption dates will exist and will involve the firms in the industry adopting sequentially. Reinganum also looks at how the market structure of the industry will affect the schedule of adoption dates in the Nash equilibrium by considering industries with different numbers of firms. She argues that if demand is linear, an increase in the number of firms will cause most firms to delay adoption.

This is a fascinating approach to the diffusion question. Essentially, the diffusion is the result of the oligopoly game. It seems however that the model implies a game that is played only once, at that date at which the new technology first appears. At that time potential users agree dates of adoption through the game. If the environment changes before adoption dates, do we expect the game to be replayed? We may also ask whether the Nash equilibrium is a likely result of the game. Might a co-operative solution be more likely? If this is the case, then the Reinganum results may no longer hold. Neither is there any indication of whether the diffusion will be sigmoid. The underlying philosophy of the model, however — that the oligopoly game is an important aspect of diffusion — should not be overlooked.

7.3 Conclusions

In this chapter we have explored a number of approaches to the explanation of inter-firm diffusion in the search for an answer to the question, why does new technology not spread to all users immediately? The explanations are of the following basic types.

1   In the Mansfield approach there is a fixed potential number of users, the proportion of whom are actual users increasing over time as epidemic learning reduces the uncertainty attached to the use of the new technology.
2   In the David/Davies approach, at any moment in time those firms for whom it is profitable to use the new technology will be users, but over time the underlying profitability of the technology changes and so does the number of users.
3   In the Bayesian learning approach, at any moment in time the number of firms that are users is determined by the number who perceive the returns to the technology to be such that they wish to use it at a significant level. Over time their perceptions change and so does the number of users.
4   In the game theoretic approach the different adoption dates (and thus the diffusion) are the results of an oligopoly game.

To some extent we must consider that each of these approaches contains ideas that are desirable in explanations of diffusion. Learning both from other firms and from experience, changing factor prices over time, and the oligopoly game would all seem to be relevant factors impinging on the adoption decision. It may thus not be a matter of choosing one theory in preference to another — perhaps we should consider that they all have something useful to tell us. However, perhaps just as important is a reminder that each of these theories is demand-orientated; the supply side is to all intents and purposes ignored. This may throw some doubt on the validity of the models in explaining real-world phenomena. Given this, however, the empirical work discussed above suggests that profitability has a major role to play in the determination of diffusion speed, as do absolute capital requirements. The role played by market structure is somewhat ambiguous, but there seems to be some indication that larger firms are earlier adopters.

References

Bain, A.D. (1964), *The Growth of Television Ownership in the United Kingdom Since the War*, Cambridge University Press.

Bonus, H. (1973), 'Quasi-Engel Curves, Diffusion and the Ownership of Major Consumer Durables', *Journal of Political Economy*, 81, 655–77.

David, P.A. (1969), *A Contribution to the Theory of Diffusion*, Stanford Center for Research in Economic Growth, Memorandum no. 71, Stanford University.

Davies, S. (1979), *The Diffusion of Process Innovations*, Cambridge University Press.

Dixon, R. (1980), 'Hybrid Corn Revisited', *Econometrica*, 48, 1451–62.

Griliches, Z. (1957), 'Hybrid Corn: An Exploration in the Economics of Technological Change', *Econometrica*, 25, 501–22.

Griliches, Z. (1980), 'Hybrid Corn Revisited: A Reply', *Econometrica*, 48, 1463–6.

Jensen, R. (1979), *On the Adoption and Diffusion of Innovations in Uncertainty*, Discussion Paper no. 410, Centre for Mathematical Economics and Management Sciences, Northwestern University, Chicago.

Lindley, D.V. (1965), *Introduction to Probability and Statistics*, Cambridge University Press.

Mansfield, E. (1968), *Industrial Research and Technological Innovation*, W.W. Norton, New York.

Nabseth, L. and Ray, G.F. (1974), *The Diffusion of New Industrial Processes: An International Study*, Cambridge University Press, London.

Norris, K. and Vaizey, J. (1973), *The Economics of Research and Technology*, George Allen and Unwin, London.

Oster, S. (1982), 'The Diffusion of Innovation among Steel Firms: The Basic Oxygen Process', *Bell Journal of Economics*, 13, 45–56.

Pyatt, G. (1963), *Priority Patterns and the Demand for Household Durable Goods*, Cambridge University Press.

Reinganum, J. (1981), 'Market Structure and the Diffusion of New Technology', *Bell Journal of Economics*, 12, 618–24.

Rogers, E.M. (1962), *The Diffusion of Innovations*, Free Press, New York.

Romeo, A.A. (1977), 'The Rate of Imitation of A Capital Embodied Process

Innovation', *Economica*, 44, 63–9.

Rosenberg, N. (1976), 'On Technological Expectations', *Economic Journal*, 86, 523–35.

Scherer, F.M. (1980), *Industrial Market Structure and Economic Performance* (2nd ed.), Rand McNally, Chicago.

Stoneman, P. (1980), 'The Rate of Imitation, Learning and Profitability', *Economics Letters*, 6, 179–83.

Chapter 8

Economy-wide and International Diffusion

8.1 Introduction

The purpose of this chapter is to consider those approaches that have been used to analyse diffusion of a new technology over the national and international economy as a whole rather than within sectors. This analysis is often restricted, for obvious reasons, to major innovations that have economy-wide applications. In many senses the work on product innovation diffusion could have been included in this chapter.

When discussing diffusion at the economy-wide level, one can use epidemic or probit models, as discussed above, but at this stage such models will not yield much in terms of further understanding. Instead, we are going to concentrate on three approaches — the Schumpeter approach, the vintage approach, and the stock adjustment approach — each of which yeilds some further insight into what has already been detailed. In the last part of the chapter international issues are discussed.

8.2 The Schumpeter approach

One of Schumpeter's (1934) objectives was to investigate and explain long swings in economic activity, the Kondratieff cycle. To this end he was particularly concerned with the impact of new technology on the economy. His hypothesis is that, while the economy is in the trough of the cycle, new inventions are made.[1] First application of these inventions is carried out by the entrepreneur, who makes excess or entrepreneurial profit from his path-breaking activities. The profit he makes acts as a signal to other economic actors that there are returns to be made from imitation, and they copy him. (These imitators are, by the way, 'Austrian' entrepreneurs; see Reekie, 1979.) As they copy, the diffusion is proceeding. However, as the new technology is used, one would expect that factor prices and product prices will change. These changes in prices will tend to force non-adopters to use the new technology as their previous technology becomes economically infeasible. The cycle is derived from this by allying the boost in gross investment from the take-up of new technology with a multiplier effect and accommodating monetary expansion. However, as the diffusion slows down, which it will as the population of non-adopters decreases, the strength of the boom will die away and the process goes into reverse.

This Schumpeter story has a number of echoes in other later work — the pull of profitability, for example. How it differs from the other models we have so

[1] Which may be in contradiction to the suggestions made in Part I of this book.

far discussed is that the profitability changes over time are induced by the new technology, and are not imposed externally. This is a result that is common to the vintage model, discussed below.

As appealing as is the Schumpeter story, it is not precise, and has not, as far as we know, been used to analyse empirically any given diffusion process.

8.3 The vintage approach

Vintage models come in a number of forms, the three basic ones being putty-putty, putty–clay, and clay–clay models. The putty/clay distinction rests upon the degree of factor substitution in the production process before and after a capital good is installed. We will discuss a putty–clay model. In this framework, once a capital good is installed it has fixed labour requirements to produce a unit of output. However, prior to installation the firm can choose the capital-labour ratio from a menu of combinations represented by a standard neoclassical production function.

Technical progress in these models is generally taken as completely embodied and exogenous. The *ex ante* production function (pre-installation) shifts over time. However, once a capital good is installed it does not experience any further technological enhancement. It is possible to allow some disembodied technological progress, but that is not important at this stage.

At any moment in time, therefore, the economy's capital stock is made up of machines of different ages (vintages) with different productive potentials. Any new investment is always in machines of the latest type. The oldest machine in existence is the one that can only just cover its operating costs (labour plus materials, etc.); in other words, machines are scrapped when they yield zero quasi-rents (to simplify matters we are assuming that machines have an infinitely long physical life, but that their economic life determines scrapping dates). We assume that installed machines are operated under constant returns to scale.

To illustrate how the model works we shall tell the story as detailed by Salter (1966). At a moment in time the economy is in a position where it has a set of vintages that make up its capital stock. Perfect competition is assumed. The price ruling for the homogeneous product produced is p, composed of operating costs *and* capital costs (including normal profits) of 'best-practice' plants (those installed in the current period). The oldest plant has operating costs alone equal to p. If for the moment we hold factor prices constant, we now argue that in the following period a new best-practice technology becomes available which at current prices can yield super-normal profits. Capacity is increased by investment in such plant until increasing output forces down prices to the level at which super-normal profits are no longer available. At this point, however, p is lower, and so the previously marginal plant is no longer covering its operating costs and it is therefore scrapped. A series of new technologies will over time lead to the gradual scrapping of old plants and their replacement by better types.

Consider now, however, that technology is not changing but that relative factor prices are. For the installed capital goods the changes in relative prices may lead to some scrapping. Moreover, if relative factor prices change then the choice of the *ex ante* capital-labour ratio will change. Thus, if the real wage increases, we expect the capital-labour ratio chosen to rise. For installed machines factor requirements are fixed. Sufficient machines are built of the desired capital-labour intensity to produce a price at which no super-normal profits exist on these new machines. Older machines with operating costs greater than this price are scrapped.

These vintage models are usually used in a macro-model context in which the usual neoclassical assumptions of perfectly clearing markets are assumed; i.e., the labour market clears in each period and savings and investment are equal in each period. These markets enable one to determine factor prices in each period.

Let us then consider the full story, and ask how an economy will react to the appearance of a new technology. Initially there is full employment of labour, savings and investment are equal, and the capital stock is made up of a set of vintages the oldest of which has an operating cost equal to the price of output. The new technology is superior to the previous technology at current prices and thus new capacity is created. This capacity requires labour to operate, and this increase in labour demand will lead to increases in wages. The equilibrium between investment and savings will yield a figure for capital rental. The level of investment will be that which yields zero super-normal profits given the wages and rentals determined. The price of output will fall as output increases. These price changes lead to scrapping decisions.

Let this new technology now remain as the best technology. Will further use of the new technology follow? Of course, if old machines physically deteriorate it will be necessary to have replacement investment, but if this were all that was going to happen it would imply that diffusion was due purely to old machines dying. (Although there is some evidence to support the view that this is a relevant determinant — see Stoneman (1976) — it is not likely to tell the whole story.) Is there any other force that will mean extensions of use? Without improvements in the new technology we must rely on factor prices. One could argue that, if no further advances in technology were made, then at the factor prices ruling in the previous period investment would be zero. However, if investment is zero and savings are positive, the rental of new machines must fall to equilibrate savings and investment. As these new rentals will not affect operating costs on old machines, their quasi-rents remain constant. However, the latest machines installed will have a more appropriate capital-labour ratio, and thus lower costs, than machines installed in the previous year. With price forced down to best-practice total costs, obsolescence of old machines results. In a similar manner, one could argue that, if the labour force is growing, then in the absence of new capacity wages will fall. New capacity with the optimal capital-labour ratio will then be installed, leading to changes in prices and obsolescence.

These factor price changes, however, are not usually considered as the driving force in the vintage models. The story told here is very much that technology will improve in every year. Thus extensions of the use of a new technology come about from improvements in that technology itself — very much as in the Davies (1979) model above.

The great strength of the vintage model is that it illustrates that it is perfectly rational for entrepreneurs to use old technology even when new best-practice techniques exist. Essentially, old machines can still yield a contribution to profits if price covers operating costs, and they are therefore worth using while this condition holds. Only as the appearance of new machines drives price below operating costs are they replaced.

The vintage model does have some disadvantages, however. For example, it argues that all investment is in machines of the latest type. This has been found to be empirically suspect (Davies, 1979; Gomulka, 1976). Second, the driving force is investment in latest machines, and the vintage model really is only as good an explanation of diffusion as its investment function. That discussed above is theoretically neat but empirically suspect. Third, the model gives us no guarantee that the diffusion curve will be sigmoid. The time-path of usage of new technology given perfect competition will depend upon

1   the existing age structure of the capital stock;
2   improvements in the new technology over time; and
3   movements in relative prices.

The combination of the three may yield sigmoid curves, but the conditions required to make them do so have not, as far as I know, been detailed.

8.4 Stock adjustment models

In the investment theory literature it is a common procedure to consider that at a moment in time an entrepreneur holds an estimate of a desired capital stock S_t^* and an actual capital stock S_t. However, his investment in time t will be only some proportion of the difference between S_t^* and S_t because of the costs of adjustment involved in changing the capital stock. It is argued that the faster the adjustment is made, the greater will be the adjustment costs involved.

These adjustment costs arise from a number of sources. One can resort to a Penrose (1959)-type argument that management has the capacity to expand a business only at a certain rate, and as expansion exceeds that rate control losses set in. Alternatively, Nickell (1978) argues that, as demand for capital goods increases, the price to be paid must also increase in order to offset rising costs in the capital goods supply industry. This is very Keynesian in nature, and for our purposes has the specific advantage of introducing supply factors into the diffusion process for the first time.

To show how we can produce diffusion curves from this approach, consider the following very simple example. Let the representative firm at time t have a capital stock S_{t-1}, left from the previous period. The firm is assumed to maximize profits by the appropriate choice of S_t subject to incurring adjustment costs as S_t diverges from S_{t-1}. We will assume that the firm is myopic in the extreme in that it expects current prices to last for ever and does not take into account that the S_t selected in this period will affect the adjustment costs it has to incur in future periods. There is no depreciation. Let the production function be

$$Y_t = \delta \log S_t \tag{8.1}$$

and let adjustment costs be written as AC. Profits may be defined as

$$\Pi = pY - rS - AC \tag{8.2}$$

where p = output price;
    r = cost of capital.
The first-order condition for profit maximization is

$$p\frac{\delta}{S_t} - r = \frac{\partial AC}{\partial S}. \tag{8.3}$$

Define the equilibrium capital stock as that S_t for which $S_t = S_{t-1} = S_t^*$; then S^* is such that

$$\frac{p\delta}{S_t^*} = r \tag{8.4}$$

and

$$S_t^* = \frac{r}{p\delta}. \tag{8.5}$$

Substituting from (8.5) into (8.3), we get

$$p\delta \left(\frac{1}{S_t} - \frac{1}{S_t^*} \right) = \frac{\partial AC}{\partial S}. \tag{8.6}$$

If adjustment costs are such that

$$\frac{\partial AC}{\partial S} = b \left(\frac{1}{S_{t-1}} - \frac{1}{S_t} \right), \tag{8.7}$$

then we may write

$$\frac{S_t - S_{t-1}}{S_{t-1}} = \frac{dS}{dt}\frac{1}{S} = \frac{p\delta}{b} \left(1 - \frac{S_t}{S_t^*} \right). \tag{8.8}$$

Equation (8.8) is of course the differential equation of the logistic curve given above with a diffusion speed dependent, *inter alia*, on the adjustment cost parameter. The adjustment cost function (8.7) has the property that, if $S_t > S_{t-1}$, then $AC > 0$, $\partial AC/\partial S_t > 0$, and $\partial^2 AC/\partial S_t > 0$, which are the required properties of an adjustment cost function (see Nickell, 1978, p. 27).

    It is of course obvious that a different adjustment cost function would have produced a different time-profile for S_t, as would the use of an alternative production function. What we wish to note however is that this stock adjustment approach does help to create a bridge between investment theory and diffusion theory. One might note that it plays a similar role when used in Chapter 6.

Most work on stock adjustment models, however, ignores the adjustment cost approach and proceeds directly to the specification of a diffusion equation from *a priori* reasoning, usually of the epidemic kind. Thus G.C. Chow states:

the rate of growth depends on the quantity of the existing stock y_t . . . partly rationalised by the idea that, the more the product has been accepted, the more prospective buyers . . . have learned about the product, [and] the difference or ratio between the equilibrium level y^*, ultimately to be reached, and the existing level y_t . . . the closer $[y_t]$ comes to the equilibrium level y^*, the smaller will be the number of prospective buyers remaining. (Chow, 1967, p. 1118)

Chow then states that this process can be represented by (8.9), where we use our terminology rather than Chow's:

$$\frac{dS}{dt}\frac{1}{S} = \alpha\,(\log S^* - \log S),\qquad(8.9)$$

the Gompertz curve, or by

$$\frac{dS}{dt}\frac{1}{S} = \gamma\,(S^* - S),\qquad(8.10)$$

which we described above as the Chow logistic. These arguments are followed also in Stoneman (1976).

The advantage of the above formulation is that S^*, the equilibrium stock, can be made endogenous by substitution into the growth equations using, for example, either

$$\log S_t^* = \beta_0 - \beta_1 \log \rho_t + \beta_2 \log x_t \qquad(8.11)$$

(where ρ_t is the relative price of the new technology and x_t is output), or

$$S_t^* = \theta_0 - \theta_1 \rho_t + \theta_2 x_t, \qquad(8.12)$$

to yield an equation that is simple to estimate. We will not go into the Chow results at this stage, for they will be discussed in Chapter 10 below. The model is however a good example of how the stock adjustment principle has been used to investigate the diffusion of process innovations.

It is in product innovations, however, that stock adjustment models have been extensively used. For example, Stone and Rowe (1957) use such a model. The principles are the same as for process innovations, although the agents are different, so we will not pursue this line any further.

8.5 The international diffusion of innovations

This literature falls into two parts:

Table 8.1 *Factors affecting the diffusion of numerically controlled machines*

	Weight	Austria Value	Austria Score	Italy Value	Italy Score	Sweden Value	Sweden Score	UK Value	UK Score	USA Value	USA Score	West Germany Value	West Germany Score
Wage level	40	+1	40	–	–	+3	120	+2	80	+5	200	+2	80
Importance of aerospace industry	10	–	–	+1	10	+1	10	+4	40	+5	50	+1	10
Quality of information system	10	+2	20	–	–	+4	40	+4	40	+5	50	+2	20
Investment financing possibilities	10	–	–	–	–	+4	40	+3	30	+5	50	+1	10
Management attitudes	5	–	–	–	–	+3	15	+2	10	+5	25	+1	5
Condition of the market	5	+1	5	+1	5	+3	15	+4	20	+3	15	+5	25
Trade union attitudes	5	–	–	–	–	–	–	–2	–10	–4	–20	+1	5
Technical factors	5	–	–	+1	5	+2	10	+3	15	+5	25	+1	5
Labour market conditions	5	–	–	+1	5	+5	25	+2	10	–	–	+4	20
Other relevant factors	5	+2	10	+3	15	+3	15	+5	25	+5	25	+5	25
Total	100		75		40		290		260		420		205

Source: Nabseth and Ray (1974), p. 56.

1   the comparison of national diffusion across countries, which may be considered an extension of the above, and

2   the transmission of technology between countries, which is really international diffusion proper — or, to take labels to extreme, inter-country or intra-world diffusion.

Within the remit of the first category, the work of Nabseth and Ray (1974) yields a number of empirical results that are worth detailing.[2] In Table 8.1 their results on the factors influencing the diffusion of numerically controlled machine tools in different countries are reproduced. In this table,

the ten probably most essential factors are listed to explain differences in actual diffusion in the various countries. These factors were selected in accordance with the findings of their inquiry . . . Using a scale ranging from +5 to —5, each factor was given a 'valuation', which when multiplied by the weight gave a score for each factor. (Nabseth and Ray, 1974, p. 57)

The results suggest that differences in wage levels were a prominent influence determining different diffusion speeds, swamping, in fact, many of the other influences. The explanation is probably that high wages imply a high return to the introduction of numerically controlled machine tools; i.e., one is returning to the influence of profitability on diffusion speeds.

Within the remit of the second category — the transmission of technology between countries — we have a new and much more fascinating question. What we are really concerned with here is why one country should 'use' a new technology before another — once innovation has taken place somewhere in the world. We do not wish to explain here why a new technology originates in a specific country, only why the imitation path in the international economy takes the direction that it does.

It is not difficult to build models to explain this latter process, although as far as we know the literature in this area is very sparse. An obvious approach is to consider that diffusion will follow information, and as information spreads use of a new technology will spread. Information may well spread via technical publications, personal contacts, etc.; i.e., we could have an epidemic approach.

Second, we might consider that information is not a barrier but that different countries have different characteristics, and above a certain threshold of characteristics adoption is profitable — a probit approach. Thus, for example, there may be a new process technology that is profitable for a real wage greater than some threshold level $\bar{w}$; then assume that w is distributed across countries with some bell-shaped density function. If this distribution shifts or $\bar{w}$ falls over time, then the use of a new technology will expand and the diffusion proceed. The country characteristics may be to do with costs, as in this example, or with skill and educational levels or similar factors. For a new product, income may be the appropriate characteristic.

---

[2] Swann (1973) has used the Mansfield methodology with diffusion parameters of fitted logistics explained by country-level explanatory variables. However, as we have already spent considerable time discussing this approach we do not pursue it here.

Finally, one cannot really consider this question of international diffusion without referring to the literature on technology transfer. Although licensing arrangements can be used, direct investment is the most important institutional arrangement by which technology is transferred, and that takes one's analysis directly to the consideration of multinational companies. It is not too unfair a characterization of this literature to state that direct investment will follow (risk-adjusted) profitability. It is then not too much of a step to explain differences in take-up dates by countries by profitability considerations, thereby yielding diffusion results.

Teece (1977) investigates the costs involved in the international transfer of technology. The value of the resources that have to be utilized to accomplish the successful transfer of a given manufacturing technology is used as a measure of the cost of transfer. Data on 26 international technology transfer projects were obtained, all transfers being by multinational firms with headquarters in the USA. The recipients of the knowledge were on average smaller and less research-intensive than the transferors; 12 were wholly owned subsidiaries, 8 were joint venture partners, 4 were wholly independent private enterprises, ·and 3 were government enterprises. Of the 26 cases, 17 were in chemicals and petroleum refining and 9 in machinery.

Transfer costs are defined as the costs of transmitting and absorbing all of the relevant unembodied knowledge. They fall into four groups:

1   the cost of pre-engineering technological exchanges;
2   engineering costs associated with transferring the relevant designs;
3   the costs of R & D personnel during the transfer;
4   the pre-start-up training costs and learning and debugging costs.

On average, transfer cost represented 19 per cent of total project costs. It is hypothesized that transfer cost as a proposition of project costs are related to:

1   the number of previous applications or start-ups that the technology has undergone by the transferor;
2   the age of the technology;
3   the number of years of manufacturing experience that the recipient has accumulated;
4   the R & D-to-sales ratio of the recipient;
5   the sales of the recipient;
6   the number of firms identified by the transferor as having similar technology;
7   the GNP of the host country of the recipient.

All effects are expected to have negative signs: in chemicals and petroleum refining significant influences arise from (1), (3), and (6) with the expected signs, whereas in machinery, (2), (3), and (6) have significant negative influences.

Teece argues that his findings indicate that more experienced enterprises have lower transfer costs, and also that transfer costs decline with each application of a given innovation — technology transfer is a decreasing cost activity. It is also argued that any firm that is moderately mature in experience, size, and R & D is a good candidate to absorb technology at minimum cost. Large size does not seem to offer any advantage over medium size.

This obviously presents some material relevant to the question of international diffusion. To fully explain that diffusion, however, one would need to consider profitability, and transfer costs are only one of the factors affecting profitability. For a very recent survey of the whole literature one if referred to Caves (1983).

8.6 Conclusions

Having built up, over the last three chapters, from the level of the individual firm to the level of the world economy, we have covered a number of approaches to the diffusion problem. However, we have not quite finished with diffusion. In the next two chapters we wish to extend the above in two directions.

1　The work detailed above concentrates almost exclusively on demand phenomena, and the supply of technologically new products has been almost ignored. In the next chapter we will consider the supply problem.
2　In the work discussed so far we have deliberately taken a theoretical-model-based approach in order to stress our belief in the idea that diffusion of technology is a positive process that has some basic underlying rationale. We have set out to explain the rationality of continuing to use old technology when new is available, and we have detailed a number of reasons why such behaviour may be rational. However, what we are missing is any sense of the detail of a diffusion process, i.e., an understanding of those forces that impinge on any particular diffusion process but are ignored in the generality of the model-building. Thus in Chapter 10 we will consider two case studies of diffusion processes – the diffusion of the steam engine, and the diffusion of the digital computer.

References

Caves, R.E. (1983), *Multinational Enterprise and Economic Analysis*, Cambridge University Press.

Chow, G.C. (1967), 'Technological Change and the Demand for Computers', *American Economic Review*, 57, 1117–30.

Davies, S. (1979), *The Diffusion of Process Innovations*, Cambridge University Press.

Gomulka, S. (1976), 'Do New Factories Embody Best Practice Technology? New Evidence', *Economic Journal*, 86, 859–63.

Nabseth, L. and Ray, G.F. (1974), *The Diffusion of New Industrial Processes: An International Study*, Cambridge University Press.

Nickell, S. (1978), *The Investment Decision of Firms*, J. Nisbet, Welwyn Garden City, and Cambridge University Press.

Penrose, E.T. (1959), *The Theory of the Growth of the Firm*, Basil Blackwell, Oxford.

Reekie, W.D. (1979), *Industry, Prices and Markets*, Philip Allan, Oxford.

Salter, W.E.G. (1966), *Productivity and Technical Change* (2nd ed.), Cambridge University Press.

Schumpeter, J. (1934), *The Theory of Economic Development*, Harvard University Press, Cambridge, Mass.

Stone, R. and Rowe, D. (1957), 'The Market Demand for Durable Goods', *Econometrica*, 25, 423–43.

Stoneman, P. (1976), *Technological Diffusion and the Computer Revolution*, Cambridge University Press.

Swann, P.L. (1973), 'The International Diffusion of an Innovation', *Journal of Industrial Economics*, 22, 61–9.

Teece, D.J. (1977), 'Technological Transfer of Multinational Firms: The Resource Cost of Transferring Technological Know How', *Economic Journal*, 87, 242–61.

Chapter 9

Diffusion and Supply

9.1 Introduction

The material covered in the previous four chapters has concentrated almost exclusively on diffusion as a demand-based phenomenon, and supply factors are either ignored or, as in the David (1969) approach, subsumed into changes in prices or quality of technologically new products. If we consider that much of the literature suggests that the demand for new products and new processes is related to profitability and prices, then we need some understanding of how new products and new processes will be priced. Moreover, it is possible that diffusion will be constrained by the inability of producers to supply new products in the required amount. One can complete the diffusion story only if the price-setting and output policies of the suppliers of technologically new products are investigated. Surprisingly, however, the literature in this area is very thin.

9.2 Epidemic diffusion and supply factors

Bass (1978) explores a fascinating model of diffusion where supply and demand are allowed to interact. Moreover, in the supply of new products there are also learning economies.

Let q_t be the output of the firm in time t of the technologically new product. Define

$$E_t = \int_0^t q_\tau \, d\tau.$$

Costs are assumed to fall with E_t as given by (9.1), with k being some constant:

$$\frac{\partial C_t}{\partial q_t} = k E_t^{-\lambda}; \qquad (9.1)$$

i.e., the marginal cost of producting the qth unit at time t will be lower the greater is the accumulated output to date.

Assume that the firm is a monopolist and behaves to maximize current profits. (This is an undesirable assumption, of course; one ought to allow present value to be maximized.) Then, defining η as the elasticity of demand (assumed constant) and p as the price of the new good, one achieves (9.2), assuming that the demand function is of the form (9.3):

$$p_t = \frac{\eta}{\eta - 1} k E_t^{-\lambda} \qquad (9.2)$$

$$q_t = f(t) \, \alpha \, p_t^{-\eta} \qquad (9.3)$$

where α is some constant and $f(t)$ is a time-based demand shifter (usefully considered as a density function of 'time to purchase' if price is fixed). Equation (9.3) is a demand diffusion relationship of the type discussed in Chapters 5–8 above. We define $F(t) = \int_0^t f(\tau)\, d\tau$.

We substitute for p_t from (9.2) into (9.3) to yield

$$E_t = K(1 - \lambda\eta)^{1/(1-\lambda\eta)} F(t)^{1/(1-\lambda\eta)} \tag{9.4}$$

where

$$K = \alpha \left(\frac{\eta}{\eta - 1} k\right)^{-\eta}. \tag{9.5}$$

Define

$$\lim_{t\to\infty} E_t = \bar{E}. \tag{9.6}$$

Then

$$q_t = \frac{\bar{E}}{1 - \lambda\eta} f(t)\, F(t)^{\lambda\eta/(1-\lambda\eta)}. \tag{9.7}$$

The demand shifter is specified by essentially Mansfield-type methods to yield

$$F(t) = \frac{1 - \exp(-\alpha - \beta t)}{1 + \beta/\alpha \exp(-\alpha - \beta t)}, \tag{9.8}$$

i.e., a logistic demand diffusion. The substitution of (9.8) into (9.7) yields the path of q_t, which is the overall diffusion path. Bass states that the result of this substitution is that q_t will initially increase with t and then decline — generating something like a diffusion curve. Price will fall continuously throughout.

Thus what we have here is that basically experience in producing the new capital good leads to falls in product cost and this is translated into price reductions. These price reductions increase demand for the good. At the same time, demand is shifting over time through learning, with the shifts at a given price generating a logistic diffusion curve. The combined effect of falling price and shifting demand generates the overall diffusion curve.

Bass tests his model on data for a number of consumer innovations, but in doing so violates his monopoly assumption; thus we will not report the results in detail. The major problem with the Bass approach is his failure to model the supply of technologically new products when there is oligopolistic rivalry rather than a monopoly position.

Glaister (1974) considers a world similar to that of Bass. Let N_t be the number of users of the new technology in time t and m_t be the number of non-users in time t.[1] By an epidemic argument Glaister suggests that the number of users is determined by

---

[1] For ease of notation we drop the time subscript where this can be done without confusion.

$$\frac{dN}{dt} = \beta(p)Nm \qquad (9.9)$$

where β is a function of price p. Obviously, (9.9) is the logistic curve if β is constant. A monopoly supplier chooses p to maximize present value so that it maximizes

$$\int_0^\infty e^{-rt} \{pq - C(q)\} \, dt \qquad (9.10)$$

where q_t is output in time t and $C(q)$ is the total cost of producing q. If demand and supply are to be equal,

$$q_t = Nh(p), \qquad (9.11)$$

in which $h(p)$ is the number of units bought by each user.

Thus we have the number of users, for a given price, following a logistic curve, based on epidemic reasoning. But as price changes the speed of diffusion β changes, and at the same time the quantity purchased by each user will change.

In an example involving constant returns to scale (i.e., Bass's learning has gone), a constant elasticity function $h(p)$, and a constant elasticity function $\beta(p)$, Glaister shows that the optimal pricing policy involves a price below cost initially, with price rising above cost as time proceeds, eventually converging on a long-run equilibrium above cost. The changing price changes β. The net result is that the growth curve of the stock of the new good is not in fact logistic but has a positive skew, very similar to the results found by Bain (1964) or Chow (1967) (e.g., a Gompertz curve). One might note that the result differs dramatically from that of Bass, where price fell over the life of the new product — a result inherently related to learning economies in new-good production. However, this model does not, like Bass, have any rivalry arguments.

This result of Glaister's is a good example of how, when one considers supply problems and associates them with a Mansfield-type epidemic model, the predictions of that model are much changed; in particular, the logistic diffusion curve is no longer predicted.

Glaister, however, takes his work further by also considering advertising policy. Holding price constant, he shows that the optimal policy is to have an early splurge of advertising which then dies out. The net result is that the growth curve of ownership then has a positive skew again. This, once again, is of particular interest, for as the epidemic results are so dependent on information transfer, it seems that an obvious extension is to consider advertising as a means of information transfer. Gould (1970) considers in some detail the optimal advertising policies of firms when information is passed by word of mouth, as in the diffusion models above, but we will not pursue that any further here.

These models by Bass and Glaister are very neoclassical in their approach. Metcalfe (1981) has approached the problem with a much less neoclassical analysis. He initially assumes that inter- and intra-firm diffusion generate a time-path for demand such that

$$\frac{dy}{dt}\frac{1}{y} = \beta(y_t^* - y_t) \tag{9.12}$$

where y_t is demand for the new good in time t. Equation (9.12) is a way of representing a logistic diffusion curve used by Chow (1967) and Stoneman (1976) and is justified by Chow on epidemic grounds. In equation (9.12) y_t^* is 'equilibrium' demand and is allowed by Metcalf (as by his predecessors) to depend on the price of the new capital good p; i.e.,

$$y_t^* = c - gp_t. \tag{9.13}$$

On the supply side Metcalf assumes that if r_t is the rate of profit in time t, Ω is the fraction of internally generated profits ploughed back in capacity expansion, and μ is the ratio of external to internal funds invested at any time, then capacity q_t will grow as follows:

$$\frac{dq_t}{dt}\frac{1}{q_t} = \Omega(1 + \mu)r_t. \tag{9.14}$$

One should note that in (9.14) Ω and μ are constant, by assumption, and moreover are not explicitly derived by reference to any maximizing or otherwise rational behaviour of the firm. If we now assume that there is a fixed coefficient technology for producing the new good with a capital–output ratio v and labour–output ratio l, with a price for labour w and depreciation at rate δ, then we can state

$$r_t = \frac{p_t - w_t l - \delta v}{v}. \tag{9.15}$$

We now allow that wages increase with labour input, so

$$w_t = w_o + \theta q_t \quad \theta > 0. \tag{9.16}$$

Then from (9.14), (9.15), and (9.16) we have

$$\frac{dq}{dt} \cdot \frac{1}{q} = \frac{\Omega(1 + \mu)}{v} p_t - \frac{(w_o l + \delta v)}{v} - \frac{\theta l}{v} q_t. \tag{9.17}$$

If we allow that demand and capacity grow in a balanced way, so that $q_t = y_t$ for all t, then from (9.12), (9.13), and (9.17) one can generate an expression for the rate of growth of usage as a function of the parameters of the model and y_t. Metcalf shows that this function is a logistic curve, with, as the diffusion proceeds, the profitability of adopting changing.

   This model of supply can be objected to on the grounds that there is no explicit consideration of the rational behaviour of firms, although it has the advantage that financial constraints on the supply industry are considered explicitly. One can also object to the analysis because the cost of producing the new good increase over time (see (9.16)), which is not really supported by the empirical analysis of new products (see for example Chow, 1967). If some learning economies or further technological advance in the supply industry were introduced, this objection might be overcome.

Spence (1979, 1981) has attempted to investigate models of the growth of new products, and his findings throw light on several parts of the analysis above. In his 1979 paper Spence investigates the firm's investment strategy in a new market. He shows that a firm financing expansion out of retained earnings, facing Penrose-type physical limits on growth and maximizing its net worth, will, given the behaviour of its competitors, invest as rapidly as possible up to some target level of capital and then stop. He also shows that a likely outcome of the oligopoly game is that each firm will behave in this way. Spence's analysis suggests that the gains to be made from leadership will encourage the fastest possible growth of output in the industry rather than the limitation of output to maximize short-term gain. This suggests fast early growth in the diffusion process and thus perhaps the skewed sigmoid again.

In his 1981 paper, Spence combines learning economies with a demand diffusion of sorts (based primarily on a learning argument) and also allows for oligopolistic rivalry. His results are not easily translated into the diffusion literature context, for he is not writing directly on this subject. A main result of his, however, is that the price of the new good will always fall over time as entry to the supplying industry occurs. This result is found without any demand diffusion. Once a demand diffusion is combined with this, one would expect to find Bass-type results.

These models tend to have a Mansfield-type, epidemic-learning-based, demand diffusion combined with a capital goods supplying sector that is monopolized. The monopolist maximizes his profits or present value given the demand function. We have shown that in such a world the diffusion curve will tend to not be logistic, as is predicted when the supply side is ignored. The skew of the diffusion curve depends upon how prices move over time, and that is related to whether firms maximize short-term profits or their present value. With price changing over time the post-diffusion level of use will depend upon the price approached as time proceeds, and the diffusion speed will depend on the time-profile of prices.

The models so far presented are restricted to the consideration of a monopolized supplying sector. It would seem advantageous to try and relax this assumption. At the same time, the models rely on an epidemic-type demand diffusion, and it would also seem reasonable to relax this assumption. Thus, in the next section we turn to a model with a probit approach to demand diffusion in which we can also make comments on a world with a non-monopolized supplying sector.

9.3 Probit diffusion and supply factors

Following David (1969), consider that a firm will be a user of a new technology in time t if its size X_i is greater than or equal to same critical level $\bar{X}$.[2] Allow that firm sizes among potential users (the industry) are distributed with the cumulative distribution function $F(X_i)$ and a probability density function $f(X_i)$, which for simplicity is assumed invariant with respect to time (which implies, somewhat unrealistically, that early use of the new technology does not affect firm size). Define D as the proportion of firms using the new technology at any moment in time (the time subscripts are dropped to simplify the notation); then

$$D = 1 - F(\bar{X}). \tag{9.18}$$

Following David, assume a new technology that requires purchase of a capital good at price p (only one capital good being required, whatever the size of the firm) and yields, in use, a saving of labour L^s per unit of output produced. Then, if the capital good does not depreciate, the wage is w, and r is the interest rate, one can define[3]

$$\bar{X} = \frac{rp}{wL^s}. \tag{9.19}$$

After substitution this yields

$$D = 1 - F\left(\frac{rp}{wL^s}\right). \tag{9.20}$$

Let N be the number of firms in the industry (assumed constant) and define x as the stock of new machines in time t; then

$$x = ND = N - NF\left(\frac{rp}{wL^s}\right). \tag{9.21}$$

If we now allow, again following David, that wages grow at a constant exponential rate λ, so that $w = w_o e^{\lambda t}$, one may write (9.21) as

$$x = N - NF\left(\frac{rp}{w_o e^{\lambda t} L^s}\right). \tag{9.22}$$

David assumes that r, p, and L^s are constant and then, assuming F is lognormal, traces out a cumulative normal time path for x_t/N, the proportion of firms using the new technology. We however wish to make p endogenous by considering the sector supplying the new capital goods. To start this invert (9.22) to yield the inverse stock demand curve facing the supplying industry, which we write as

$$p = g(x, t). \tag{9.23}$$

This tells us, for any t, at what price the next unit of new machines can be sold

[2] The analysis presented here is based upon Stoneman and Ireland (1983).
[3] Assuming completely myopic price expectations. Despite the desirability of introducing expectations more explicitly (see Chapter 7), this proved impractical.

given the existing stock of machines. Assume that the supplying industry has knowledge of (9.23).

Turn now to consider the supplying industry. The suppliers of capital goods are assumed to maximize their discounted stream of profits subject to (1) the industry demand function (9.23), (2) an initial stock of the new technology $x(0)$, (3) the condition that $dx/dt \equiv \dot{x} = q$, where q is the current output of all capital goods suppliers, (4) the reaction of rival producers, and (5) the cost conditions.

Write (9.24) as the general form of the cost function facing capital goods producer i with rivals j:

$$C_i = C(x_i, x_j, q_i, t) \qquad (9.24)$$

where x_i and x_j are the cumulative outputs of the firm and its rivals respectively and $x = x_i + x_j$. In (9.24) learning economies are allowed for through the influence of x on C, reductions in costs through further technological advance are included through the term in t, and the effect of current output on costs is allowed for through the term in q_i. With this cost function we proceed first to analyse a monopoly case and then to consider oligopoly. In the monopoly model it is assumed that $C_t = 0$ (subscripts are used to indicate partial derivatives), and this is allied with an assumption that λ, the rate of wage increase, is zero. These assumptions imply that the diffusion path is generated internally to the model rather than being the result of exogenous forces.

In the oligopoly model time is allowed directly to affect both the cost function and the inverse demand function, and the implications arising from different market structures are explored.

Consider then the case of a monopoly supplier. In this case one may drop the i and j subscripts and note that $\dot{x} = q$. Assume $C_t = 0$, $\lambda = 0$, and also that the cost function has the simple form $C(x, q) = c(x) \Psi(q)$, where $\Psi(0) = 0$, $\Psi_q > 0$, and $\Psi(q)/q$ has a minimum at the efficient scale of production $q^* > 0$. Thus, for $q > q^*$, marginal cost is rising.

The monopolist's problem is to choose the time path of q so as to maximize

$$V = \int_0^\infty e^{-rt} \, \Pi dt \qquad (9.25)$$

subject to $x(0) = 0$, $\dot{x} = q$, where r is the discount rate and

$$\Pi = g(x)q - c(x) \Psi(q). \qquad (9.26)$$

In Stoneman and Ireland (1983) it is shown

1   that the spread of the new technology will continue until $x = x^*$, where x^* is determined as that x at which, even if production costs are at a minimum ($q = q^*$), the price at which an extra unit of x can be sold is below average cost;

2   that, assuming $r > 0$, if $\Pi(x, q)$ is declining in x for a given q, then $\dot{q} < 0$ for all t, i.e., output of the new good, and therefore $\dot{x}$ declines over the whole diffusion period;

3   that, assuming $r > 0$, if $\Pi(x, q)$ has an interior maximum for some x^{**}, $0 < x^{**} < x^{*}$, then $\dot{q} > 0$ for low values of t and $\dot{q} < 0$ for higher values of t; i.e., one obtains a sigmoid diffusion path. From the definition of $\Pi(x, q)$, if $c_x = 0$ then Π_x has the sign of g_x, which is negative, and thus $\dot{q} < 0$ for all t. Thus, if this model is to generate a sigmoid diffusion path it is necessary that $c_x \neq 0$; i.e., learning economies are necessary to generate sigmoid diffusion. In this model it is not sufficient for a sigmoid diffusion curve that g_{xx} changes sign as x increases, which is the basis for most sigmoid diffusion models of this kind. One must, in fact, concentrate on the $\Pi(x, q)$ function and not the demand function;

4   that, if the diffusion curve is sigmoid, the point of inflexion on the curve and thus its skew will depend not only on the inverse demand function but also on the cost function;

5   that the price of the new capital good will decline for all t.

The logic of the model is that the monopoly producer faces a conflict between (1) a positive discount rate, making early profits attractive, and (2) rising marginal cost, making high levels of output unattractive, the whole conflict overlaid by costs reducing as experience accumulates. By resolving the conflict between time preference and rising marginal costs, output for each time period, and thus a price, is determined. As experience accumulates output can be produced at lower costs and the price clearing the market will be lower. Eventually so little of the market is left that the price at which the next unit can be sold will not cover even minimum average costs and so no further output is produced. The model predicts that every unit will be sold at a price below that at which the previous unit was sold. One should note that in this version of the model there are no exogenous forces generating the diffusion. Wages are not rising, nor is technology changing (the factors driving David's (1969) and Davies's (1979) diffusion models).

Let us turn now to consider an oligopoly model. Allow there to be a number of competing oligopolists supplying the capital good, and introduce demand growth by allowing the user wage to increase at rate $\lambda > 0$ per period. The simplifications introduced to obtain a tractable model are that

1   a firm only learns from its own accumulated experience;
2   costs are proportional to output;
3   all firms have the same cost function which for firm i is:

$$C(x_i, x_j, q_i, t) = c(x_i, t)\, q_i, \quad \text{all } i \tag{9.27}$$

One implication of the form of (9.27), which implies a profit function linear in current output, is that, without further modelling, one is able to discuss only 'steady-state' solutions in this model, whereas in the monopoly model the complete time-paths could be characterized.

The oligopoly game is modelled, following Spence (1981), by using the open-loop equilibrium concept, whereby each firm selects its optimal path given the paths of competitors, which are also optimal and are correctly anticipated by the individual firm. If one allows that there are n producers, and models a symetric equilibrium, so that all firms are the same, then

$$\frac{dx}{dt} = nq_i.$$

The problem of the ith firm is thus[4] to maximize

$$\int_0^\infty \Pi_i e^{-rt}\, dt$$

where

$$\Pi_i = \{e^{\lambda t} g(x) - c(x_i, t)\} q_i$$

subject to

$$\dot{x}_i = q_i;$$

$$x = x_i + x_j;$$

$$x(0) = 0.$$

The nature of the solution (see Stoneman and Ireland, 1983) is that there is an initial stage in which x adjusts to a level consistent with a steady-state path to yield $\dot{x}(0)$, and the solution then follows the steady-state path.

To illustrate the nature of the steady-state path, consider the case where $c_t = c_{x_i} = 0$ so $c(x, t) = c$, $\lambda > 0$, $r > 0$ and $n > 1$.

One may then show that

$$p = e^{\lambda t} g(x) = \frac{rcn}{nr - \lambda} \tag{9.28}$$

so that price is a constant on the steady state path. As (9.28) is decreasing in n, the more producers, the lower will be the price on the steady-state path; also, the lower will be the price cost margin, $(p - c)/p = \lambda/rn$. On the other hand, both price and price cost margin on the steady-state path are increasing in λ.

Given that price is constant, the spread of use of the new technology is driven by the wage rising at rate λ. To illustrate more precisely the implications of these results for the diffusion path, make the assumption that user firm sizes are distributed log-logistically; i.e., that

$$F(X_i) = [1 + \exp\{-(\log X_i - \alpha)/\beta\}]^{-1} \tag{9.29}$$

where α is the mean of the distribution and $\beta = \sigma\sqrt{3}/\Pi$ where σ^2 is the variance of the distribution. With this distribution one may derive that

$$p = e^{\lambda t} g(x) \tag{9.30}$$

---

[4] In this problem we have allowed $p = g(x, t)$ to be written as $p = e^{\lambda t} g(x)$, which will hold for certain distributions but not for others. It does hold for the distribution considered below.

where

$$g(x) = \frac{w_o L^s e^\alpha}{r} \left(\frac{x}{N-x}\right)^{-\beta}.$$

(9.31)

Given that

$$p = \frac{rcn}{rn - \lambda},$$

one may solve for x as

$$x = \frac{N}{1 + \exp\left\{\log\dfrac{N - \hat{x}(0)}{\hat{x}(0)} - \dfrac{\lambda}{\beta}t\right\}}$$

(9.32)

where

$$\log\frac{N - \hat{x}(0)}{\hat{x}(0)} = \beta \log\left\{\frac{cr^2 n}{L^s}\frac{1}{(rn - \lambda)w_o e^\alpha}\right\}.$$

(9.33)

Following Mansfield (1968), define the diffusion speed as

$$z = \frac{dx}{dt}\frac{1}{x}\frac{N}{N-x}$$

(9.34)

and from (9.32)

$$z = \frac{\lambda}{\beta}.$$

Thus the diffusion speed is the greater the faster wages rise and the smaller is the variance of the log of user firm sizes. It is independent of the number of producers. However, $d\hat{x}(0)/dN$ and $d\hat{x}(0)/dn$ are both positive (from 9.33).

Thus, the greater is N or n, the greater is $\hat{x}(0)$ and the greater will be the level of use of the new technology on the steady-state path. Furthermore,

1   x will be greater for all t the larger the number of producers and/or users of the new capital good;
2   prices and price–cost margins will be lower in capital goods production the larger the number of producers;
3   the diffusion speed z is invariant with respect to n and N.

Although this oligopoly analysis is not totally satisfactory, it does give some indication of the role that supply factors can play in the diffusion process. Objections to the analysis may be listed.

1   The oligopoly model is really concerned only with steady states.
2   The structure of the supplying industry is invariant with respect to time. It does not develop as, for example, Gort and Klepper's (1982) analysis suggests it should (see Chapter 16).
3   The model assumes perfect information and foresight on the behalf of producers (but not users), which is rather unrealistic.

One interesting deduction one can make from the analysis concerns patenting and its influence on diffusion. This is referred to elsewhere; and in, for example, the study of steam engines in the next chapter, it is argued that Watt's patent slowed down diffusion. This model would suggest that, if the patent laws affect the number of producers of the capital good so that n is low, the effect will be:

1   a higher price for the new capital good on the steady-state path;
2   a higher price–cost margin for capital goods producers on the steady-state path;
3   an unaffected diffusion speed; but
4   a lower level of use for all t on the steady-state path through a lower $\hat{x}(0)$.

The patent protection will thus reduce the level of use of the new technology but will increase the capital good producer's profitability.

Before drawing some conclusions, one can use the current model to represent a version of Spence's (1981) paper. If one assumes

1   a monopoly supplier;
2   given wages and no endogenous demand growth;
3   no endogenous learning;
4   continuous exogenous technological change in capital goods production,

then one can set up the firm's problem as to choose q to maximize

$$V = \int_0^T \{g(x)q - C(t,q)\}e^{-rt}\,dt$$

subject to $x(0) = 0$, $\dot{x} = q$, which is a problem of exactly the form analysed by Spence. It can be shown that, if $r > 0$, then the solution to the problem is that, if marginal cost decreases with time but increases with q, then $\dot{q} > 0$ for all t. If $r > 0$ but $C_{qt} = 0$, then $\dot{q} < 0$ for all t. Although it is possible to conceive of an example where q follows a path such that initially $\dot{q} > 0$ and then $\dot{q} < 0$, it is difficult to generate one. This does however represent another version of the model, and illustrates once again how changes in capital goods supply can affect the diffusion path.

9.4 Conclusions

Much of the literature on diffusion is directed towards explaining empirical observations of real-world diffusion processes. To date the majority of the studies attempting to model these processes have concentrated on demand-side phenomena. We have argued that this involves a significant omission, in that the time-path of usage will depend on the interaction of the supplying industry and the using industry. In this chapter we have attempted to rectify the omission by explicitly modelling the interaction of demand and supply. The models we have constructed assume that the market clears in each time period by the establishment of an equilibrium price. This is not necessary, however. One could construct models where queues build up for a new good (see for example the discussion of computer diffusion in Stoneman, 1976). However, whatever approach one takes, it is important to realize that supply also matters.

We have shown that the market structure of the supplying industry, the objectives of the firms in that industry, the degree of learning economies, and the extent of any financial constraints can be important influences on the diffusion path. It would be particularly informative if we had some empirical work to support the theoretical reasoning. We do not. From the next chapter, however, it is clear that supply factors can be important when one attempts to analyse real-world diffusion phenomena.

References

Bain, A. (1964), *The Growth of Television Ownership in the U.K. since the War*, Cambridge University Press.

Bass, F. (1978), *The Relationship between Diffusion Rates, Experience Curves and Demand Elasticities for Consumer Durable Technological Innovations*, Working Paper, Kranner Graduate School of Management, Purdue University, Illinois.

Chow, G. (1967), 'Technological Change and the Demand for Computers', *American Economic Review*, 57, 1117–30.

David, P. (1969), *A Contribution to the Theory of Diffusion*, Stanford Center for Research in Economic Growth, Memorandum no. 71.

Davies, S. (1979), *The Diffusion of Process Innovations*, Cambridge University Press.

Glaister, S. (1974), 'Advertising Policy and Returns to Scale in Markets where Information is Passed between Individuals', *Economica*, 41, 139–56.

Gort, M. and Klepper, S. (1982), 'Time Paths in the Diffusion of Product Innovations', *Economic Journal*, 92, 630–53.

Gould, J.P. (1970), 'Diffusion Processes and Optimal Advertising Policy', in E.S. Phelps *et al.*, *Microeconomic Foundations of Employment and Inflation Theory*, 338–68, W.W. Norton, New York.

Mansfield, E. (1968), *Industrial Research and Technological Innovation*, W.W. Norton, New York.

Metcalfe, J. (1981), 'Impulse and Diffusion in the Study of Technical Change', *Futures*, 5, 347–59.

Spence, A. (1979), 'Investment, Strategy and Growth in a New Market', *Bell Journal of Economics*, 10, 1–19.

Spence, A. (1981), 'The Learning Curve and Competition', *Bell Journal of Economics*, 12, 49–70.

Stoneman, P. (1976), *Technological Diffusion and the Computer Revolution*, Cambridge University Press.

Stoneman, P. and Ireland, N. (1983), 'The Role of Supply Factors in the Diffusion of New Process Technology', *Economic Journal*, Supplement, March, 1983, 65–77.

Chapter 10

Steam Power and Computerization:
Two Case Studies in Diffusion

10.1 Introduction

At first sight if may seem odd to pair steam power and computerization to illustrate how diffusion has been studied. There is however a certain logic in the choice.

1   Steam power is often considered, rightly or wrongly, to be the major innovation or *sine qua non* of the first industrial revolution, whereas the computer is considered to play a similar role in the second industrial revolution (also known as the 'information revolution').
2   The choice of two historically disparate technologies enables one to illuminate how similar principles of analysis can be applied to quite dissimilar epochs, while at the same time illustrating how the historian must overcome problems not faced by the researcher on contemporary phenomena, such as data shortages.
3   Steam power and computerization have a number of similarities in their development; for example, one can identify generations of steam engines as one can computers.

10.2 Computerization in the UK and USA

In this section we investigate the spread of digital computer use in the USA to 1965 as analysed in Chow (1967) and in the UK to 1970 as discussed in Stoneman (1976). The 1970 date is particularly appropriate, 1971 being the year that the first micro-processor was launched.

The first electronic digital computers appeared in the UK and West Germany during the Second World War and were used mainly for code-breaking. The first commerical machines were not really available until 1950. In the years that followed the basic technology underwent numerous improvements whereby the price–efficiency ratios of the machines for sale and in use improved dramatically. These improvements have often been represented by discussing computer generations. One can label the generations as follows:

1   the zeroth-generation machines — mainly primitive prototype machines appearing up to 1955;
2   the first generation, consisting of valve-based technology machines, dated 1956–60;
3   second-generation machines, using transistor-based technologies and appearing around 1960;

4   third-generation machines, based on integrated circuits and appearing around 1964–5;
5   fourth- and later-generation machines, appearing from the early 1970s and using large- or very large-scale integration in their circuitry.

The dramatic changes in the computer over time mean that

1   a head count of the numbers of computers in existence will not give a good estimate of the extent of computer usage, for it will take no account of the differences in quality between computers installed; and
2   in order to generate a series on computer prices one needs to take into account the differing quality of machines being sold.

Before one can proceed with any analysis of diffusion it is necessary to tackle this problem of quality change directly. The method that has been used carries a number of labels, the two most popular labels being the 'characteristics approach' and the 'hedonic price approach'.

The basis of the characteristics approach is similar to Lancaster's (1966) approach to consumer theory as first detailed by Ironmonger (1972). A good is considered to be a bundle of characteristics. By valuing the characteristics in the good we get an estimate of the good's 'quality'. By summing qualities over all machines in existence we will get an estimate of the total stock adjusted for quality. By dividing an individual machine's price by its quality we get an estimate of quality-adjusted price. By averaging over all machines installed in a given year we can generate a price, quality adjusted, each year, and thus produce a quality-adjusted time-series on prices. The basic principle therefore is that, if we define P_{it} as the price of machine i in time t and $\mathbf{X_{it}}$ as the vector of characteristics embodied in machine i in time t, then

$$P_{it} = f(\mathbf{X_{it}}, e_{it})$$

where

$$\mathbf{X_{it}} = (x_{1it}, x_{2it}, \ldots, x_{nit}) \tag{10.1}$$

and e_{it} is some random error term.

The hedonic approach to the treatment of quality differences has been extensively criticized by, among others, Deaton and Muellbauer (1980). Although it may be possible rigorously to validate the procedures used in these studies (cf. Deaton and Muellbauer, 1960, pp. 263–4), no attempt is made to do so in the studies themselves.

The problems that arise in applying the hedonic procedure are twofold:

1   defining the appropriate characteristics;
2   estimating the shadow prices.

In Chow's work the appropriate characteristics are considered to be (1) multiplication time, M, (2) memory size, S, (3) access time, A. In Stoneman a data set covering twenty-one characteristics was assembled, and using empirical performance as the yardstick the set was reduced to (1) cycle time, Z, (2) floor area, F, and (3) maximum memory size, $\bar{S}$. The main problem with both sets of characteristics in that software aspects are nowhere reflected.

In both Chow and Stoneman the function $f(X_{it}, \epsilon_{it})$ is considered to be linear in logs; i.e., the function is of the form

$$\log P_{it} = a_t + b_{1t} \log x_{1it} + b_{2t} \log x_{2it} + \ldots \epsilon_{it}, \qquad (10.2)$$

this choice being made on empirical grounds.

The estimates, $\hat{a}_t, \hat{b}_{jt}$ ($j = 1, \ldots, n$) were obtained by applying OLS regression techniques (although weighted least squares might have been better — thus giving more weight to those machines most frequently purchased) to (10.2) for different t. Thus, for example (10.2) was estimated across all machines introduced in a given year or a given period, or across all machines of a given generation. Tests were then made (the Chow test) to see if the $\hat{b}_{jt}$ varied significantly across time. In Chow the P_{it} were measured by rentals and in Stoneman they were measured by the average purchase price. Chow finds that the $\hat{b}_{jt}$ do not vary significantly across his sample although the intercept term does. Chow's sample covers 1960-5. Stoneman finds that, looking at 1955-70, there are significant changes in the $\hat{b}_{jt}$. Given the constancy found in Chow's results, he selects his $\hat{b}_{jt}$ estimates from a pooled regression covering 1960-5 (using the 1960 intercept) to use as his shadow prices. He then generates his quality of machine i, $\hat{P}_i$ by

$$\log \hat{P}_i = -0.1045 + 0.5793 \log S - 0.0654 \log M - 0.1406 \log A.$$
$$\qquad\qquad (0.00354) \qquad (0.0284) \qquad (0.0293)$$

(standard errors of estimates in parentheses); i.e., the quality of a machine is measured by what its rental would have been had it been introduced in 1960. Stoneman, somewhat arbitrarily, given the change in coefficients over time, chooses coefficients appropriate to 1963 to generate

$$\log \hat{P}_i = 1.269 - 0.191 \log Z + 0.425 \log F + 0.247 \log \bar{S}$$
$$\qquad\quad (-3.10) \qquad\qquad (6.22) \qquad\qquad (5.06)$$

(*t*-statistics of estimates in parentheses), in which case quality is measured by what a machine would have cost had it been introduced in 1963.[1]

In generating the quality-adjusted price series, Chow does not weight machines by sales, whereas Stoneman does. The net result of these procedures is to yield the data in Table 10.1. The price and quantity data are then used by Chow to estimate a Chow logistic diffusion curve and a Gompertz curve from time-series data on national computer use, the rationalization of the curves being as detailed in Chapter 8. As the Gompertz yields the better fit, we will only report the results on that curve. The Gompertz may be written as

---

[1] Horsley and Swann (1981) have recently reworked the data for the UK.

$$\frac{dS_t}{dt} \frac{1}{S_t} = \alpha(\log S_t^* - \log S_t).$$

(10.3)

The post-diffusion level of use is given by

$$\log S_t^* = \beta_0 + \beta_1 \log P_t + \beta_2 \log Y_t$$

(10.4)

where P_t = computer prices relative to the GNP deflator;
Y_t = output (GNP).

Table 10.1 *Computer quantities and prices, quality-adjusted, UK and USA**

| | USA | | UK | |
| | Quantity t | Price | Quantity t | Price |
Year (t)	Quantity $t-1$	index	Quantity $t-1$	index
1954		3.25		1.63
1955	2.678	2.96	2.194	2.05
1956	2.410	2.53	2.013	2.00
1957	2.129	2.32	3.669	1.73
1958	1.644	2.03	1.914	1.96
1959	1.501	1.58	1.537	1.62
1960	1.520	1.07	2.073	0.94
1961	2.006	0.90	1.787	1.03
1962	1.682	0.68	1.706	0.97
1963	1.489	0.57	1.587	1.00
1964	1.428	0.42	1.472	0.65
1965	1.419	0.34	1.686	0.49
1966			1.813	0.40
1967			1.507	0.41
1968			1.409	0.55
1969			1.377	0.51
1970			1.192	0.47

* The UK and US price series are not comparable with each other; neither are the quantity series. For units of quantity see the text.
Source: Chow (1967), Stoneman (1976).

From (10.3) and (10.4) one obtains an estimating equation:[2]

$$\log S_t - \log S_{t-1} = \alpha\beta_0 + \alpha\beta_1 \log P_t + \alpha\beta_2 \log Y_t - \alpha \log S_{t-1} + e_t \quad (10.5)$$

and the estimate of (10.5) is given by

$$\log S_t - \log S_{t-1} = 10.45 - 0.5924 \log P_t - 1.160 \log Y_t - 0.2828 \log S_{t-1}$$
$$(0.4056) \qquad (1.845) \qquad (0.0906)$$

$$R^2 = 0.843 \qquad (10.6)$$

(standard errors in parentheses). As can be seen, the estimates of $\alpha\beta_1$ and $\alpha\beta_2$ are not significantly different from zero. Although Chow does try other methods for

[2] In which e_t is the error term, which is assumed to have the standard properties for OLS regression. The justification for the error term is not made explicit (cf. Chapter 6).

obtaining significant estimates of the price and output coefficients, in general the effect of these two variables on the growth of the computer stock is difficult to isolate.

In Stoneman a similar exercise is performed. In this case the 'stock of computer data' is adjusted to allow for unfilled orders so that the S_t variable will better reflect demand. This is an illustration of supply-side effects (see Chapter 9). Once again, different types of diffusion curves are estimated, although the Gompertz works best. The problem of separating out the effect of P_t and Y_t again arises and is overcome by reducing P_t to a generation dummy variable and by using external information to argue that $\beta_2 = 0.7$. This then yields the following estimates:

$$\log S_t - \log S_{t-1} = 1.652 - 1.304D_1 - 0.772D_2 - 0.438D_3$$
$$(6.672)\ (-3.93)\quad (-3.22)\quad (-2.96)$$

$$+\ 0.234$$
$$(5.386)\quad (0.7 \log Y_t - \log S_{t-1})\ .\ R^2 = 0.847\ DW = 1.88\quad (10.7)$$

where D_1, D_2, D_3 are generation dummies and t-statistics are given in parentheses. This result yields an estimate of $\alpha = 0.234$, which suggests that S_t will go from $S_t/S_t^* = 10$ per cent to $S_t/S_t^* = 90$ per cent in a period of twenty-three years.

Table 10.2   *Prime reasons for installing computers*

Reasons	Installations
Difficulty in recruiting staff	22
To replace worn out equipment	197
To reduce cost of processing data	83
To speed processing of data	97
To provide better service to management	325
To provide better service to customer	75
Other	27
Total number of installations	806

Source: Stoneman (1976)

Table 10.3   *Reasons for rejecting computers*

	All	Computers Used	Not used
Number sampled	95	24	71
Computerization not considered	62	8	54
Computerization considered but:	33	16	17
Insufficient work	13	3	10
Too expensive	11	3	8
No technical staff	1	—	1
Intend to; no time yet	4	4	—
Other	11	2	9

Source: Stoneman (1976)

Table 10.4   *Practice with respect to evaluation methods*

	Number using	
	Sample no. 1	Sample no. 2
Discounted cash flow	24	12
Payback	14	4
Book return	2	5
Other	16	26
Total	72	56

Source: Stoneman (1976)

In Stoneman's work the analysis is extended to look at diffusion in individual industries and also the diffusion of certain computer applications. However, rather than follow that here we will consider other data provided in this study, those that consider computerization directly rather than looking at aggregate diffusion curves. In Stoneman's Chapter 6, the results of surveys of firms' decision-making procedures with respect to computerization are discussed. Three samples of the data are detailed in Tables 10.2, 10.3, and 10.4.

It is argued, on the basis of this and similar data, that one can throw doubt on a theory of technique choice that depends upon maximization, for the data indicate that in many cases no evaluation takes place, and in the cases where it does take place the methods used are of a satisficing and not optimizing (or maximizing) nature. Moreover in Table 10.3 the data indicate that a number of potential users never even considered computerization, so any theory of technique choice that considered that techniques were continually compared is not acceptable.

An interpretation of the data on the lines of behavioural theory (cf. Cyert and March, 1963) is provided.[3] With a bias towards the present context, this theory says that prior to a change in technique a problem is required, or at least an expected problem or a pet solution seeking a problem. The problem stimulates a search for a solution, which to be accepted must satisfy certain standard rules. Once the decision is made to implement the solution a further problem must arise before another solution will be considered. To support this argument, the data in Table 10.2 are interpreted as follows. An initial important observation is that, among the reasons given for installing computers are some that imply the inoperability of the existing technique. These reasons are listed as the replacement of worn out equipment, savings in office space, savings in manpower, and difficulty in recruiting staff. It may well be that office and staff considerations are just indicators that the firm is not willing to operate its old technique at higher prices, but they still indicate that the firm is being forced to reconsider its technique.

---

[3] One can argue that in the long run behavioural theory and maximization are not incompatible. The problem is whether the economy is ever in the long run.

It is argued that in the sample 27–40 per cent of machines were installed primarily because the previous technology was no longer operable. The conclusions derived from these arguments state that the *timing* of the technique choice decision is determined for the firm by the interaction of its goal formation and goal achievement characteristics, leading to problems and the search for solutions. The actual *composition* of the technique choice decision depends on the nature of the search and evaluation procedures. The search procedures are influenced by the number of machines installed in comparable organizations and, if new tasks are being considered, whether a machine is already installed. The evaluation procedures are simple-minded; satisficing rather than optimizing is the rule; non-quantifiable benefits are considered as parameters; feasibility is important; and short-run considerations dominate. Moreover, it is illustrated that the conception of the costs and benefits is biased by the views of the evaluator.

Although in Stoneman's own empirical study of data on computer stocks very little of the richness of this approach is reflected, the information provides a neat contrast to the formality of the diffusion models discussed above. However, arising out of it also is, first, a lead into the behavioural-based work on technical change of Nelson and Winter discussed in several chapters below and, second, a reference back to Chapter 7, where Davies's model of diffusion is specifically based on a behavioural approach to technique choice.

In this study of computerization we have had occasion to consider not only diffusion models but also quality adjustment and survey data. This detail adds more flesh to the bones of the diffusion theories discussed above. We now turn to steam engines for detail of diffusion in an historical context.

10.3 The diffusion of steam power to 1860

This section draws overwhelmingly on the work of Von Tunzleman (1978). I have selected those parts of his study that refer to the diffusion of steam power, and it is that material that is presented here. We shall have occasion in Chapter 15 to consider other parts of his work, but as our purpose here is to discuss diffusion rather than steam power *per se*, the selectiveness seems justified.

The steam engine went through essentially three generations:

1   the atmospheric engines of Savery (introduced in 1695) and Newcomen (introduced in 1712), which, because of their required alternate heating and cooling of the cylinder, were always wasteful of fuel;
2   the Watt engines (patented in 1769), which achieved greater fuel economy by the use of separate condensers;
3   the high-pressure engines of the early nineteenth century (associated with Trevithick), which yielded even further gains in fuel economy.

Across the three generations, progress was illustrated by the achievement of greater efficiency in the use of fuel, but at each stage capital intensity increased.

However, not only did technology improve by changes in generations, but each generation also experienced improvements in performance. One measure of performance is the increasing horsepower of the most powerful engines. In Table 10.5 we illustrate how this changed for Newcomen (first-generation) engines.

Table 10.5   *Cylinder diameters and power of largest extant Newcomen engines*

Date	Locality	Estimated horsepower	Cylinder diameter (in.)
1712	Dudley	5	21
1720s	–	–	15–24
1733	Heaton	15	33
1734	Heaton	–	42
1755	Horsehay	30	48
1757	Horsehay	45	60
1763	Walker Collier	52	74
1810	Whitehaven	100+	82

Source: Von Tunzleman (1978), p. 25

To study diffusion we need some estimate of the usage of new technology at different times, but we do not have good data on the number of engines installed in toto. In Table 10.6 we detail horsepower installed in a number of major industrial towns. Von Tunzleman suggests that installed horsepower grew at about 6 per cent per annum between 1800 and the mid-1820s with slower growth thereafter, with a total installed horsepower of 239,000 in 1838 for the British Isles.

Table 10.6   *Approximate horsepower installed, 1800–1840*

	Early 1800s	Mid-1820s	Late 1830	Growth rates 1800–25	1825–40
				%	%
London	1355	5460	–	7.3	–
Manchester	c. 650–750	4760–4875	9925	7.7–8.4	6.3
Bolton	–	1604	5251	–	10.4
Leeds	c. 450	2318–2330	–	6.8	7.6
Birmingham	–	1000–1262	3436	–	5.7
Glasgow	–	4480	–	–	–

Source: Von Tunzleman (1978)

The lack of detailed data obviously limits the scope for a detailed econometric analysis; however, Von Tunzleman carries out five major diffusion exercises, covering:

1   the shift from atmospheric to Watt engines at the end of the eighteenth century;
2   the shift to high-pressure engines in manufacturing in the first half of the nineteeth century;

3   the spread of steam power using techniques in the cotton industry in the
    nineteenth century as illustrated by
    (a)   the self-acting mule;
    (b)   the power loom;
4   the spread of high-pressure engines in Cornish mining in the first half of
    the nineteenth century.

The underlying basis of Von Tunzleman's diffusion analysis is the David
(1969) model detailed in Chapter 7. In analysing the shift from atmospheric
to Watt engines at the end of the eighteenth century, it is observed that the
Watt engine had higher capital costs but lower fuel inputs. Von Tunzleman
thus proceeds to calculate the threshold price of coal at which it would be
profitable for an entrepreneur to use a Watt engine rather than an atmospheric
engine. He estimates that prior to the late 1790s the thresholds were:

1   7*s*. 10*d*. per ton between a Newcomen engine and a Watt rotative engine,
    on a comparison of total costs;
2   14*s*. per ton between a Newcomen engine and a Watt rotative engine on a
    comparison of Watt's total costs and Newcomen operating costs as is appro-
    priate if a Newcomen is already installed;
3   5*s*. 10*d*. per ton between a Newcomen and Watt reciprocating engine on a
    total cost comparison.

After the late 1790s the threshold price fell to 1*s*. 1*d*. per ton in case (1) and
6*s*. 6*d*. per ton in case (2). This change is due to the expiry of Watts patents.
Prior to their expiry Watt had charged various kinds of premia — in the case of
reciprocating engines, the premium came from a policy of charging users one-
third of the savings in fuel costs that the engine effected, whereas for rotative
engines the premium was capitalized, setting it usually at £5 per horsepower
per annum (£6 in London). David's threshold approach would suggest that such
pricing behaviour would considerably slow down the diffusion process. It is a
good example of how supply-side forces and the patent system can affect a
diffusion process (cf. Chapter 9).

These calculations suggest that the use of the Watt engine would be very
responsive to the price of coal. Of course, other factors mattered; for example,
the Newcomen engine was too erratic in motion for delicate work in a textile mill.

The threshold approach is also used to look at the spread of high-pressure
engines, which were used extensively in mining in Cornwall but spread slowly
in manufacturing. High-pressure engines were more fuel-efficient than earlier
engines but the degree of fuel saving depended on the speed at which they
were run. In Cornwall fuel economy was maximized by slow running, but in
Lancashire, for example, the need for high speeds to run delicate machinery
precluded the achievement of the same economies. The differences in diffusion
speed may thus well be due to differences in running costs. Von Tunzleman
calculates that, if coal requirements were cut by three-quarters as a result of
installing a Cornish engine in Manchester, annual costs would have fallen by £185.

However, the capital costs of achieving this would be £230 per year. This apparent lack of profitability could well have been a major factor generating slow diffusion. It is suggested that coal prices above 12*s*. per ton would have been required to make the Cornish engine worthwhile.

The upshot of this discussion is that the location of steam engines should be very responsive to coal prices. In Table 10.7 the data on the location of textile mills in relation to coal prices in 1838 are given. These data support the basic hypothesis.

Table 10.7   *Textile mill location, coal prices and steam power use*

	Total steam horsepower	< 10*s*./ton	% in locations with coal at: between 10*s*. and 20*s*./ton	> 20*s*./ton
		%	%	%
Cotton	40,783	96	4	< 0.5
Worsted	5,863	86	12	2
Woollen	10,887	81	19	1
Flax	3,134	69	29	2
Silk	2,320	54	33	12

Source: Von Tunzleman (1978), p. 66

For his study of steam power in the cotton industry, Von Tunzleman concentrates on the diffusion of self-acting mules and the power loom. He states:

the diffusion of machinery and power in the cotton industry cannot be understood if factors of yarn and cloth quality are not taken into account. But whereas in the first few years of the nineteenth century it was the finer qualities that bore the brunt of the impact of steam power, twenty and thirty years later it was the coarse grades. Since coarse spinning was so 'power intensive' any improvement in the application of power should have had a disproportionately large effect on the costs of production. (Von Tunzleman, 1978)

If we consider the self-acting mule, this was available in 1830 but by 1834 only about 3 per cent of the whole spindleage was of this type. The new technology made little headway in fine yarns until about 1860, but in the second quarter of the nineteenth century coarser yarns were affected. Von Tunzleman's figures suggest that, in the 1830s,

1   for a low-grade yarn an entrepreneur could expect a return on total costs 9 per cent higher from self-actors than from hand mules;
2   on medium-grade yarn the hand mule was superior on both capital and running costs;
3   there was no economic incentive to replace working hand mules by self-acting mules;
4   a conversion of a hand mule could yield a 7 per cent rate of return on low-grade yarn. However, a rise in coal prices could easily have removed this gain.

The argument then is that, for the 1830s, self-acting mules were profitable only for coarser yarns. However, with falls in power costs in the 1850s finer spinning on self-actors became profitable. Again, it is argued that the slow diffusion is related to profitability, although in addition one could isolate other influences.

Table 10.8 *Power looms installed in the UK, 1813–1849*

Year	Number installed
1813	2,400
1820	18,140
1825	40,000
1830	80,000
1835	108,210
1849	249,627

Source: Von Tunzleman (1978).

The power loom was patented in 1803, but only in 1813 was the first workable machine available. Numbers installed are shown in Table 10.8. Despite the apparently rapid spread of power looms, however, it was nearly 1850 before the hand loom weavers were threatened with extinction. The reasons for this slow diffusion by period are that in the 1820s the hand loom weavers took lower wages, thereby reducing the gain to be derived from power looms, and moreover the power looms were still limited in performance. The limitations on performance were partly relieved in the 1830s, although the power loom was still unable to compete economically with either fine or figured work produced by hand looms. However, there was a further succession of improvements to power looms in the 1840s that tilted the balance towards the power loom even for finer products.

These two investigations of the introduction of new technology into the cotton industry are supplemented by the study of the introduction of the stationary high-pressure engine in Cornish mining. Von Tunzleman considers that one can identify a high-pressure engine by its 'duty'. Duty is defined as the number of pounds raised 1 ft high by consuming 84 lb of coal. An engine producing 35 million of duty is considered to be a high-pressure engine. He then argues that, following much of the diffusion literature, one can attempt to approximate the diffusion process by a logistic curve. His reasoning for the logistic relationship is as follows:

Initially the new technique is in an experimental stage. The scope for its application can only be known at some cost. There may be considerable resistance, perhaps out of ignorance but also because its cost in relation to the advantage it brings is high. . . . if its adoption does, however, lead to significant economic gain other firms in different cost subsections or less favourably disposed to innovation will be induced to follow suit. . . . More and more firms introduce it, until diminishing returns set in. Gradually a ceiling is approached at which all those firms that can benefit will have done so. (Von Tunzleman, 1978)

This heuristic explanation does not really give us a rational for an S-curve, let alone a logistic; it just gives a story of why diffusion is not immediate.

However, the logistic is estimated.

D is defined as the percentage of all engines that are of 35 million or over in duty.

$$D^* = \lim_{t \to \infty} D_t.$$

D^* is (apparently arbitrarily) set to 91 per cent. It is estimated that, for the period 1816–38,

$$\log \frac{D_t}{91 - D_t} = -4.402 + 0.2516t \quad R^2 = 0.93. \tag{10.8}$$

We are not given any t-statistics. The predictions from (10.8) understate the rate at which the improved practices were diffused during the central years of 1825–32. In fact, the estimated logistic has a point of inflexion in 1826, whereas the actual curve was more skewed to the right, with a point of inflexion around 1830.

Von Tunzleman explains the faster-than-predicted diffusion on the grounds that many of the technical advances that supported the high-pressure engine were disembodied in character. He argues that on the logistic curve the rate at which high-duty engines increase is governed by the number of engines already performing at high duty; however, if advances are disembodied, the rate of diffusion can proceed at a rate unencumbered by past history.

We find this to be a rather strange argument in the context of the logistic curve, which we have argued above is justified primarily on epidemic grounds. Von Tunzleman's arguments, plus his previous preference for David's model, suggest he might have been better occupied in estimating a lognormal diffusion curve, David's justification for which explicitly takes account of the degree of embodiment of the new technology (capital costs). Perhaps the most interesting point, however, is that Von Tunzleman believes that the methods applied to study diffusion in a modern context can be fruitfully applied to an historical example even if the application is subject to some criticism.

10.4 Conclusions

In this chapter we have considered two separate technologies and their diffusion. Their historical disparateness was however noticeable only in the minimal influence it had on the analysis of the processes. The same principles were governing the study of both diffusion processes. The main differences are related to the data problems that are involved in historical research and thus the lower level of quantification in the study of steam power. The wealth of data on computers, on the other hand, enabled one to study diffusion processes not only by estimating curves but also by considering the technique choice decision directly. Moreover, it enabled one to investigate and quantify quality changes over time. Even if it is only on these grounds, the material in these chapters justifies inclusion; but the other purpose — to illustrate the nature of real-life diffusion processes — was the main reason for presenting the material.

With this chapter we now have come to the end of the second part of this book. Our starting point for this part was that, through invention and innovation, a new technology was available. We went on to consider those forces that will influence the rate at which the new technology is accepted or diffused. Our material is both theoretical and empirical, the present chapter being a major contribution to the empirical section. It may appear that there is something of a contradiction between the work in this chapter and the formal models discussed above, for example on the question of firm behaviour. The computerization study suggests non-profit-maximizing behaviour, the models to a large extent relying on maximizing behaviour. This should not be particularly surprising. Every student eventually is asked to write on whether firms actually profit-maximize and whether it really matters. We do not intend to try and resolve the issue here. It is sufficient for us to state that the maximization assumption in the models may not be crucial. If one just compares David's (1969) and Davies's (1979) models of inter-firm diffusion, it is clear that in terms of prediction and performance they differ very little, but one is based on profit maximization and the other on rule-of-thumb decision-making procedures. That the behavioural approach has a greater grounding in reality, and thus greater intuitive appeal, is not disputed, but this does not invalidate the insight that one can obtain by the use of the more formal analysis based on profit-maximizing assumptions.

The discussion of these last five chapters merits some conclusions. We provide these by considering the forces that we have isolated as impinging on the diffusion process.

1   New technology and its application involves risk and uncertainty, and the attitude of economic actors to such factors needs to be considered. The degree of uncertainty involved may be related to the level of use of the technology and how learning proceeds.

2   The learning process impinges at a number of points. It can involve learning about existence of a new technology or learning about its true characteristics. It can also involve learning about ways of producing technologically new products that reduces their cost of production. For a given initial state of knowledge, the faster that learning occurs, the faster diffusion is likely to proceed.

3   During a diffusion process, learning may not be the only factor changing. The good itself may be improving, as we have illustrated with steam engines and computers. This improvement may have a two-edged effect on diffusion: a direct effect, stimulating greater use, and an indirect effect, whereby expectations of future advances may lead to the postponement of adoption.

4   We have argued that to a large degree the adoption decision for the firm will be related to expected profitability, which in turn will be dependent upon a number of factors. Thus differences between firms will be important, as may the behaviour of the industry supplying any new goods. The market structure of user and supplying industries has also been considered relevant, as has the oligopoly game. The existence of a patent system and the ease of imitation can

impinge here also. However, perhaps as worthy of attention as the general point is that it may be unrealistic to consider profitability as exogenous to the diffusion process. By looking at the interaction of supply and demand forces we have stressed one dimension of its endogeneity. It may be that profitability varies across the whole time-profile of the diffusion process.

5 Finally, we can return to a question that was raised at the beginning of Part II. Why, rationally, might not all firms/users adopt a new technology immediately? Our answers are now numerous: they cover lack of knowledge, misconceptions of returns available, positive quasi-rents on existing technology, expectations of future advances, insufficient expected return, and the nature of the oligopoly game. Each has its own role to play, and each has been discussed in detail above.

References

Chow, G.C. (1967), 'Technological Change and the Demand for Computers', *American Economic Review*, 57, 1117–30.

Cyert, R. and March, J. (1963), *A Behavioural Theory of the Firm*, Prentice-Hall, Englewood Cliffs, NJ.

David, P. (1969), *A Contribution to the Theory of Diffusion*, Stanford Center for Economic Growth, Memorandum no. 71.

Davies, S. (1979), *The Diffusion of Process Innovations*, Cambridge University Press.

Deaton, A. and Muellbauer, J. (1980), *Economics and Consumer Behaviour*, Cambridge University Press.

Horsley, A. and Swann, G.M. (1981), 'Computer Price Functions', Working Paper, ICERD, London School of Economics.

Ironmonger, D.S. (1972), *New Commodities and Consumer Behaviour*, Cambridge University Press.

Lancaster, K. (1966), 'A New Approach to Consumer Theory', *Journal of Political Economy*, 74, 132–57.

Stoneman, P. (1976), *Technological Diffusion and the Computer Revolution*, Cambridge University Press.

Von Tunzleman, G. (1978), *Steam Power and British Industrialisation to 1860*, Clarendon Press, Oxford.

Part III
The Impact of Technological Change

Chapter 11

The Impact of Technological Change on Output and Employment: A Microeconomic Approach

11.1 Introduction

This chapter is the first in a series of chapters in which we investigate the impact of technological change on the economy. We will look at effects on output, employment, income distribution, investment, market structure, etc., to see what theoretical work suggests will arise as a reaction to technological change, and to what extent the theory is supported by empirical work.

The tendency in the chapters that follow is to consider that a new technology has appeared though forces largely discussed in Part I, i.e., that innovation has already occurred, but that only as the diffusion process proceeds will the impacts start to make themselves felt. However we will not, to any great degree, be reflecting the endogeneity to the economic process of the invention, innovation, and diffusion process; to consider the endogeneity while looking at the impacts would be desirable, but is only to a limited degree attainable.

In this chapter we start our analysis by looking at the impact on output, prices, and employment of technological change at the firm and industry level. More than with most of the chapters that follow, the theoretical analysis is rather neoclassical, although the empirical work is less so. We consider less neoclassical approaches in the next chapter.

11.2 Theory: prices, output, and employment at the firm level

We begin by considering an individual firm producing output q^1 of an homogeneous product. Let a new process technology be introduced that shifts the firm's marginal cost curve from MC_0 to MC_1 (see Fig. 11.1). Assume that the firm is a profit-maximizer. Allow first that the firm is in a perfectly competitive industry and does not gain monopoly power through the innovation. In this case the price received for the product will be the same before and after the innovation, say p_0, and the profit-maximizing firm will increase its output (from that making $MC_0 = p_0$) to bring MC_1 equal to p_0. Thus (see Fig. 11.1) the technological change will lead the profit-maximizing firm to increase its output, if all other things remain constant.

If we move away from the perfectly competitive firm, then the profit-maximizing firm will set marginal cost equal to marginal revenue (MR); i.e., after the technological change we have

---

[1] Through this chapter we use a lower-case q for output of the firm and upper-case Q for industry output.

$$MC_1 \ = \ MR_1$$

whereas prior to the change we had

$$MC_0 \ = \ MR_0.$$

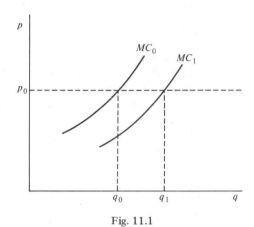

Fig. 11.1

In Fig. 11.2 let DD' be the firm's demand curve, with associated marginal revenue curve MR; then technological change will mean that output increases from q_0 to q_1, with price falling from p_0 to p_1.

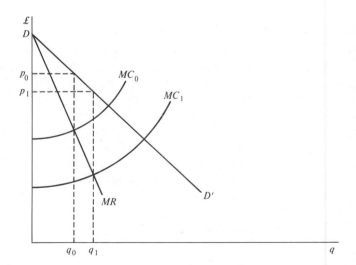

Fig. 11.2

From such diagrams it is clear that, with an upward-sloping marginal cost curve, any technological change that lowers the marginal cost curve for all levels of output will generate higher levels of output and lower prices as long as the demand curve is not vertical. By similar diagrammatic techniques we can argue that a product innovation that shifts the demand curve and the marginal revenue curve outwards will also yield higher output. In both cases, profit maximization and an invariant market structure are being assumed.

To make the conditions for these results more explicit and to investigate impacts on factor employment, we will consider a more formal model.

Let us start with the simplest case – a firm with a Cobb–Douglas production technology using two inputs, x_1 and x_2, to produce a homogeneous product q. To simplify further let x_2 be fixed at some level k. If factor prices are given as w_1 and w_2 and the price received for output is p, then, with the production function given by

$$q = A x_1^a k^b \qquad\qquad a > 0, b > 0, \qquad\qquad (11.1)$$

one can proceed to derive the firm's cost function.

The cost function, to be derived from the Cobb–Douglas (see Varian, 1978) when one input is predetermined, can be written as

$$C(w_1, w_2, q, k) = w_1 q^{1/a} A^{-1/a} k^{b/a} + w_2 k, \qquad (11.2)$$

from which it is clear that one can write the marginal cost curve as

$$MC = B q^\alpha \qquad\qquad (11.3)$$

where

$$B \equiv \frac{1}{a} w_1 A^{-1/a} k^{b/a} \qquad\qquad (11.4)$$

$$\alpha \equiv \frac{1}{a} - 1. \qquad\qquad (11.5)$$

We will investigate the impact of changes in A, which from Chapter 1 we know are Hicks-, Harrod, and Solow-neutral. For reasons that will become apparent below, we can note that in terms of (11.3) a change in A will affect only B and not α.

If we now specify that the firm faces the demand function

$$TR = H q^{1-1/\eta}, \qquad\qquad (11.6)$$

then the derived demand for input one is given by

$$x_1 = \left(\frac{w_1}{a(1 - 1/\eta) H k^{b-b/\eta} A^{1-1/\eta}} \right)^{1/(a-1-a/\eta)}, \qquad (11.7)$$

which will be positive if $a(1 - 1/\eta) > 0$. If $x_1 > 0$, then from (11.7) we can derive that

$$\frac{dx_1}{dA} \frac{A}{x_1} = \frac{1 - \eta}{a\eta - \eta - a},$$ (11.8)

and from (11.8) we can derive that

$$\frac{dq}{dA} \frac{A}{q} = 1 + a\left(\frac{dx_1}{dA}\right)\left(\frac{A}{x_1}\right) = 1 + \frac{a(1 - \eta)}{a\eta - \eta - a}.$$ (11.9)

In the perfectly competitive case where $\eta \to \infty$, these reduce to

$$\frac{dx_1}{dA} \frac{A}{x_1} = \frac{-1}{a - 1}$$ (11.10)

and

$$\frac{dq}{dA} \frac{A}{q} = 1 - \frac{a}{a - 1},$$ (11.11)

and if $a < 1$, then both output and employment will increase in reaction to the change in A. In this model, with $x_2 = k$, the size of a indicates the scale economies available, and if $a < 1$, there are decreasing returns to scale.

In the general case, if $a > 0$, x_1 will be positive only if $\eta > 1$. Thus, assuming $\eta > 1$, $(dx/dA)(A/x) > 0$ if $0 < a < \eta/(\eta - 1)$. However, if $\eta > 1$, this condition is satisfied if $0 < a < 1$. Also, if $a < 1$, then output will be increasing. However, the extent of any change in output and employment will be positively related to η. Thus if there are decreasing returns to scale, both employment of the non-fixed factor and output will increase in reaction to a neutral technological change, the size of the increase being the greater the larger is the elasticity of demand (assuming profit maximization and an invariant market structure).

These results have been derived allowing for a neutral technological change. What if the change is non-neutral? In terms of the Cobb–Douglas, we can represent a non-neutral change by allowing a or b to change. If prices are fixed, one can derive

$$\frac{dx_1}{db} \frac{1}{x_1} = \frac{1}{1 - a} \log k,$$ (11.12)

which, if $0 < a < 1$, is positive; and we can also derive

$$\frac{dx_1}{da} \frac{a}{x_1} = \frac{1}{1 - a}(a \log x_1 - 1),$$ (11.13)

which for higher values of x_1 will also be positive if $0 < a < 1$. Thus, if new technology increases a or b, x_1 will increase. A new technology that reduces a but increases b may lead to a fall in x_1, and thus it is possible for a non-neutral change to reduce the desired input of the non-fixed factor. One may show that $d \log x_1 < 0$ with fixed prices if

$$da < \frac{-a \log k}{a \log x_1 - 1} db.$$

From our diagramatic apparatus above, however, we can argue that, as long as *MC* will be lower for all levels of output after the change than before, output will be higher whatever the bias in the technological change.

As we have been holding one factor fixed, we might consider that these results refer to a 'short run'. In the longer run we expect both factors to be variable. For variety, and to move away from the unit elasticity of substitution assumption of the Cobb–Douglas production function, we analyse this case using the CES production function. This may be written as

$$q = \gamma \{\beta x_1^{-\rho} + (1 - \beta) x_2^{-\rho}\}^{-\nu/\rho} \tag{11.14}$$

and again will yield a marginal cost curve of the constant-elasticity type (see Varian, 1978). With the CES function, ν represents returns to scale and changes in γ are neutral (that is, neither labour-saving or -using) in the Hicksian sense. The elasticity of substitution of x_1 for x_2, σ, $= 1/(1 + \rho)$.

Murray Brown (1966) has worked through the CES example at some length. His most general example assumes

$$p = H_0 q^{-1/\eta} \tag{11.15}$$

$$w_1 = H_1 x_1^{1/\eta_1} \tag{11.16}$$

$$w_2 = H_2 x_2^{1/\eta_2} \tag{11.17}$$

which is a constant-elasticity demand curve with factor prices increasing with demand. We concentrate on the demand for x_1, and consider a change in γ.

In the Appendix we show that, if $1 + (1/\eta_1) = 1 + (1/\eta_2) = \epsilon > 0$, then, given $\eta > 1$,

$$\frac{dx_1}{d\gamma} \frac{\gamma}{x_1} = \frac{1/\eta - 1}{\nu \{1 - (1/\eta)\} - \epsilon} \tag{11.18}$$

If the firm operates in perfectly competitive markets then $\epsilon = 1, 1 - (1/\eta) = 1$, and thus

$$\frac{dx_1}{d\gamma} \frac{\gamma}{x_1} = \frac{-1}{\nu - 1}, \tag{11.19}$$

and thus, with decreasing returns to scale ($\nu < 1$), a neutral technological change will lead to an increase in the demand for x_1. Given A11.7 in the Appendix, x_2 will also increase (given constant factor prices, the condition tells us that the ratio of x_1/x_2 is a constant). If x_1 and x_2 are increasing output is also increasing.

In the imperfectly competitive case given, $\epsilon > 0$ and $\eta > 1$ (11.18) holds, from which it is clear that employment of x_1 (and thus x_2) will increase if

$$\nu < \frac{\epsilon}{1 - (1/\eta)},$$

and as should be obvious, if x_1 and x_2 increase so will output. From (11.18) it is also clear that the impact on x_1 (and thus x_2) of a change in γ will be the greater the larger is the elasticity of demand.

Thus again we have conditions relating scale economies to demand elasticities that tell us of the relationship between factor demands and a neutral technological change. We can also look at the impact of a product innovation.

If we consider an increase in H_0 to represent product innovation, it is simple to show that

$$\frac{dx_1}{dH_0} > 0 \quad \text{if } v < \frac{\epsilon}{1 - (1/\eta)}$$

(given $\eta > 1$).

The case of non-neutral technological change is algebraically very messy. However, one can simply conceive of a technological change that leads to an increase in output but changes the desired ratio of x_1 to x_2 to such an extent that the demand for either input may fall. Changes in β can have such an effect. Thus, if technological change is Hicksian non-neutral, one could find the demand for either input falling; although as long as the marginal cost curve is lower for all levels of output, output will be higher.

In essence, what is happening in this framework is that technological change increases output per unit of input. If output remains the same, then obviously the derived demand for at least one input must fall. However, because the higher productivity produces lower costs, more output will be produced at a given price under profit maximization. This increase in supply will mean that the market will clear at a lower price and higher output than previously. The extent of the increase in output is dependent on the extent of the cost reduction and on the elasticity of demand. The increase in output produced must be compared with the increase in productivity to see if total factor demand will increase. If the technological change is biased towards saving a specific factor, then the increase in output may not be sufficient to increase the demand for that factor. If the technological change is neutral, then the effect on the demand for both factors of the increase in output will depend on the extent of scale economies. Thus it is that our conditions above centre on inequalities relating the degree of scale economies to the elasticity of demand.

11.3 Theory: prices, output, and employment at the industry level

Thus far our analysis has considered only one firm, the innovating firm, ignoring all others. What we must now investigate in order to obtain industry-level results is whether one firm introducing new technology and increasing output will lead to other firms changing their outputs. After considering outputs we will then discuss employment. We will assume that we have an industry with N firms, each with initial marginal costs given by

$$MC_0 = B_0 \, q_0{}^{\alpha} \tag{11.20}$$

where q_0 is the output of a firm using the old technology. Let a new technology be in the process of being diffused with marginal costs

$$MC_1 = B_1 q_1{}^{\alpha} \qquad\qquad B_1 < B_0. \qquad\qquad (11.21)$$

These constant-elasticity marginal cost curves correspond exactly to those generated by the production functions discussed above. We will assume $\alpha > 0$, which is equivalent to assuming decreasing returns to scale. We have represented our technological change by the reduction of B from B_0 to B_1, which as we have shown above is equivalent to having a neutral process innovation.

If we let N_0 be the number of firms in the industry using old technology and N_1 be the number using new, then, writing Q as industry output and defining $N = N_0 + N_1$, we have

$$Q = N_0 q_0 + N_1 q_1. \qquad\qquad (11.22)$$

Let the industry demand curve be of the form

$$p = HQ^{-1/\hat{\eta}} \qquad\qquad \hat{\eta} > 0. \qquad\qquad (11.23)$$

Consider now that each firm is a price-taker, but the price is set to clear the market in each period. Thus, as output changes in reaction to new technology, so will price.

A price-taking firm using either technology will choose output so that price equals marginal cost; thus we will have

$$p = B_0 q_0{}^{\alpha} = B_1 q_1{}^{\alpha}, \qquad\qquad (11.24)$$

which yields

$$q_1 = q_0 \left(\frac{B_0}{B_1}\right)^{1/\alpha} \qquad\qquad (11.25)$$

and, given $B_0 > B_1$, $\alpha > 0$, we have $q_1 > q_0$, and a firm switching from old to new technology will produce higher output. Using the definition of Q (11.22), we can derive

$$Q = N_1 q_0 (B_0/B_1)^{1/\alpha} + (N - N_1) q_0. \qquad\qquad (11.26)$$

From (11.24) we get

$$q_0 = \left(\frac{p}{B_0}\right)^{1/\alpha}, \qquad\qquad (11.27)$$

and from (11.27) we obtain

$$\frac{dq_0}{dN_1} = \frac{1}{\alpha} \left(\frac{p}{B_0}\right)^{(1/\alpha)-1} \frac{dp}{dN_1}. \qquad\qquad (11.28)$$

From (11.23) and (11.26) we may derive

$$\frac{dp}{dN_1} = -\frac{1}{\hat{\eta}}HQ^{(1/\hat{\eta})-1}\frac{dQ}{dN_1} = -\frac{1}{\hat{\eta}}HQ^{(1/\hat{\eta})-1}.$$

$$\left\{ q_0\left(\frac{B_0}{B_1}\right)^{1/\alpha} + \left(\frac{B_0}{B_1}\right)^{1/\alpha} N_1\frac{dq_0}{dN_1} - q_0 + (N-N_1)\frac{dq_0}{dN_1} \right\}. \quad (11.29)$$

Substituting into (11.28) from (11.29), solving for dq_0/dN_1, and assuming $\alpha > 0$, $\hat{\eta} > 0$, we obtain that $dq_0/dN_1 < 0$; i.e., as the number of users of the new technology increases, the output of old technology users decreases. However as $dq_0/dN_1 < 0$ then from (11.28) $dp/dN_1 < 0$; thus from (11.29) $dQ/dN_1 > 0$ and industry output is increasing. Thus, as more firms use the new technology, each firm switching from old to new technology experiences an increase in output, remaining non-users suffer declines in output, but total industry output increases and price declines. One can take the story further to argue that, if when all firms have innovated there is still excess profits in the industry, then new entry will lead to further increases in output, in the absence of entry barriers.

The simplicity of the above case arises through firms being price-takers. In an oligopoly one cannot maintain this assumption. There is however no reason to believe the results will be dramatically changed. Rather than consider the oligopoly case in greater detail, however, let us consider the case of a multi-plant monopolist.

Let the monopolist have N plants, with N_0 as the number of plants using old technology, N_1 as the number using new. A monopoly firm, to minimize total costs, given N_0 old plants and N_1 new, will set the output on the plants, q_0 and q_1, respectively such that MC is equal on the two plants. As the number of new-type plants increases (intra-firm diffusion proceeds), the firm's marginal cost curve shifts down, and at the point of profit maximization ($MR = MC$) output must be higher. This higher output will be associated with a lower price. As the lower price will generally mean lower MR, increases in N_1 must mean that MC is lower on old plants, and thus the output on these plants must be lower. Thus, replacing old plant with new means a higher level of output for the firm but declining output on remaining old plants.

The effects on industry output and the output of users of old technology shown for atomistic price-taking industry can thus be confirmed for the monopoly case when we have process innovation. What of product innovation? As the innovating firm expands its share of the market, the price received by non innovators should fall, if the cross-price elasticity of demand is non-zero, and their desired output will decrease. However, it does not make a lot of sense to add together what will now be products in a differentiated market, and thus we cannot really say much about total industry output.

This discussion has centred on output. What can we say about employment? We have argued in section 11.2 that, if factors are in perfectly elastic supply and the firms' own price elasticity of demand is greater than one, then a neutral technological change will generate increases in the demand for both factors if there are decreasing returns to scale. By analogy, if we think of an industry with

N firms facing similar decreasing returns and an industry elasticity of demand greater than unity, we may argue that industry employment will increase in reaction to a neutral technological change. However, we have shown that as new technology spreads the users of old technology will suffer reductions in output; they must therefore reduce employment. The users of the new technology must thus increase employment. Thus as diffusion proceeds, if there are decreasing returns to scale, users will increase output and employment but non-users will suffer declines in output and employment. Once again, however, we should state that these results refer to a world in which firms profit-maximize, market structure is not changing, and technological change is neutral. If technological change is not neutral the employment of a factor may well be reduced although we expect our results on output to be maintained.

The outcome of our analysis is thus that (1) the introduction of new technology will lead to increases in the level of output of the profit-maximizing firm; (2) if the technological change is neutral and there are decreasing returns to scale, factor demand will also increase; (3) these results can be extended to an industry level; and (4) as diffusion proceeds non-users will suffer reductions in output and employment. If technological change is non-neutral, then the results on employment will not necessarily hold. There is thus no guarantee that technological change will not lead to technological unemployment, although the basic results of the analysis are reasonably optimistic. Even so, the model has certain limitations and we must thus consider a number of factors that have so far been ignored.

1 We have treated our inputs as homogeneous. In reality, this is invalid. If the new technology requires, for example, different skills than the old, then although the overall demand for the labour input may increase, the old type of skills will become redundant. The owners of these old skills may consider themselves technologically unemployed.

2 We have argued that as the new technology spreads inputs will be released by non-users although more may be demanded by users. If the location of non-users and users are disparate, this may create pools of regional unemployment. This problem may be particularly important when the innovators are importers and the laggards are domestic firms.

3 On a similar line to (1) and (2), the sex composition of the demand for labour may also change.

4 Our analysis has implicitly assumed that the new technology is disembodied and that inputs can be costlessly transformed into the new type. If the new technology is embodied in new capital equipment our results may have to be amended. A full amendment requires consideration of inter-industry effects and is thus considered in the next chapter, but at the firm level we must argue that (a) any constraints on the supply of investible funds will limit the firm's ability to adopt new technology and thus firms unable to change may reduce employment and perhaps eventually disappear, and (b) if technology is embodied in new capital equipment then technological advance may imply capital losses

for the owners of old machines. These capital losses may have impacts on firm behaviour that have been ignored above.

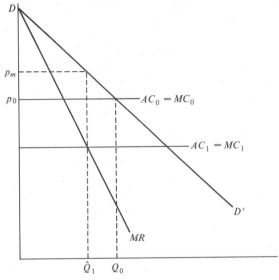

Fig. 11.3

5   Finally, we have not considered the monopoly-creating power of technological change. To illustrate, consider Fig. 11.3, where DD' is the demand curve and $MC_0 = AC_0$ are initial costs and $MC_1 = AC_1$ post-invention costs. Allow initially that the industry is perfectly competitive; then the initial price would be $p_0 = MC_0 = AC_0$. Allow now that one firm has a monopoly on the new technology. By pricing at a small margin below p_0, this firm will be able to drive out all other firms, although Q, output, will hardly change. If there are barriers to re-entry to the industry the monopoly may then price at p_m to maximize profits with output $\hat{Q}_1$, and output will fall and thus the employment of at least one factor will fall. The point is that if new technology creates monopoly power, this power may lead to reductions in output and consequently to reduced demands for inputs. Similar examples could be constructed for the case of product innovation.

11.4   Empirics

The purpose of this section is to investigate the impact that technological change has had on employment, output, etc., at the firm or industry level. The theory above, if correct, would argue that technological change will have definite impacts on employment, output, and prices, the actual extent of the impact depending among other factors on the nature of the technological change, demand elasticities and scale economies.

An obvious place to start is to consider what sort of impact technological change has had on productivity, and the classic approach to this is to include a technological change term in a production function. The majority of the work that uses this approach is macroeconomic, and thus the literature is considered in greater detail in the next chapter. There seems to have been only a limited amount of work using this approach at the micro-level. One extensive study for the UK, however, is that by Wragg and Robertson (1978), whose declared aim is to update the work of Salter (1969). Assuming that one can define a production function at the industry level that is Cobb–Douglas with exponents a and b, then total factor productivity is defined by

$$\frac{\dot{A}}{A} = \frac{\dot{Q}}{Q} - a\frac{\dot{L}}{L} - b\frac{\dot{K}}{K} \tag{11.30}$$

where $\dot{A}/A$ = rate of growth of total factor productivity;
$\dot{Q}/Q$ = rate of growth of output;
$\dot{L}/L$ = rate of growth of labour input;
$\dot{K}/K$ = rate of growth of capital input;

and a, b will equal the shares of labour and capital in income. An estimate of $\dot{A}/A$ will indicate the extent of the shift in the production function over time (not necessarily, of course, all due to technological change, although if this is the only change allowed the assumption is that all changes are neutral). It is not uncommon to use estimates of $\dot{A}/A$ as rough indicators of the change in technology, which, subject to the reservations detailed in the next chapter, we do. In Table 11.1 we reproduce, in the first column, Wragg and Robertson's estimates of total factor productivity growth in UK manufacturing industry, 1963–73.

Our analysis above indicates that increases in productivity (technological change) will be associated with, at the industry level, (1) reductions in price and increases in output; and (2) effects on the level of inputs dependent upon scale economies, the elasticity of demand, and the bias of technological change. Thus in Table 11.1 we have included Wragg and Robertson's estimates of above (+) and below (−) average rates of growth for each industry of total factor productivity, output, employment, capital stock, and prices for the same 1963–73 period.

Let us begin to relate the predictions to performance. Fitting a simple OLS regression to their data (summarized in the table), Wragg and Robertson find that the elasticity of prices with respect to total factor productivity is − 0.43 (their regression having, however, an R^2 of only 0.23), and there is thus some reason to support the view that faster technological advance will lead to lower relative prices (a view also given support by Salter, 1969). With a downward-sloping demand curve this must imply increases in output.

However, we cannot really argue that all the effects of the productivity increase are passed on in lower prices. The effects may be taken as higher wages or higher profits rather than lower prices. Wragg and Robertson can find no significant relation cross-sectionally between earnings and productivity growth

Table 11.1   *Manufacturing industry performance, 1963–73*

	UK average annual rate of growth of total factor productivity	Deviations from average growth rates of				
		Output	Employment	Capital stock	Total factor productivity	Prices
Radio, computers, etc.	5.6	+	+	+	+	–
Man-made fibres	3.8	+	+	+	+	–
Spirit distilling	0.7	+	+	+	–	–
Chemicals	5.2	+	+	+	+	–
Surgical & other instruments	4.2	+	+	+	+	–
Miscellaneous paper and board	2.9	+	+	+	+	+
Carpets	3.7	+	+	+	+	–
Dressmaking	5.4	+	+	+	+	+
Telegraph & telephone appliances	4.8	+	+	+	+	–
Hosiery	4.2	+	+	+	+	–
Office machinery	6.0	+	–	–	+	–
Construction equipment	3.6	+	–	+	–	+
Glass	4.1	+	+	+	+	–
Toilet preparations	2.5	+	+	+	–	+
Fertilizers	4.1	+	+	+	+	–
Engineering tools	2.5	+	+	+	+	+
Mechanical handling equipment	2.9	+	+	+	+	+
Industrial plant	2.6	+	+	+	–	+
Electrical appliances	5.9	+	–	+	+	–
Brewing	1.6	+	–	+	–	–
Overalls	4.2	+	+	+	+	–
Vegetable fats	0.5	+	+	–	–	+
Metal cans & boxes	2.5	+	+	+	–	–
Explosives	3.5	+	–	+	+	+
General mechanical engineering	3.1	+	+	–	+	–
Agricultural machines	2.4	+	+	–	+	+
Asbestos	2.8	+	–	–	+	–
Paint	3.9	+	+	–	+	–
Cement	1.2	+	+	+	+	–
Fruit & vegetable products	1.6	+	+	+	–	+

| | + | + | | + +   + + | + | | | | + | + + | + | | | + + + | + + + | |

| + | + + | + + + + + | + + + + | + + + + | | | | | + | | | | | + + +

+ | + | | | | + | | + | | | | | + | | | | | + | | + | | | | | | | | | |

| + + | | + | | | + | + + + | | + | | | | | + | + + | | | + + | | + | | |

+ + + + + | | | | | | | | | | | | | | | | | | | | | | | | | | | | | | | |

Industry	Value
Electricity	0.5
Corsets	3.5
Pumps, valves, etc.	2.2
Shipbuilding	4.5
Textile machinery	5.9
Motor vehicles	1.5
Rubber	3.1
Brick & fireclay	4.9
Soap & detergents	4.3
Narrow fabrics	3.0
Cocoa confectionery	3.1
Animal & poultry foods	1.7
Women's wear	3.5
Leather goods	2.9
Footwear	4.2
Linoleum	3.8
Lubricating oil	-3.8
Men's wear	3.9
Iron & steel	2.8
Textile finishings	4.3
Industrial engines	3.8
Hand tools	1.7
Sugar	1.3
Miscellaneous metals	0.6
Brushes & brooms	2.5
Paper & board	1.0
Motor cycles	2.8
Nuts & bolts	1.5
Steel tubes	1.6
Biscuits	0.1
Grain milling	0.9
Metal working tools	2.4
Wire & wire manufacturing	0.9
Electrical machinery	4.5
Lace	3.7
Outer wear	3.7

Table 11.1 (*cont.*)

	UK average annual rate of growth of total factor productivity	Deviations from average growth rates of				
		Output	Employment	Capital stock	Total factor productivity	Prices
Non-ferrous metals	−0.8	−	−	+	−	+
Gloves	3.4	−	−	+	+	−
Margarine	−0.3	−	−	+	−	+
Leather and fellmongery	2.8	−	−	−	+	+
Canvas	−0.3	−	−	−	−	+
Insulated cables	1.6	−	−	−	−	+
Spinning & doubling	3.1	−	−	−	+	+
Woollen & worsted	3.0	−	−	−	+	+
Weaving	3.3	−	+	−	+	+
Bread & flour confectionery	−1.6	−	−	−	−	+
Hats & caps	3.2	−	−	−	+	+
Jute	1.7	−	−	−	−	−
Rope	0.7	−	−	−	−	+
Railway vehicles	2.1	−	−	−	−	+
Coal mining	−1.1	−	−	−	−	+
Coke ovens	−5.4	−	−	−	−	+

Source: Wragg and Robertson (1978).

(and neither did Salter), suggesting that the benefits have not gone directly into higher wages. There is however some evidence that firm profits and technological progressiveness are positively related. For example, for the USA, Grabowski and Mueller (1978) argue that progressive firms from high-research industries earn significantly above-average returns to R & D (16.7 per cent calculated at the sample mean). Firms in industries that are not so classified do not earn such high rates. They find also, across an 86-firm sample, that there is a significant relationship between profitability and R & D intensity. This suggests that faster technological advance means higher profits, and thus not all productivity gains are passed on in prices. (We say more in Chapter 15 on the returns to be achieved from R & D.)

As to the effect of total factor productivity on the inputs of factors, Wragg and Robertson find no statistical relation from total factor productivity to the inputs of labour and capital (which they suggest might imply that productivity growth is no threat to employment). They argue, on the lines of our theoretical section, that the impact of technological change on employment would depend among other things on demand elasticities. They find that, although after 1950 there is a positive relationship between the growth of productivity and employment in retail distribution, it did not exist in manufacturing, and overall they could not find a statistically significant relationship between growth rates of productivity and employment.

Even in terms of above- or below-average growth rates of employment, they find that, of the 45 industries with below-average productivity growth, 23 had above-average growth of employment and 22 below-average. Of the 18 industries that increased employment between 1963 and 1973, 9 had above-average productivity growth and 9 below-average.

These results on employment can be said to reflect the above theory, which states that the effects depend on elasticities, biases, and scale economies, and can be signed only once these are known. It must also be considered that other factors, such as income elasticities, which have so far been ignored, will be significant influences on employment.

However, one does not have to restrict oneself to a neoclassical approach to see how technological change will affect output, employment, etc. The most common alternative approach is via case studies. A number of these are investigated in Rothwell and Zegveld (1979). The advantage of a case study approach is that one can range more widely than in a strict neoclassical framework. Thus in this work the effect on skills, job satisfaction, etc., are also discussed. In Table 11.2 a summary of the results is reproduced. As one can see from this table, an output expansion effect is common although the effects on the labour force are mixed. It is clear that the skill structure of the labour force is also strongly affected by new technology.

11.5 Conclusions

We have approached the theoretical analysis of the impact of technology on

Table 11.2 *Summary of the quantitative and qualitative impacts of technical change on employment in a number of industries*

	Agriculture	Coal mining	Canadian railways	Textile machinery industry	Textile industry	Cement industry	Steel industry	Metal working industry	NC machine tools	Computer-aided design	Automation
Reduction in labour force	✓	✓									
Increased output with same or reduced labour force (jobless growth)	✓	✓	✓	✓	✓		✓	✓			
De-skilling or making certain skills redundant	✓	✓	✓		✓		✓	✓	✓	✓	✓
Generation of the need for new skills	✓	✓	✓	✓	✓	✓		✓	✓	✓	✓
Reduction in job satisfaction					✓	✓			✓	✓	✓
Requirement of higher-level management skills	✓	✓		✓	✓	✓		✓	✓	✓	✓
Displacement of specialist skills outside the factory					✓	✓		✓	✓	✓	✓
Job loss owing to lack of technical change competitiveness							✓		✓		

Source: Rothwell and Zegveld (1979).

employment at the firm and industry level in a neoclassical way. Our theory predicts that prices should fall as technological advances occur, that the output of the firm and the industry should increase as new technology is introduced, and moreover that the employment of factors will be affected. The direction of movement in factor demand will depend on the bias in technological change, the elasticities of demand and factor supply, and economies of scale.

Empirical work suggested (1) that technological advance is associated with reductions in relative price but not with higher earnings, (2) that some of the gain from higher productivity will be taken in increased profits, (3) that price reductions should mean higher output, (4) that we cannot isolate any particular direction of effect of technological change on factor employment. We also illustrated that skills and job satisfaction may well be influenced by technological change.

Appendix

The purpose of this Appendix is to derive the effect on the demand for a factor of a technological change when the production function in CES (see 11.14) and the demand and factor supply functions are of the constant elasticity type (11.15–11.17).

Define total cost TC as

$$TC = w_1 x_1 + w_2 x_2 \tag{A11.1}$$

which after substitution from (11.15) and (11.16) yields

$$TC = H_1 x_1^{1+(1/\eta_1)} + H_2 x_2^{1+(1/\eta_2)} \tag{A11.2}$$

If we minimize total cost subject to the production function we obtain

$$\frac{x_1^{-\rho-1-(1/\eta_1)}}{x_2^{-\rho-1-(1/\eta_2)}} = \frac{1-\beta}{\beta} \frac{H_1}{H_2} \frac{1+(1/\eta_1)}{1+(1/\eta_2)}. \tag{A11.3}$$

To proceed further we assume $\eta_1 = \eta_2$ such that $1+(1/\eta_1) = 1+(1/\eta_2) = \epsilon$; therefore

$$\left(\frac{x_1}{x_2}\right)^{-\rho-\epsilon} = \frac{1-\beta}{\beta} \frac{H_1}{H_2}. \tag{A11.4}$$

Thus the cost-minimizing firm will choose x_2 such that

$$x_2 = x_1 \left(\frac{\beta}{1-\beta} \frac{H_2}{H_1}\right)^{-1/(\rho+\epsilon)} \tag{A11.5}$$

Define

$$L \equiv \left(\frac{\beta}{1-\beta} \frac{H_2}{H_1}\right)^{-1/(\rho+\epsilon)} \tag{A11.6}$$

Then

$$x_2 = Lx_1. \tag{A11.7}$$

We may now write the cost function as

$$TC = H_1 x_1^\epsilon + H_2 L^\epsilon x_1^\epsilon = (H_1 + H_2 L^\epsilon) x_1^\epsilon \equiv M x_1^\epsilon \tag{A11.8}$$

(where $M = H_1 + H_2 L^\epsilon$) and the production function as

$$q = \gamma \left\{ \beta x_1^{-\rho} + (1 - \beta) x_1^{-\rho} L^{-\rho} \right\}^{-\nu/\rho} \qquad (A11.9)$$

$$= \gamma x_1^{\nu} \left\{ \beta + (1 - \beta) L^{-\rho} \right\}^{-\nu/\rho}$$

$$= \gamma z x_1^{\nu}$$

where $z = \left\{ \beta + (1 - \beta) L^{-\rho} \right\}^{-\nu/\rho}$. To maximize profits the firm will choose x_1 to maximize

$$\Pi = pq - TC \qquad (A11.10)$$

which can be rewritten (by use of (11.15) as

$$\Pi = H_0 q^{1-(1/\eta)} - TC.$$

After substitution from (A11.8) and (A11.9) we have

$$\Pi = H_0 (\gamma z)^{1-(1/\eta)} x_1^{\nu \{1-(1/\eta)\}} - Mx_1^{\epsilon}. \qquad (A11.11)$$

Maximizing (A11.11), we obtain

$$\frac{d\Pi}{dx_1} = \nu \left(1 - \frac{1}{\eta} \right) H_0 \ \gamma^{1-(1/\eta)} z^{1-(1/\eta)} x_1^{\nu \{1-(1/\eta)\}-1} - \epsilon Mx_1^{\epsilon-1} = 0 \qquad (A11.12)$$

yielding

$$x_1 = \left[\frac{\epsilon M}{\nu \{1 - (1/\eta)\} H_0 \ \gamma^{1-(1/\eta)} z^{1-(1/\eta)}} \right] [\nu \{1 - (1/\eta)\} - \epsilon]^{-1}. \qquad (A11.13)$$

Define

$$R = \left[\frac{\epsilon M}{\nu \{1 - (1/\eta)\} H_0 z^{1-(1/\eta)}} \right] [\nu \{1 - (1/\eta)\} - \epsilon]^{-1}. \qquad (A11.14)$$

Then

$$x_1 = R \ \gamma^{-(1/\eta)} \left[\frac{1}{\nu \{1 - (1/\eta)\} - \epsilon} \right]. \qquad (A11.15)$$

To find the effect on x_1 of a neutral technological change, we consider $(dx_1/d\gamma)$ (γ/x_1), and from (A11.15) we get

$$\frac{dx_1}{d\gamma} \frac{\gamma}{x_1} = \frac{(1/\eta) - 1}{\nu \{1 - (1/\eta)\} - \epsilon}. \qquad (A11.16)$$

References

Brown, Murray (1966), *On the Theory and Measurement of Technical Change*, Cambridge University Press.

Grabowski, H.G. and Mueller, D.C. (1978), 'Industrial Research and Development, Intangible Capital Stocks and Firm Profit Rates', *Bell Journal of Economics*, 9, 328–43.

Rothwell, R. and Zegveld, W. (1979), *Technical Change and Employment*, Frances Pinter, London.

Salter, W.E.G. (1969), *Productivity and Technical Change* (2nd ed.), Cambridge

University Press.
Varian, H. (1978), *Microeconomic Analysis*, W.W. Norton, New York.
Wragg, R. and Robertson, T. (1978), *Post War Trends in Employment, Productivity, Output, Labour Costs and Prices by Industry in the UK*, Research Paper no. 3, Department of Employment, London.

Chapter 12

Technological Change, Output, and Employment: A Macroeconomic Approach

12.1 Introduction

In the previous chapter we explored the implications of technological change for output and employment at the firm and industry level, using essentially a neoclassical framework. In this chapter we extend to an analysis at the macro-level. The justification for this is that

1 the macro-literature includes approaches that are less optimistic in terms of price flexibility, and we show that price inflexibility can be a major influence on the impact of technological change;
2 the impact of technological change may not be constrained to one industry; inter-industry linkages may be important in determining its effects;
3 much of the analysis of the impact of technological change has been conducted in macroeconomic frameworks.

It is not our intention to provide a guide to the history of economic thought on technological change and its impact. We are concerned only with the more recent work in this area. Thus we will not, for example, be discussing Smith or Ricardo. For the interested reader the survey by Gourvitch (1966) on the literature to 1940 is invaluable. What is noticeable from that survey is how many of the issues that are currently at the forefront of the debate on technological change and employment were also key issues in earlier discussions. We do not feel therefore that a modern perspective will make our discussion incomplete.

In this chapter we constrain ourselves basically to closed-economy analysis. Open-economy aspects are discussed in Chapter 17. The plan of the chapter is initially to consider a simple Keynesian-type macro-model to illustrate some basic points in the debate. We then move to consider other models that lay stress on particular aspects of importance in the debate. A general conclusion from the theoretical discussion is provided before moving to empirical work. We should note at the outset, however, that nearly all the discussion below refers essentially to changes in process technology. Modelling the impact of changes in product technology at the macro-level is an area that is much less analysed — not because of irrelevance, but because of its difficulty.

12.2 Some basic observations

Consider the standard aggregate supply/aggregate demand analysis that can be found in any macroeconomics textbook. We have an aggregate demand curve

$$p = F(Y, \bar{G}, \bar{M}) \qquad F_1 < 0, F_2 > 0, F_3 > 0, \qquad (12.1)$$

where p = price level;
$\quad Y$ = GNP;
$\quad \bar{G}$ = government expenditure (assumed fixed exogenously);
$\quad \bar{M}$ = nominal money supply (assumed fixed exogenously);
which is derived rom the demand and supply of money functions

$$\left. \begin{array}{l} M^D = M^D\,(p, Y, r) \\[2mm] M^S = \bar{M} \\[2mm] M^S = M^D \end{array} \right\} \qquad (12.2)$$

and the product market equilibrium conditions

$$Y = C + I + G \qquad (12.3)$$

$$G = \bar{G} \qquad (12.4)$$

$$I = I(r, Y) \qquad (12.5)$$

$$C = C(Y, r) \qquad (12.6)$$

where r = interest rate, C is aggregate consumption expenditures, and I is aggregate investment.

The aggregate supply curve,

$$p = G(Y, w) \qquad G_1 > 0, G_2 > 0, \qquad (12.7)$$

where w is the money wage, is derived as the inverse of the notional demand for labour curve. Given the production function

$$Y = H(\bar{K}, L) \qquad H_2 > 0, H_{22} < 0 \qquad (12.8)$$

(where $\bar{K}$ is the given capital stock), the profit-maximizing firm employs labour to the point where

$$\frac{\partial Y}{\partial L} = \frac{w}{p} \qquad (12.9)$$

and from (12.8) and (12.9) we can write the aggregate supply curve.

In Fig. 12.1 we have a three-sector Keynesian diagram with the aggregate supply (SS') and demand curves (DD') in sector A, the production function in sector B, and the labour market in sector C. LL' is the notional demand for labour curve and NN' is the supply of labour curve. A full-employment equilibrium is assumed at an initial position (p_0, Y_0, w_0, L_0) given $\bar{G}, \bar{M}, \bar{K}$.

Now postulate that a disembodied technological change occurs.[1] This will have three initial effects:

---

[1] For a similar analysis on these lines see Sinclair (1981).

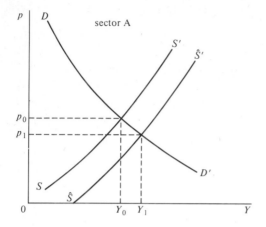

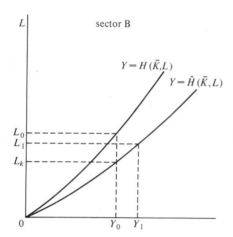

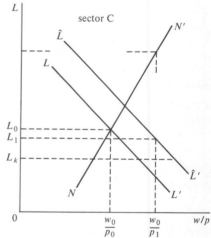

Fig. 12.1

1   the production function will shift, so that for any given labour input more output can be produced;
2   the labour demand curve will shift: let us allow that for any real wage the technological change means a higher labour demand; i.e., the marginal product of labour is higher for any L;
3   the aggregate supply curve will shift to the right, firms being willing to supply more for any price, given w.

Let the curves with a caret superflex represent the new curves, the aggregate supply curve being drawn assuming $w = w_0$. To predict what will happen from here consider the following scenarios.

1 Prices and wages remain fixed at p_0 and w_0, respectively. In this case the price level is such that aggregate demand is less than aggregate supply and firms cannot sell all the output they wish to produce. We allow, in a temporary equilibrium frame of mind, that the short (demand) side of the market dominates, in which case firms produce the output they can sell at price p_0, i.e., Y_0. Thus output has not changed. However the labour required to produce this output will fall to L_k (k for Keynesian), yielding an effective demand for labour at w_0/p_0 equal to L_k. At wage w_0/p_0 labour supply is still L_0, and we thus have unemployment $L_0 - L_k$.

2 Either wages or prices react to excess demands. Consider the case where the wage is fixed at w_0 but price is flexible. Then at p_0 the notional excess supply of products will force down the price level to p_1 and output is now Y_1. With the new production function employment is L_1, and given that the aggregate supply curve and the notional demand curve for labour are different ways of drawing the same relationship, the real wage is equivalent to that read off the notional demand for labour curve at employment L_1.

With this degree of price flexibility, we see that employment will be higher than with rigid wages and prices; however, there is no guarantee that the new position will involve full employment.[2] We have drawn the new position as one of unemployment with labour supply greater than labour demand at prices w_0/p_1.

3 In this third scenario consider both w and p flexible in response to excess demand. Now starting from $(Y_1, L_1, w_0/p_1)$, the excess supply of labour will, through some Phillips curve-type mechanism, lead to reductions in the money wage. This shifts the aggregate supply curve to the right (firms are willing to produce more at any given price), yielding a lower p, higher Y, higher L, and lower unemployment. As long as unemployment remains, the wage continues to fall until all unemployment is removed. At the new full-employment equilibrium, compared with the pre-change postion, p will be lower, Y higher, L higher and w/p higher. If wage and price responsiveness is infinitely fast there will not even be any transitional unemployment.

The point we are trying to establish here is that the responsiveness of all prices to excess demand is a crucial factor in determining the impact of technology on employment. Complete price flexibility is a characteristic of neoclassical or Walrasian approaches to economics, and, as has successfully been argued by Hahn (1982) and others, can be theoretically supported only if the whole Walrasian apparatus (of an auctioneer, price adjustment in response to notional concepts rather than quantity adjustment in response to effective concepts, no trading out of equilibrium, and complete futures markets) is assumed to exist. In the absence of this apparatus we have price rigidities, and this leads us to very different conclusions as to how the economy behaves. These price

---

[2] From our analysis in the last chapter we can state that employment will depend on scale returns in production, the elasticity of the aggregate demand curve, and the bias in the change in technology.

rigidities can be justified on the grounds of the absence of a Walrasian apparatus or through oligopolistic pricing, fixed-term contracts, etc. (see Casson, 1981).

We have then that, if prices and wages are sufficiently flexible, there is no unemployment resulting from the introduction of new technology. In general however we would not expect prices to be sufficiently flexible, and thus unemployment may well result.

In some ways the models discussed here are somewhat 'old fashioned'; perhaps more representative of the price adjustment model is the New Classical framework (see, for example, Parkin and Bade (1982)). In this view of the world, prices and wages always adjust to clear markets but price expectations enter the picture. The aggregate demand curve may be carried over from above, but the aggregate supply is constructed rather differently.

Given the production function, the demand for labour is still determined by the marginal productivity condition

$$\frac{\partial Y}{\partial L} = \frac{w}{p}$$

but labour supply is related to wage rates and expected prices (p^e) such that

$$L^S = L^S(w, p^e) \qquad L_1^S > 0, L_2^S < 0. \qquad (12.10)$$

L^S and L^D for given p^e came into equality through changes in w, and thus using the production function we may solve for Y as a function of p and p^e, this yielding the aggregate supply curve. At the full-equilibrium position aggregate supply and demand are equal and $p = p^e$. Full employment always holds (given p^e), but only at full equilibrium is the economy at its natural rate of unemployment.

In this framework a disembodied technological change, of the type discussed above (i.e., one that increases $\partial Y/\partial L$ for all L), will increase the demand for labour at any given real wage. The labour market-clearing money wage will thus be higher and employment and aggregate supply will be higher for any given p and p^e. The technological change will thus shift the aggregate supply curve outwards (for the given p^e) from SS' to SS'', (see Fig. 12.2). Initially we are at $(p_0\, Y_0)$, and then the technological change occurs. The product market now clears at prices p_1 with output Y_1. Thus the technological change leads to an increase in Y and a reduction in p (holding price expectations, p^e, constant). However, we now have $p < p^e$, which leads to expected prices falling, in turn leading SS'' to shift further to the right, inducing further falls in p and p^e until $p = p^e$ at, say, point Y_2 in Fig. 12.2 (where $S_2\, S_2'$ is drawn assuming $p^e = p_2$). At this point we are again at the natural rate of unemployment.

In this framework, therefore, the technological change creates no involuntary unemployment, for the labour market always clears; however, we are away from the natural rate of unemployment for the whole of the period during which p^e is adjusting to p. If p^e adjusts to p with infinite speed (through, say, a rational expectations mechanism), the economy would jump from one full equilibrium directly to the next without the intermediate stage and the economy would always be at the natural rate of unemployment. Again, therefore, if technological change is to be associated with a divergence from equilibrium in the labour market (here defined as unemployment at the natural rate), some price inflexibility (here inflexibility in p^e) is required.

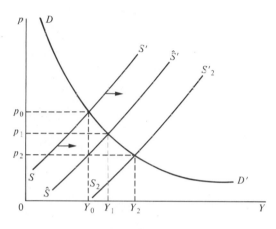

Fig. 12.2

These results are generated for one technological change with a fixed capital stock. If we wish to extend to continuous technological change and a variable capital stock we must consider the growth models literature. This literature cannot be surveyed adequately here, but by picking out the two most basic and simplest models one can illustrate how price flexibility assumptions reflect in predictions as to how technological progress affects employment. The two simplest models are those that appear in most basic macro-texts, the Harrod–Domar and neoclassical models. Let us assume that in both cases we are discussing a world with Harrod–neutral technical progress at rate g.

In the Harrod–Domar model we assume

Labour supply:
$$L^S = L_0 e^{nt} \tag{12.11}$$

Production function:
$$Y = \frac{1}{v_r} K = \frac{1}{u} L e^{gt} \tag{12.12}$$

Savings:
$$S = s_w Y \tag{12.13}$$

Investment:
$$I = \frac{dK}{dt} \tag{12.14}$$

where v_r is the desired capital–output ratio and v will be used to represent the actual capital–output ratio.

In this world full employment will be assured if output grows at the natural rate $n + g \equiv G_n$. Desired savings and investment will always be equal if output grows at the warranted rate, $S_w/v_r \equiv G_w$, but in general output grows at the actual rate $G_a = s_w/v$. It is argued that if $G_a \gtrless G_w$, G_a will deviate further from G_w and that in general there is no reason to expect $G_w = G_n$ or $G_a = G_w$. Thus an increase in the rate of technical progress will increase G_n, but this will not affect G_a or G_w, and thus, if initially $G_a = G_w = G_n$, the increase in G_n will only lead to increasing unemployment. Ways of bringing G_w into equality with G_n that have been suggested (see for example Kaldor, 1955; Passinetti, 1962) generally rely on some wage or price flexibility, which is absent from the basic Harrod–Domar story.

In the neoclassical growth model complete wage and price flexibility is assumed to clear all markets at every moment in time. Then

$$L = L^S = L_0 e^{nt} \tag{12.15}$$

$$\frac{dK}{dt} = sY \tag{12.16}$$

and, assuming a Cobb–Douglas production function, we have

$$Y = A\,K^\alpha\,(Le^{gt})^{1-\alpha}. \tag{12.17}$$

Defining

$$r = K/Le^{gt}, \tag{12.18}$$

we follow Solow (1956) to generate the fundamental differential equation:

$$\frac{dr}{dt} = sAr^\alpha - nr. \tag{12.19}$$

On the steady-state growth path $dr/dt = 0$, and thus

$$r = \left(\frac{n}{sA}\right)^{1/(\alpha-1)} \tag{12.20}$$

and if $dr/dt = 0$,

$$\frac{dK}{dt}\frac{1}{K} = \frac{d\,(Le^{gt})}{dt}\frac{1}{(Le^{gt})} = n + g$$

and with constant returns to scale, $(dY/dt)\,(1/Y) = n + g$.

Thus, an increase in g will lead to an increase in the rate of growth of Y and K, a constant value of r, and full employment will be maintained throughout. The neoclassical predictions are thus somewhat different from those of the Harrod–Domar fixed-price world.[3]

---

[3] We ought to state that these predictions are long-run, the adjustment times required being measured in terms of years rather than months (see Sato, 1963).

Given the similarity of the conclusions to be drawn from models with continuous technological change to those drawn from models with the change
represented by one-off shifts, we will be restricting ourselves below to those of
the latter type. We find however that we must get away from the sorts of models
discussed above, for various reasons.

1   The models above consider the technological change as disembodied. It
     seems to be reasonable to argue that to a large extent new technology is
     embodied in new capital goods, and that to change technology some positive
     gross investment is required.
2   By assuming disembodiment the models above are restricted to a comparative
     statics analysis. Essentially, one is comparing an economy using an old
     technology with one using a new. However, in Part II of this book we
     argued that the take-up of new technology is a time-intensive process.
     With their comparative statics emphasis, the models above are not really
     suited to analysing the time-path of the economy during the long period
     in which the economy is making the transition from using the old to using
     the new technology. In the model we discuss below these two restrictive
     assumptions can be relaxed.

12.3 A two-sector model

This approach to the analysis of the effect of technological change is designed
partly to overcome the deficiences of the above approach and partly to incorporate into the analysis a number of what are termed 'compensation effects'.
In essence, our whole discussion to date has centred on these compensation
effects implicitly, but now is the time to consider them explicitly. The compensation principle concerns how indirect effects resulting from the introduction
of a factor-saving new technology can compensate for direct reductions in factor
demand. Compensation effects fall into three main classes.

1   *Technology multiplier effects.* This can be exemplified as the effect on
     factor demands in industry *j* resulting from the introduction of a new
     technology in industry *i* (for a given final demand vector). An example
     of such an effect is the effect on the demand for new capital goods resulting
     from the introduction of a new embodied technology.
2   *Income effects.* If new technology leads to changes in income levels or
     distribution, then the level and/or pattern of final demand may change with
     a consequent effect on the demand for factor inputs.
3   *Price effects.* If new technology affects prices, this may well affect demand
     and thus factor employment. Moreover, as we have shown, if prices react
     to notional excess demand and supply, then this will also influence the
     employment of factors.

We should state initially that these compensation effects need not always
(or ever) have a positive effect on the employment of factors. They may in fact

reinforce a direct negative effect. Moreover, the nature of these effects may mean that, if any direct reduction in factor demand is offset by an indirect increase, then the increase may be 'distant' in a geographic, industrial, or skill sense from the original reduction. However, even so, one cannot properly analyse the effects of new technology and ignore these effects.

In Stoneman (1982) a model is constructed that is designed to reflect these compensation effects. As has already been shown, complete price flexibility will mean that there will be no technological unemployment, and thus such flexibility is not assumed. Compensation effects are introduced by using a two-sector model with embodied technological change, thus allowing for technology multiplier effects; income effects are introduced by linking aggregate demands to wages and profits; and price effects are introduced by linking prices to costs. Moreover, the observation that new technology takes time to spread is reflected by the incorporation of a specific diffusion profile for new technology.

The model is set out in some detail in the Appendix to this chapter. The important points to note about the model are as follows.

1   The new and old technologies are both considered to be of fixed-coefficient type, the new being superior to the old at all wage rates.
2   The output of capital goods of each type is just sufficient to meet demand in each period. In general, when a new technology is first introduced both new and old capital goods will be produced and used side by side. However, as time proceeds and the demand for old capital goods falls, their production will cease although their use will continue for some time. The date at which the industry producing the old capital goods 'dies' is endogenous.
3   It is assumed that all wages are consumed and all profits are saved and invested.
4   The diffusion process is considered as either of the epidemic or the probit type.
5   Wage rates are considered to be determined independently of the demand for labour.

In the Appendix it is shown that the time-path of employment will depend on

1   the time-path of consumption, which in turn depends on the time-path of wages;
2   the time-path of the diffusion process; and
3   the technical coefficients of the two technologies (i.e., the factor-saving bias).

In Stoneman (1982) the time-path of employment under a variety of assumptions regarding (1)-(3) above is analysed. In particular, a model with a fixed wage w is analysed combined with an epidemic approach to diffusion, and a model with a growing wage and a Davies/David type probit diffusion approach is analysed.

Starting from full employment, it is shown that under the first set of assumptions the employment ratio will increase for the whole of the diffusion process, the jobs lost from the old process being more than adequately compensated for by the new. There is also some indication that a higher diffusion speed will increase employment, at least in the early years of the transition.

It is also shown that, if an epidemic approach to diffusion is combined with a wage rising over time, at a rate conditioned by the diffusion speed, from that appropriate to full-employment balanced growth using technology 1 to that appropriate to full-employment balanced growth using technology 2, then if the new technology is more capital-intensive than the old, the employment ratio will fall for all t, whereas if the old technology is more capital-intensive, then the employment ratio will either increase for all t or will initially increase and then decrease as time proceeds.

Within a probit diffusion model, an exponentially increasing wage is combined with a Pareto-distributed reservation wage across entrepreneurs and employment can be shown to fall over time.

The effect of technology on employment, then, even when taking into account compensation effects, cannot be signed without some knowledge on the behaviour of other key economic variables. In particular, this model reflects that the diffusion path and the wage path are important factors in defining the impact of technology.

Essentially, the operation of the model generates time-paths for employment by having endogenous compensation that offsets direct reductions in labour demand. As the diffusion proceeds, with the new technology being less labour-intensive than the old, a given level of output can be produced by fewer workers. However, as the new technology is introduced, the time-path of profit is being changed. As profits change so investment is affected, and this investment affects total labour demand. The time-path of wages and diffusion interact to generate profit per unit of output. The diffusion path generates employment per unit of output. The total level of output and employment is generated by the all-profits-invested, all-wages-consumed assumption.

The model can be criticized on a number of grounds. The most important of these is the assumed savings and investment relationship. This assumption means that any extra profits realized from the use of the new technology are always invested, and this investment creates employment opportunities and multiplier effects. If the extra profits are not invested, but are retained, then we may well get a 'realization' problem, whereby the increase in productive potential is not matched by any increase in demand, and thus employment falls.

In many ways this two-sector analysis could be considered to be in the tradition of Schumpeter's trade cycle analysis — the take-up of a new technology and its consequent multiplier effects leading to booms and slumps in the economy. However, Schumpeter's analysis is concerned primarily with output. The upward phase of a trade cycle is associated with increasing output as the new technology is innovated and imitated. The repercussions for labour demand will depend, as in the above model, on the diffusion path and wage movements, and also on the labour saving–using bias of the new technology. Expansion in output need not imply increases in labour demand if the new technology is sufficiently labour-saving. The Schumpeterian approach to the impact of new technology on labour demand has been pursued by Freeman, Clark, and Soete (1982), and we review their work further below.

In the discussion that follows we will also refer to other theoretical approaches to the question of how new technology will affect employment, for we should note that we have not to any extent covered the rich diversity of different models that have been used to investigate the link between technology, output, and employment. In particular, we have not presented any work in a vintage model framework, nor have we so far investigated input–output or neo-Austrian models. The justification for this is that the analysis already undertaken has been sufficient for us to isolate theoretically those factors that will determine how technology will affect output and employment, and the use of other models will not yield much greater theoretical insight. The way we must proceed is to look at some empirical work to see how the different factors have (or will) interact to generate time-paths for output and employment. In this work we will introduce some of the, so far neglected, theoretical frameworks.

Before proceeding to this, it is worth summarizing our theoretical results so far, both from this and the previous chapters. First, we have argued that, in a Walrasian world in which prices and wages react to notional excess demands and supplies, there will not be unemployment as a result of technological change; factor and output prices will, however, change, and so will output. The direction of movement of prices will depend on the factor-saving bias of the technology, so that, for example, a labour-saving technology will imply a new equilibrium in which the relative price of labour is lower. Given that technological change must mean an increase in total factor productivity, the new equilibrium must have higher output.

In the absence of price behaviour of a Walrasian kind we have argued that the impact of technological change on output and employment will depend on the strength of compensation effects and their time-profile. Through the models above we have argued that (1) comparative statics is an unsatisfactory approach for looking at the impact of new technology; (2) the transition path that the economy follows depends upon links between costs and prices and incomes and demands; (3) the time-path of factor prices and the speed of diffusion are important determinants of the impact during the transition phase; (4) compensation effects are not always positive; and (5) the impact of new technology on factor demands is related to the bias in technological change.

We have also argued that, even if compensation effects do act to offset indirect reductions in factor demands, one would still expect to find the indirect impact to be 'distant' — e.g., in other industries — from the direct impact. This in itself may have further implications. For example, we have an implicit assumption so far that labour is homogeneous. We have not really considered the problems of, say, the skill structure of labour demand. This can be particularly relevant in that new technologies may require different skills from the old, making labour market-clearing somewhat less likely. Moreover, if during the transition there is an excess demand for certain skills, this may slow down the diffusion, which in turn may impact on the aggregate demand for labour. We shall have more to say however on the skill problem in section 12.4 below.

One final comment before we move to the empirical section. So far we have discussed process innovations. How would the analysis be affected if product innovations dominated? The effect of a product innovation would be essentially to shift the composition of demand away from all other products to the new. If the new is produced in a less labour-intensive way per unit of expenditure than the other products, then aggregate labour demand should fall unless compensation, through for example price changes, increased expenditure, or increased gross investment acted to reverse the direct impact. However, the analysis and conclusions should not be dramatically different from the process innovation cases studied above.

12.4 Empirics

The purpose of this section is to investigate empirical estimates of the impact of technological change. We are concerned primarily with the impacts on output and labour employment. The empirics of investment are discussed in Chapter 13. We start by looking at a neoclassical approach before moving on to an Austrian approach, and finally consider an input–output example.

As should be clear, the question we wish to ask can be stated in two ways; for example, with respect to output we can ask (1) what is the impact of technological change on output? or (2) what proportion of a realized change in output is due to technological change? This latter approach is more common in the neoclassical literature, which is where we start.

12.4.1 The neoclassical approach

The essence of the literature on the neoclassical approach is to assume the existence of an aggregate production function. Generally two inputs, capital K and labour L, are assumed, with technology represented by a multiplicative shift parameter A. We write the production function as

$$Q = AF(K, L)$$

and from the production function it is clear that we can write

$$\frac{dQ}{Q} = \frac{dA}{A} + \left(\frac{LF_L}{Q}\frac{dL}{L} + \frac{KF_K}{Q}\frac{dK}{K}\right) \tag{12.21}$$

or

$$\frac{dA}{A} = \frac{dQ}{Q} - \left(\frac{LF_L}{Q}\frac{dL}{L} + \frac{KF_K}{Q}\frac{dK}{K}\right). \tag{12.22}$$

The term in brackets is a weighted sum of the rate of growth of the inputs.

Thus, dA/A equals the rate of growth of output minus this weighted sum, i.e., output growth minus input growth. Generally termed the residual, dA/A is taken to be a measure of the rate of technical change, and A itself is termed total factor productivity. It is often implicitly assumed in this literature that dA/A yields a measure of the proportion of output growth due to technological change. Thus one may argue that, if one can estimate the rate of growth of total factor productivity, one can get an estimate of the contribution of technical change to output growth.

Let us consider the classic work in this area, that by Solow (1957). Assuming a constant returns to scale production function of the form above, and allowing factors to be paid their marginal products, we can say that LF_L/Q and KF_K/Q are the shares of labour and capital in national income which must add to unity. Let w_k be the share of capital; then we may write that

$$\frac{dQ}{Q} - \frac{dL}{L} = \frac{dA}{A} + w_k \left(\frac{dK}{K} - \frac{dL}{L} \right) \qquad (12.23)$$

and from series on output per man, capital per man, and the capital share one can obtain estimates of dA/A for any particular years. This is what Solow does for the non-farm private sector of the USA for 1909–49, obtaining estimates for dA/A and thus $A(t)$ (with $A_{1909} = 1$). $A(t)$ rises from 1.000 in 1909 to 1.809 in 1949, i.e., 81 per cent over forty years. Total factor productivity is 81 per cent higher in 1949 than in 1909, and Solow suggests that A has risen at about 1½ per cent per year. Solow's conclusion is that 'gross output per man hour doubled over the interval with 87½% of the increase attributable to technical change and the remaining 12½% due to increased use of capital.'

There is now an extensive literature on the estimation of total factor productivity (much of it critical of Solow's approach), which has been surveyed by both Kennedy and Thirlwall (1972) and by Nadiri (1970).[4] This literature itself overlaps significantly with the literature on estimating production functions.

Given that these surveys already exist, we shall not go into great detail on the problems identified in the literature. A number are however worthy of mention.

1   Solow's analysis essentially assumes that factor shares are constant, which implies that the production function is of Cobb–Douglas form. Numerous attempts have been made to move away from the Cobb–Douglas assumption.
2   Much effort has been expanded on the appropriate measurements of the inputs, for any error in this will lead to bias in the estimates of the rate of productivity gain.
3   Attempts have been made to introduce embodied instead of disembodied technical change into the models.
4   Attempts have been made to make the rate of productivity change a function of variables earlier in the technological progress chain, e.g., R & D.

---

[4] The latest survey, and in many ways the most interesting, is that of Nelson (1981).

For details on all these modifications one may look at a standard text on estimating production functions or at Nadiri (1970). We shall concentrate solely on the last of these modifications, for it seems, potentially at least, to have the most interesting implications. In Mansfield (1972) the literature on the inclusion of R & D in production function studies is admirably reviewed. However, the recent work of Schott (1978) for the UK extends one stage beyond the studies summarized by Mansfield. In her model R & D contributes to technical knowledge, which is represented by the shift parameter in the production function. The extension however is that the level of technological knowledge is determined endogenously as a function of its price relative to other factor prices. Knowledge is shown to respond positively to expected output, wages, and unemployment but negatively to the user costs of capital and knowledge. If this approach is a correct one, then productivity increase is merely a mechanism whereby factor prices affect production rather than an independent factor affecting production. In many ways the endogeneity does reflect our discussion in Part I of the book, and is also an alternative, equilibrium way of reflecting some of the characteristics of the Nelson and Winter (1974) model below.

One must agree to some extent with the Schott view that technological advance is endogenous to the economic system. Whether it can be considered as determined as a derived demand using an aggregate production function is somewhat more questionable. However, if we accept the model, the estimates show that an increase in the expected user cost of knowledge leads to an increase in employment (but a reduction in hours) and an increase in capital and capital utilization. This suggests that, as the stock of knowledge is inversely related to its own user cost, technological advance is associated with lower factor demands.

One alternative approach that has merited some investigation is that, for non-'technological-leader' countries, one can consider that the technological advances possible at any moment in time are reflected in their technology gap relative to the leader country; i.e., for non-leader countries productivity can be increased by catching up technologically. A primary advocate of the technology gap approach is Gomulka (1971, 1982). Looking at the period 1950–70, he regresses the rate of growth of output per head for country i on a measure of the technological gap (output per head in USA relative to output per head in country i) for fifteen advanced capitalist countries, and finds a significant relationship indicating that the growth rate is higher the greater the technology gap. The conclusion is that greater technological opportunity yields faster growth in output per head. But this does not tell us of what happens to output or employment as technology advances.

These studies, however, are just extensions to the basic model, and this basic model is still subject to a number of objections. Assume first of all that the rate of growth of total factor productivity has been correctly estimated. What exactly can we derive from the estimates? Say, for example, it is calculated that $\dot{A}/A$ is 2½ per cent per annum: what does this tell us of how output will grow as technological change occurs? We would seem to be on reasonable grounds if we argued that, in general, it does not mean that, even in the absence of changes induced by non-technological change-related phenomena, output will grow by 2½ per cent per year. The reasons for this are two-fold.

1   If productivity does increase, the increase in output will at least partially be related to the demand for the product. If effective demand is static, there seems no good reason why output should grow. This is what we have discussed in the early part of this chapter.
2   Even if we overcome this objection, there is no reason why output should grow at the rate of increase of A; for one must realize that in neoclassical analysis the inputs of K and L are dependent on A. Our analysis in the previous chapter at the micro-level indicated that the demands for inputs are related to the rate of technological change. Thus the increase in output resulting from technological change (assuming all the output can be sold) will be equal to the increase in output directly induced by the change in A plus the increase induced by the changes in K and L occurring in reaction to the changes in A. In many ways this is similar to the point stressed by Hulten (1979), who argues that Solow-type estimates are in error, for some of the increase in the capital stock results from the extra savings generated by the increase in output brought forth by the technological change. The point to stress however is that one cannot tell what output increase will arise from a productivity increase just by looking at the productivity increase.

Despite the considerable effort that has been expended, the Solow-type approach is still subject to considerable criticism. Perhaps the most basic objection concerns whether an aggregate production function can actually be defined. The 'Cambridge capital controversy' of the 1960s would suggest that in general it cannot (see Harcourt, 1972). If this is so, then all the effort is being expended to no great avail. Of particular interest on this point are the results of Nelson and Winter (1974) (N & W), who not only criticize the above approach but also are able to 'explain' the same data as 'explained' by Solow without requiring an aggregate production function, or in fact any aggregate analysis at all.

N & W essentially argue that to represent technological change by a 'time' term in a production function, and to concentrate on equilibria, is completely unsatisfactory, for 'the essential forces of growth are innovation and selection, with augmentation of capital stocks more or less tied to these processes.' Their underlying view is Schumpeterian, whereby the generation and application of new technology stimulates profit-making, changes in techniques, growth and decline in firms, and the development of the economy, rather than technological changes being some 'additional factor' imposed on the top of the 'normal' growth process of an economy.

N & W construct a model, which is then simulated, in which firms act in a behavioural manner. They satisfice rather than maximize, and in response to unsatisfactory performance they undertake localized search for improved technology. This local search can lead to imitation (generating a diffusion process similar to that observed by Stoneman, 1976 — see Chapter 10).

Technically advanced firms make profits, re-invest and expand driving up the wage rate facing other firms. Firms with low rates of return look for better techniques, sometimes finding them, sometimes not . . . Imitation helps to keep the technical race fairly close, but at any given time there is considerable cross-section dispersion in factor ratios, efficiency and rates of return.

The results derived on the economy's performance are thus the result of technological change and competition at the micro-level, which is then aggregated up to 'explain' macro-data.

A technique is represented by a capital–labour ratio, and a set of feasible techniques (100 in all) was selected from the range of K/L ratios found in Solow's data set. Initial conditions were set so that firms' initial techniques were in the vicinity of 1909 aggregate input coefficient values and capital stocks were set to a convenient size. Initial labour supply was set to yield approximately the 1909 wage rate. Then labour supply was allowed to grow at 1.25 per cent per year, depreciation was allowed at 4 per cent per year, and the critical rate of return in the satisficing mechanism was set at 16 per cent. The simulation of the competitive process among firms was then allowed to run for forty periods. The results are that 'The historically observed trends in the output–labour ratio, capital–labour ratio and the wage rate are all visible in the simulated data.'

The point to be made is that the generally chaotic world of Schumpeterian competition can yield smooth aggregate time-series. The macro-phenomena can be generated through simulating micro-processes that reflect the real nature of technological change. One does not have to revert to inserting t into a macro-model to 'explain' macro-data. Moreover, the results being derived by 'growth accounting' exercises may be considered misleading, for they give no consideration to the role that the Schumpeterian competitive struggle to achieve superior technological performance by the firm plays in generating changes in conventionally measured capital and labour inputs. At its simplest level, it is another way of saying that capital and labour inputs not only change in addition to technology changing, but also change because technology is changing. The whole development of the economic system is the result of the technological change process. To summarize, 'the diversity and change that are suppressed by aggregation, maximisation and equilibria are not the epiphenomena of technical advance. They are the central phenomena' (N & W, 1974, p. 103).

The proximity of much of the N & W model to the microeconomics of technological change discussed in Parts I and II of this book make it a very attractive approach to the analysis of the role of technological change in economic development. We have little to add to their own view of its implications.

Given the N & W reference to Schumpeter, this seems an appropriate place to pick up our reference to the work of Freeman, Clark, and Soete (1982) on Kondratieff cycles, technology, and employment. Again, the basic view is that technology is not an 'additional' factor in growth; it is the 'driving' factor in economic development. The argument is that

The diffusion of clusters of technical innovations of wide adaptability is capable of imparting a substantial upthrust to the growth of the economic system, creating many new opportunities for investment and employment and generating widespread secondary demands for goods and services. Over *time*, however, these new 'technological systems' mature and their investment and employment consequences tend to change. The combination of standardization, growing capital intensity and scale economies means that the employment generated per unit of investment tends to diminish whilst profitability is eroded during the diffusion process. The 'compensation effects' which might mitigate these tendencies operate only imperfectly and often with long delays. (Freeman, Clark, and Soete, pp. 189–90)

One might consider this statement to reflect the operation of an embellished version of the two-sector model discussed above. However, it is not unreasonable to argue that the authors have only circumstantial evidence to support their view that the recent movements in unemployment are due to the operation of such mechanisms (and, moreover, there are many other suggestions as to why unemployment is currently high). It is felt therefore that, although the link between technology and employment could vary such that in the first steps of a Kondratieff cycle technological advances lead to greater employment, with the process being reversed in later stages, we feel that the case is unproven. We turn now to consider a different approach to the link between technology, employment and output, the neo-Austrian approach.

12.4.2 The neo-Austrian approach

This approach has, unfortunately, not received the attention it deserves. In a paper (Hicks, 1970) and a book (Hicks, 1973), the model has been used at some length to explore the basic issue of how technological change will affect output and employment.

The technology is represented by a time-profile of inputs and outputs. A unit operation consists of the application of a flow of labour inputs, $a(t)$ (t, being measured from the start of the operation) which yields a flow of outputs, $b(t)$.

We let $x(T)$ be the number of operations started in calendar time T, and let u be the total length of the operation. We may thus state that, for an economy using this technology,

1 total employment $\equiv A(T) = \int_0^u x(T-t)\,a(t)\,dt$
2 total output $\equiv B(T) = \int_0^u x(T-t)\,b(t)\,dt.$

We define w as the wage in terms of the product and define

$$b(t) - wa(t) = q(t)$$

so that $q(t)$ is the surplus of output over wages earned on a unit operation at time t from its start. We can similarly write

$$B(T) - wA(T) = Q(t)$$

where $Q(T)$ is the surplus of production over wages that accrues to economy as a whole in calendar time T.

We let ρ be the rate of discount (rate of interest) in the economy, and it is argued that with perfect capital markets

$$k(0) \equiv \int_0^u q(t)e^{-\rho t}\, dt = 0.$$

This is termed the 'Fisher condition', and states that ρ must be such that at the start of an operation that operation will have zero present value; if this were not the case, announcement of a start, without any actual production or investment, could yield a positive present value. Given the technology and the wage, the Fisher condition will determine the rate of interest.

To look at the impact of new technology we start at calendar time $0\,(T=0)$ in an economy using an old technology, the coefficients of which we represent by (a^*, b^*, u^*). We assume that the economy had been growing using this technology at a constant full-employment rate. At time 0, a new technology (a, b, u) is introduced. This new technology is assumed to be able to yield a higher ρ for any w than the old technology and is thus superior. Then at any calendar time t we may write

$$A(T) = \int_0^u x(T - t)\, a(t)\, dt + \int_0^{u^*} x^*(T - t)\, a^*(t)\, dt$$

and

$$B(T) = \int_0^u x(T - t)\, b(t)\, dt + \int_0^{u^*} x^*(T - t)\, a^*(t)\, dt$$

where $x(T)$ and $x^*(T)$ are the number of starts on new and old processes respectively (assuming that no old processes are discontinued at time zero because of the appearance of the new technique). As can be immediately seen, the time-path of output and employment will depend upon the technical coefficients of the two technologies and on the number of starts — which is in fact no more than the diffusion path.

If we are to trace out the time-path of output and employment we need some way of determining $x(T)$ and $x^*(T)$, and we do this by following Hick's fix-wage path analysis. Here it is assumed that w remains constant during the whole period of the switch from old to new technology. If w remains constant, then for the Fisher condition to hold on the new technology ρ must rise, and with an increase in ρ the capital value of a starting old process $(k^*(0))$ will be negative. We may thus argue that, for $T > 0$, $x^*(T) = 0$. It is obvious that, for $T < 0$, $x(T) = 0$ for the old technology did not exist. Thus we may write that

$$A(T) = \int_0^T x(T - t)\, a(t)\, dt + \int_T^{u^*} x^*(T - t)\, a^*(t)\, dt$$

and similarly for $B(T)$.

We may also define

$$Q(T) = B(T) - wA(T).$$

Hick's derives all his results relative to a reference path upon which only the old technique is used, and on which employment would have been

$$A^*(T) = \int_0^{u^*} x^*(T - t)\, a^*(t)\, dt$$

where $x^*(T)$, for $T > 0$, refers to starts that would have been made. Then we may write that

$$A(T) - A^*(T) = \int_0^T x(T - t)\, a(t)\, dt - \int_0^T x^*(T - t)\, a^*(t)\, dt$$

where $A^*(T)$ is employment on the reference path. We may similarly define $B^*(T)$ and $Q^*(T)$. Now to determine $x(T)$ Hicks argues that it is illuminating to consider $x(T)$ as determined by the condition that $Q(T) = Q^*(T)$.

The final product of our economy, $B(T)$, is a flow of consumption goods. $wA(T)$ is the total wage bill in terms of consumption goods. Thus $Q(T) = B(T) - wA(T)$ must equal the consumption goods produced over and above the wage bill, and thus be equal to consumption out of profits. Thus an assumption that $Q(T) = Q^*(T)$ is equivalent to assuming that capitalist consumption remains constant over time. The idea is thus that out of a total amount of output some is taken for consumption by capitalists, the rest is used to employ labour at wage w, and the total employment is conditioned by the fact that all the output must be used up. It is a sort of wage fund theory of employment. Given that a number of processes are already operating using up the wage fund determines the number of starts.

In a general way we can say that, at time zero, given that the new technology does have a period of non-zero length before producing output, and in reaction to the new technology no old processes are stopped, the introduction of the new technology will not change the sum of outputs available to meet consumption. If $Q(T)$ is the same as on the reference path and w stays constant, the number of starts on the new process will be the number that makes $A(T) = A^*(T)$. Thus $x(0)/x^*(0) = a(1)/a^*(1)$; in other words, workers are just transferred from building old machines to building new.

As time proceeds, however, and as the introduction of new technology starts to affect the output flow, the effects on employment could start to build up. Consider the following.

1   The new technology has a longer constructional period than the old. Thus, as old technologies end their useful life (of length u^*) the flow of consumption goods will start to fall below that on the reference path. Given $Q(T) = Q^*(T)$, the wage fund will be lower and employment at the given wage must reduce, for there is insufficient 'gross investment' to employ all workers in constructing new machines.

2   The new technology has the same construction period but $b(t) < b^*(t)$ (although $u > u^*$ and thus the new technology is more profitable), or, alternatively, $b(t) = b^*(t)$ but $a(t) > a^*(t)$ and $u > u^*$. Again, we may have a reduced wage fund and thus reductions in employment.

Both examples may be said to be examples of a shift to a more capital-

(time)-intensive technique, and employment is reduced because insufficient funds are available to employ all workers. However, if one moves to less capital-(time)-intensive techniques the reverse will hold and employment will always rise. Even in the increased capital intensity case, the higher productivity of the new technology must eventually lead to higher output and thus a greater wage fund, and so employment will eventually rise.

The 'wage fund' approach is really very classical, and the unemployment results of this model rely extensively on the fixed-consumption-out-of-profits assumptions. If the consumption out of profits becomes an endogenous variable, our results may change considerably. Moreover, if the assumption of the automatic investment of non-consumed profits is relaxed, then again the results may be considerably modified. Hicks himself has relaxed the fixed-wage assumption. The story the model tells, however, is to reaffirm that the time-path of wages and diffusion are important determinants of employment and output on the transition path, and moreover, because there are lags in getting new technology into production, one should not expect its impact to be immediate.

In an attempt to assess the impact of computers on labour demand Stoneman (1976) uses an essentially Austrian representation of technology combined with an actual diffusion path of computer usage and generates a series of job losses and gains resulting from computer use in the UK. It is argued that a computer took two years to build, one year to install, and had a life of seven years. Relative to the replaced technology there were extra construction costs in years -2 and -1 (measured from an installation year base) of twenty-five and twenty-three man-years, respectively. The user made no labour savings in the installation year but then saved twenty-seven man-years of labour in each year of the computer's seven-year life. This generated for each computer installed a profile of net labour use:

Year (t)	-2	-1	0	1	2	3	4	5	6	7
Net labour use per computer used	$+25$	$+23$	0	-27	-27	-27	-27	-27	-27	-27

Using this representation of the technology, two calculations are then made.

1   What would the labour saving be in the 1970 UK economy if its actual level of output had been produced with a technology that had complete diffusion of computer technology?
2   What labour-saving had been achieved during the diffusion process 1952–70 as computers spread across the economy, assuming the economy's output had not been affected by computer use?

In answering (1), it is argued that the post-diffusion stock of computers would have been 10,142 in 1970. In a fully diffused economy one-eighth of these would have to be replaced each year, meaning employing 60,852 man-years of labour extra compared with the previous technology, but this number of computers would save 239,604 man-years of labour in use, showing a net saving of 178,752 man-years or 0.79 per cent the 1970 labour force. In other words, the 1970 level of output could have been produced by 0.79 per cent fewer men than under the previous technology if the diffusion had been complete in 1970.

In answering (2), one takes the profile of net labour-saving detailed above and defining S_T as the computer stock and x_T as additions to the computer stock in year T; then for any calendar date T, labour-saving is given as

$$(S_T - x_T) \times 27$$

and increased labour use is given as

$$\sum_{t=-2}^{0} x_{T-t}\, a_t$$

where $a_{-2} = 25$, $a_{-1} = 23$, and $a_0 = 0$.

By summing over labour-saving and use, one obtains the figures in Table 12.1. These figures illustrate that, as the economy underwent its diffusion, extra labour was needed in the construction of new capital goods. This extra labour was offset only partly by the savings in computer use for a number of years.

Table 12.1   *The transition path for computerization in the UK*

Year	Stock at end of year (1)	Net additions to computer stock (2)	Gross additions to computer stock (3)	Net labour reqts in computer construction (man-years) (4)	Net labour-saving in computer usage (man-years) (5)	Balance, (4)−(5) (6)
1952				300	0	300
1953				551	0	551
1954	12	12	12	678	0	678
1955	23	11	11	1,566	324	1,242
1956	40	17	17	2,931	621	2,310
1957	87	47	47	3,177	1,080	2,097
1958	161	74	74	3,507	2,349	1,158
1959	220	59	59	5,103	4,347	756
1960	306	86	86	8,450	5,940	2,510
1961	417	111	125	12,379	7,884	4,495
1962	620	203	223	16,495	10,719	5,776
1963	875	225	290	21,139	15,795	5,344
1964	982	107	393	26,757	15,903	10,854
1965	1,424	442	484	36,950	25,380	11,570
1966	1,956	532	625	49,019	35,937	13,082
1967	2,595	639	903	57,540	45,684	11,856
1968	3,522	927	1,130	66,801	64,584	2,217
1969	4,319	797	1,262	69,353	82,539	−13,186
1970	5,470	1,151	1,151	69,557	106,893	−37,336

Source: Stoneman (1976).

The diffusion was fast, and thus the capital goods sector had to expand quickly to supply the users' demands, this resulting in a net gain in employment. As the computer stock built up, however, the labour-saving gradually begins to dominate.

These figures illustrate how an analysis of the transition path can yield results so different from comparative statics analysis. Even though computerization could lead to labour-saving comparing computerized with non-computerized economies, this does not mean that those labour-savings will be realized immediately, or that during the transition stage that there will be a net reduction in labour demand.

This exercise is open to one basic criticism, however. The output of the economy is assumed to be independent of the use of the new technology, and this is probably unrealistic.

Table 12.2 *Posts replaced and created per computer used for administrative purposes, 1964 and 1969*

Occupation	Change in number of posts	
	1964	1969
Managers and executives	} −1.0	−1.0
Supervisors		−5.0
Clerks	−47.3	−49.0
Typists	−3.4	−7.0
Calculating machine operators		−12.0
Other office-machine operators	−20.7	−5.0
Other office workers		−1.0
Data processing managers	+1.1	+1.3
Systems analysts	+3.7	+5.4
Programmers	+5.6	+7.4
Machine operators	+19.1	+20.1
Other computer staff		+5.1
Net reduction	42.9	40.7

Source: Stoneman (1976).

In the course of the analysis of the impact of computer use, some useful data on the effect on skill patterns of the labour force are generated. In Table 12.2 the reported data are reproduced. As can be seen, the majority of posts lost were in the low-skill category, the posts being gained having a different skill requirement and probably higher skills. Now, one cannot generalize from this one example as to the effect of technological change on skills, but it is not unreasonable to say that the desired skill mix of the labour force will change as new technology is introduced. This in fact is also reflected in the results discussed in Chapter 11.

12.4.3 The input–output approach

Burmeister (1974) has shown that Hick's neo-Austrian model is really a special case of the Von Neumann approach to modelling production processes. Rather than deal with the complicated Von Neumann model however we will discuss how input–output analysis has been applied to consider the effect of new technology on employment. The great advantage of the input–output approach is the stress that it puts upon the technology multiplier effect.

Essentially, one considers technology to be represented by an input–output matrix **A**. Let the old technology matrix be $\mathbf{A_0}$ and the new $\mathbf{A_1}$. Assume a given final demand vector **x**. The essence of the approach is to compare the total requirements of primary and intermediate inputs for the generation of **x**, using technology $\mathbf{A_0}$ and technology $\mathbf{A_1}$. As can be seen, this stresses the multiplier compensation effects, but ignores any impact that the new technology may have on the level or pattern of demand — and it is often this that is stressed in the other approaches discussed above.

Carter (1970) uses an input–output framework to analyse the impact of structural change on the US economy. Despite the fact that the analysis is performed in the context of a fixed bill of commodities, thus preventing a more realistic analysis of technological change allowing for the changing nature of the product input and output mix, her results are most informative. Her conclusion is that 'with 1958 technology the economy was capable of delivering 1958 final demand with a 22% lower total factor cost than with 1947 technology in all sectors' (Carter, 1970, p. 171). This gives one a fair idea of the overall impact of technological change. One might note that the implied annual growth rate of productivity does not differ substantially from Solow's estimates discussed above.

We can criticize input–output models on the grounds that in general they failed to take account of the demand impact of new technology. To illustrate the importance of the inclusion of this in such models we shall now consider the work of Whitley and Wilson (1982) using an augmented input–output model to look at the possible effects of microelectronics on employment and output.

The model has equations explaining consumption, employment, exports, imports, prices, investment, and wages and an input–output sector that determines intermediate demands. There are forty-nine employing activities in the model, and each element of demand is considered to be a function of prices, incomes, and trend terms representing changes in tastes and technologies. Labour supply growth is assumed exogenous but labour demand is determined by output and technology. In the presence of disequilibrium in the labour market, the wage rate changes, which has feedbacks on prices. However, wages and prices are not infinitely flexible in the Walrasian sense so unemployment can exist.

A simulation exercise is performed to illustrate the impact on employment of a faster rate of take-up of microelectronics in the UK economy. Comparison is made with a projection of past technological trends for 1990. The technological changes are represented by faster labour productivity growth in 'best-practice' techniques. This affects actual output per man with the actual differing from best practice. The labour-displacing effects of this are offset partly by increases in demand through resulting cost and price reduction. These offsetting

Table 12.3 *Employment impact of faster take-up of microelectronic technology*

	Improvement in best-practice output per man by 1990 (1)	Best-practice potential displacement (2)	Actual potential displacement (3)	First-round compensation (4)	Other compensation (5)	Total effect (4)+(5)−(3) (6)
	%	('000)	('000)	('000)	('000)	('000)
Mechanical engineering	5	34	26	11	34	19
Instrument engineering	5	6	4	1	2	−1
Electrical engineering	5	31	26	8	16	−2
Textile fibres	11.5	1	1	0	1	0
Textile n.e.s.	8	23	22	6	9	−7
Leather, clothing, etc.	4	13	15	9	5	−1
Papers & Board	5	2	3	2	1	0
Printing & Publishing	5	23	17	6	7	−4
Rubber	6	4	4	2	1	−1
Manufacture n.e.s.	5	10	9	3	5	−1
Communications	8	37	33	10	16	−7
Distribution	3	94	66	22	34	−10
Insurance	3	46	62	27	13	−22
Professional services	3	18	23	8	3	−12
Others	–	–		29	13	130
Whole economy	–	342	311	144	4	81

Source: Whitley and Wilson (1983).

effects are calculated at the aggregate level to approximate 50 per cent of any initial direct reduction in employment. In addition to these first-round compensation effects, the authors argue that further increases in demand and thus employment and output will come from 'second-round' compensation because of

1   extra investment demand for installing new technology;
2   changes in intermediate demands;
3   improved trade performances from improved non-price competiveness.

The sum of these three effects, they calculate, will more than offset the gap between first-round indirect effects and the direct effect.

In Table 12.3 we present data on their simulation. Column (2) gives the potential displacement of jobs that could arise from the productivity improvement detailed in column (1). The fact that not all the productivity improvement is realized leads to some reduction in the number of displacements in column (3). In columns (4) and (5) the impact of first-round and second-round compensatory increases in demand on the economy are shown. The overall effect is shown in column (6). We offer the following comments.

1   The calculation that about 50 per cent of direct displacement can be offset by first-round compensation has been found in other applications of input–output models to these types of questions; see Stoneman *et al.* (1982).
2   The ranking of the total effect is not the same as the ranking of rates of potential productivity gains across industries. This illustrates that compensation may occur at a distance from the original productivity gain, and also reflects the results of Wragg and Robertson (1978) discussed in the previous chapter.
3   To limit the calculation of the impact of new technology to looking at only direct effects can be most misleading.

The Whitley and Wilson exercise also yields the following overall effects by 1990 from their projected faster rate of technological change compared with the base run: output per man, +1.8 per cent; consumption, +1.2 per cent; gross domestic fixed capital function, +8.7 per cent; exports, +3.6 per cent; imports, +4.8 per cent; prices, −2.4 per cent; gross output, +2.3 per cent; employment, +81,000 men.

As a simulation exercise the whole process is rather illuminating. There are however problems. In generating second-round compensation effects the assumptions are somewhat arbitrary, and it is worrying (see Table 12.3) that the major growth of employment occurs in 'others', i.e., non-specific industries. However, as an exercise it does reflect the power of compensation effects.

12.5 Conclusions

Over these last two chapters we have considered one of the major questions in economics at the present time. How will new technology affect output and factor demand? At the macroeconomic level we have argued that if the economy is demand-constrained, then only if technological change leads to increases in demand will output increase and thus the direct impacts on factor demand be at least partially offset. The mechanisms by which demand might increase depend upon price movements and other compensation effects. At the level of the firm, even if the economy is demand-constrained an individual firm may increase its employment of inputs by taking market share from rivals; however, if the overall size of the market does not change this must mean that factor employment in the industry must be reduced (for both factors if the advance is neutral, for at least one of if it is biased). In terms of the analysis in chapter 11, this is the same as saying that the impact on factor employment is related to the elasticity of demand. In a demand-constrained world one firm's addition to factor employment is likely to be more than offset by another firm's reduction. This may be particularly important if the innovator is an importer.

We have argued that with perfect price flexibility there will be no demand constraints; however, we consider this assumption to be unrealistic. In a world with price inflexibility we have essentially stated that demand can still be affected by new technology, and in such a world the impact of new technology on employment will depend upon

1   the nature of the new technology;
2   the diffusion path;
3   movements in factor prices;
4   the strength of various compensation effects.

We have drawn these conclusions by the investigation of a number of different modelling frameworks, some neoclassical, some more behaviourally based. On a personal level I am inclined to consider the world as better represented by the latter approach. The choice of approach is crucial if one wishes to answer the question, 'to what extent have actual increases in output been due to technological change?' The neoclassical approach considers capital growth as essentially exogenous to the technical change process (although our next chapter will explore this more fully), and it can thus be treated as a separate influence on output growth. The Nelson and Winter approach considers that technological advance is essentially the *sine qua non* of economic development. I tend to support this view.

In our discussion we have at several times made references to the limitations of our analysis. We have largely relied on models where labour demand and supply are homogeneous, and although we have pointed it out, it is worth stating again that changes in the skill composition of labour demand and supply may well result from technological change and this may make market-clearing less likely. We have also had occasion to refer to the possible impact on market structures of technological change (which we explore in Chapter 16), and although we have pointed out how the acquisition of monopoly power may reduce output and demand, our results have not really reflected this in any detail.

Finally, our analysis is largely confined to a closed economy. The open economy is explored in Chapter 17.

To summarize, economic theory guides us in isolating those factors that will influence how output and employment react to technological change and can also tell us under what conditions there will be no technological unemployment. However, to isolate the impact of any one technological change or technological change in general we must turn to empirical analysis. There we find a mixture of results, some optimistic, some pessimistic. It seems that a reasonable conclusion is that many technological changes will reduce directly the demand for factor inputs; however, because of compensation effects the direct effect may be partly offset through indirect reactions. It is the strength of these indirect reactions that is in dispute. Elsewhere (Stoneman *et al.*, 1982) I have argued for compensation in excess of 60 per cent of the direct effect. I have no grounds for changing my view on this. As far as output is concerned, an overview of the last two chapters seems to be that technological change encourages increased output. If nothing else, we can state that technological change yields the opportunity to increase economic welfare. As to who gets the benefits, who suffers the costs, and how much of the potential is realized, we are not in a position to be definitive.

Appendix: A two-sector model

The two-sector model has a consumption good and a capital good sector. We consider only process innovations, and we assume that all process innovations are wholly embodied. Thus to change technology a whole new set of capital goods is required.

We consider two technologies, labelled 1 and 2, that are the old and new respectively. Both capital goods are made wholly by labour (although circulating capital could be allowed), but consumption goods are made by capital and labour. The technology of capital goods production is considered in this manner for it yields considerable simplification; however, the omission of a capital goods input from capital goods production will bias any compensation effects we find. We represent the quantity relations of the two technologies as follows.

$$K_{it} = \alpha_i C_{it} \qquad\qquad i = 1,2 \qquad\qquad (A12.1)$$

$$L_{it} = \beta_i C_{it} + b_i x_{it} \qquad\qquad i = 1, 2 \qquad\qquad (A12.2)$$

$$x_{it} = \frac{dK_{it}}{dt} \qquad\qquad i = 1, 2 \qquad\qquad (A12.3)$$

where K_{it} = stock of capital good i in time t;
  C_{it} = output of consumption goods on capital good i in time t;
  L_{it} = labour employed producing or using capital good i in time t;
  x_{it} = output of capital good i in time t.
We consider that L_{it} is measured in man-hours. Let E_t be the number of men employed and h_t be the number of hours worked by each man (the same for each man); then, defining $L_t = L_{1t} + L_{2t}$, we have

$$L_t = h_t E_t. \tag{A12.4}$$

We now allow that the working population E_t^s grows at rate n so that $E_t^s = E_0 e^{nt}$. We consider that there is a standard working day $\bar{h}$, and full employment is defined by

$$L_t = \bar{h} E_t^s. \tag{A12.5}$$

We further argue that, should the demand for labour be greater than the supply, then h will increase, so that employers are never constrained in the labour market; however, should the demand for labour fall below $\bar{h} E_t^s$, the number of men employed will fall; that is, workers will work not less than the standard day but will work more. This device allows us to consider a world in which labour demand grows faster than labour supply without introducing all the problems resulting from constraining the producer on the labour market.

We now assume that capital has an infinitely long physical life and there is thus no depreciation. However, capital can become economically obsolescent. Consider that at time t the economy is in the process of making the transaction from using technology 1 to using technology 2, and at any moment in time the two technologies are used in consumption goods production in the proportions $1 - \theta_t : \theta_t$. Allow now that at any time t the output of the relevant capital good is just sufficient to meet the demands of the consumer goods industry subject only to $x_{it} \geqslant 0$. Thus we have for $i = 1, 2$, using a $\cdot$ to represent a time derivative,

$$\begin{aligned} x_{it} &= \alpha_i \dot{C}_{it} \quad \text{if } \alpha_i \dot{C}_{it} \geqslant 0 \\ &= 0 \qquad \quad \text{if } \alpha_i \dot{C}_{it} < 0 \end{aligned} \tag{A12.6}$$

This yields, for $x_{it} > 0$,

$$x_{1t} = \alpha_1 \{(1 - \theta_t) \dot{C}_t - C_t \dot{\theta}_t\} \tag{A12.7}$$

and

$$x_{2t} = \alpha_2 \{\theta_t \dot{C}_t + C_t \dot{\theta}_t\} \tag{A12.8}$$

where $C_t \equiv C_{1t} + C_{2t}$. Defining $g_t \equiv \dot{C}_t/C_t$, we may state from A12.7 and A12.8 that

$$x_{1t} > 0 \quad \text{if } g_t > \frac{\dot{\theta}_t}{1 - \theta_t}$$

$$x_{2t} > 0 \quad \text{if } g_t > -\frac{\dot{\theta}_t}{\theta_t}$$

We may therefore argue that, if $g_t < \dot{\theta}/(1 - \theta_t)$, then the old capital good is in excess supply and no more will be produced. If $g_t > \dot{\theta}/(1 - \theta_t)$, then additions to the stock of the old capital good are still required and it will continue to be produced. If we define T as the date at which $g_T = \dot{\theta}_T/(1 - \theta_T)$, then time T is the date at which the industry producing the old capital good dies. Prior to time T, the economy will be producing and using new and old capital goods side by side; after time T, only the new capital good is being produced but the old (if it has a positive quasi-rent) and new will both be used.

Given the quantity relations we will now consider price relations. Consider first the case where machines are not in excess supply, i.e., $x_{it} > 0$, $i = 1, 2$.

We assume that, because there is no capital used in capital goods production, machines are priced at their labour cost. Let w_t be the wage rate per hour, which we will assume to be the same for all workers no matter what they are producing or which machines they are using or how many hours they are working. Let p_{it} be the price of machine i in time t; then we have

$$p_{it} = w_t \, b_i \qquad\qquad i = 1, 2. \qquad\qquad \text{(A12.9)}$$

Defining r_{it} as the rate of profit in consumption goods production from the use of machine i in time t, and letting the consumption goods be the numeraire with a price of unity, then

$$1 = r_{it} \, p_{it} \, \alpha_i + w_t \, \beta_i, \qquad\qquad i = 1, 2. \qquad\qquad \text{(A12.10)}$$

For the new technology to be superior, and therefore for the entrepreneur's change of technique to be rational, technology 2 must be more profitable than technique 1. The superiority of technique 2 can be local or global; i.e., it can apply for one, some, or all wage rates. We will assume global superiority to avoid switchbacks from new to old technologies with changing wage rates. For technology 2 to yield a higher profit rate (r_{2t}) than technology 1 (r_{1t}), for all w_t we require that

$$b_1 \, \alpha_1 \geqslant b_2 \, \alpha_2 \text{ and } \beta_1 \geqslant \beta_2 \qquad\qquad \text{(A12.11)}$$

with at least one strict inequality holding. We will be assuming that both inequalities hold; that technology 2 is labour-saving relative to technology 1, requiring less labour in both direct and indirect use. This would seem to be the most pessimistic assumption for unemployment implications. We have considered the behaviour of the model when technology 2 is only locally superior but the results add little.

If a machine is in excess supply, which situation arises with respect to type 1 machines if $g_t < \theta_t/(1 - \theta_t)$, then there is no reason to expect that it will be valued at its construction cost. We have considered a case where p_{1t} falls to zero when $x_{1t} = 0$ (which we do not report in detail here), but in general we are able to proceed without being explicit as to what happens to p_{1t}. Implicitly, we assume that when capital good 1 is in excess supply $p_{1t} = (1 - w_t \beta_1)/ (\alpha_1 \, r_{2t})$ with the prospect of capital losses guaranteeing the superiority of technique 2.

Given the price and quantity relations, we now close the model by specifying the demand environment. We assume that savings can be represented by a classical savings function with all wages consumed, all profits saved. We have experimented with a proportional savings function, but as Hicks (1965) states, it sits much less happily in this framework. Thus we may write

$$C_t = w_t \, L_t. \qquad\qquad \text{(A12.12)}$$

Define γ_t as the ratio of the full-employment workforce employed in time t; then

$$\gamma_t \equiv \frac{L_t}{\bar{h} E_0 e^{nt}}. \qquad\qquad \text{(A12.13)}$$

The rest of this section is primarily concerned with the time-path of γ_t as the economy makes the transition from using technology 1 to using technology 2.

From A12.13 we may derive

$$\frac{\dot{\gamma}_t}{\gamma_t} = \frac{\dot{L}_t}{L_t} - n. \qquad\qquad \text{(A12.14)}$$

From A12.12 we have

$$g_t = \frac{\dot{L}_t}{L_t} + \frac{\dot{w}_t}{w_t} \qquad \text{(A12.15)}$$

and thus

$$\frac{\dot{\gamma}_t}{\gamma_t} = g_t - \frac{\dot{w}_t}{w_t} - n. \qquad \text{(A12.16)}$$

Define

$$z_t \equiv \beta_1 (1 - \theta_t) + \beta_2 \theta_t$$

$$q_t \equiv \alpha_1 b_1 (1 - \theta_t) + \alpha_2 b_2 \theta_t$$

$$\Delta\beta \equiv \beta_1 - \beta_2 = \frac{-\partial z_t}{d\theta_t} \geqslant 0$$

and

$$\Delta A \equiv \alpha_1 b_1 - \alpha_2 b_2 = \frac{-dq_t}{d\theta_t} \geqslant 0.$$

Then z_t may be defined as the direct labour requirement per unit output of the consumption good and q_t the indirect labour (or capital) requirement, both varying over time as θ_t changes.

Using these definitions we can derive from A12.2, A12.7, A12.8, and A12.12 that if $x_{1t} > 0$,

$$\frac{C_t}{w_t} = L_t = C_t (z_t + g_t q_t - \dot{\theta}_t \Delta A) \qquad \text{(A12.17)}$$

and if $x_{1t} = 0$,

$$\frac{C_t}{w_t} = L_t = C_t \{z_t + \alpha_2 b_2 (g_t \theta_t + \dot{\theta}_t)\}. \qquad \text{(A12.18)}$$

From which it is clear that, if $x_{1t} > 0$,

$$g_t = \frac{1 - w_t z_t + w_t \Delta A \dot{\theta}_t}{w_t q_t} \qquad \text{(A12.19)}$$

and if $x_{1t} = 0$,

$$g_t = \frac{1 - w_t z_t}{w_t \alpha_2 b_2 \theta_t} - \frac{\dot{\theta}_t}{\theta_t}. \qquad \text{(A12.20)}$$

From (A12.16), (A12.19), and (A12.20) we can thus state that the time-path of employment depends upon

1   the time-path of consumption, which in turn depends on the time-path of wages;
2   the time path of θ_t, the diffusion process; and
3   the technical coefficients of the two technologies (i.e., the factor-saving bias).

References

Burmeister, E. (1974), 'Neo-Austrian and Alternative Approaches to Capital Theory', *Journal of Economic Literature*, XII, 413–56.

Carter, A.P. (1970), *Structural Change in the American Economy*, Harvard

University Press, Cambridge, Mass.

Casson, M. (1981), *Unemployment: A Disequilibrium Approach*, Martin Robertson, Oxford.

Freeman, C., Clark, J. and Soete, L. (1982), *Unemployment and Technical Innovation*, Frances Pinter, London.

Gomulka, S. (1971), *Inventive Activity, Diffusion and the Stages of Economic Growth*, Institute of Economics, Aarhus.

Gomulka, S. (1982), 'Kaldor's Sylized Facts, Dynamic Economics of Scale and Diffusional Effect in Productivity Change', London School of Economics, mimeo.

Gourvitch, A. (1966), *Survey of Economic Theory on Technological Change and Employment*, A.M. Kelley, New York.

Hahn, F. (1982), *Reflections on the Invisible Hand*, Warwick Economic Research Paper no. 196, University of Warwick.

Harcourt, G.C. (1972), *Some Cambridge Controversies in the Theory of Capital*, Cambridge University Press.

Hicks, Sir J.R. (1965), *Capital and Growth*, Oxford University Press, London.

Hicks, Sir J.R. (1970), 'A Neo-Austrian Growth Theory', *Economic Journal*, LXXX, 257-81.

Hicks, Sir J.R. (1973), *Capital and Time*, Oxford University Press, London.

Hulten, C.R. (1979), 'On the "Importance" of Productivity Change', *American Economic Review*, 69, 126-36.

Kaldor, N. (1955), 'Alternative Theories of Distribution', *Review of Economic Studies*, 22, 83-100.

Kennedy, C. and Thirwall, A.P. (1972), 'Technical Progress: A Survey', *Economic Journal*, 82, 11-72.

Mansfield, E. (1972), 'Contribution of R & D to Economic Growth in the United States', *Science*, 175, 477-86.

Nadiri, M.I. (1970), 'Some Approaches to the Theory and Measurement of Total Factor Productivity: A Survey', *Journal of Economic Literature*, 8, 1137-77.

Nelson, R. (1981), 'Research on Productivity Growth and Productivity Differences: Dead Ends and New Departures', *Journal of Economic Literature*, 19, 1029-64.

Nelson, R. and Winter, S. (1974), 'Neoclassical vs Evolutionary Theories of Economic Growth: Critique and Prospectus', *Economic Journal*, 84, 886-905.

Parkin, M. and Bade, R. (1982), *Modern Macroeconomics*, Philip Allan, Oxford.

Pasinetti, L. (1962), 'Rate of Profit and Income Distribution in Relation to the Rate of Economic Growth', *Review of Economic Studies*, 29, 267-79.

Sato, R. (1963), 'Fiscal Policy in a Neoclassical Growth Model: An Analysis of Time Required for Equilibriating Adjustment', *Review of Economic Studies*, 30, 16-23.

Schott, K. (1978), 'The Relations Between Industrial R & D and Factor Demands', *Economic Journal*, 88, 85-106.

Sinclair, P. (1981), 'When Will Technical Progress Destroy Jobs?', *Oxford Economic Papers*, 33, 1-18.

Solow, R.M. (1956), 'A Contribution to the Theory of Economic Growth', *Quarterly Journal of Economics*, 70, 65-94.

Solow, R.M. (1957), 'Technical Change and the Aggregate Production Function', *Review of Economics and Statistics*, 39, 312-20.

Stoneman, P. (1976), *Technological Diffusion and the Computer Revolution*, Cambridge University Press.

Stoneman, P. (1982), 'Technology, Diffusion, Wages and Employment', Warwick

Economic Research Paper no. 190, University of Warwick.

Stoneman, P., Blattner, N. and Pastre, O. (1982), 'Information Technology, Productivity and Employment', in *Microelectronics, Robotics and Jobs*, ICCP, no. 7, OECD, Paris.

Whitley, J. and Wilson, R. (1982), 'Quantifying the Employment Effects of Micro-electronics', *Futures*, 14, 486–96.

Wragg, R. and Robertson, T. (1978), *Post War Trends in Output, Employment, Productivity, Output, Labour Costs and Prices by Industry in the United Kingdom*, Research Paper no. 3, Department of Employment, London.

Chapter 13

Technological Change and Investment

13.1 Introduction

In Chapter 11 we investigated the impact that technological change might have on the demand for factor inputs, one of which may be capital. In Chapter 12 we stressed the importance of increased investment expenditures as a compensatory factor offsetting direct reduction in labour demand. In this chapter we explore further the link between technological change, the demand for capital, and investment.

The underlying rationale for this chapter is the view that technological change is an important neglected influence on the level of investment expenditures in an economy. Helliwell (1976, p. 20) states that 'investment equations typically ignore technological change', and to a large extent we agree with this view. In this chapter however we attempt to illustrate some of the ways in which technological change has been treated in the literature when included in investment equations, and in section 13.6 we suggest an alternative approach. This approach is based on the realization that studies of diffusion processes are really studies, in many cases, of the demand for individual capital goods, and thus by extension one can approach the analysis of the determinants of investment through the summation over diffusion processes.

We should note initially, as Nickell (1978) states, that firms may invest in 'training for their employees, in goodwill via advertising expenditures, in knowledge via research and development activities, in stocks of finished good or raw materials or work in progress and finally in fixed capital stock such as plant and machinery, office blocks or vehicles'. Following Nickell, it is investment in this last category of goods with which we are concerned. This also means that we rule out household investment in durables and housing and concentrate on firms. However, what we say with respect to firms may well, suitably modified, be applied to households.

13.2 The neoclassical approach

In a series of papers Jorgenson (e.g., 1963) has developed what is termed the neoclassical theory of investment. Assume a frictionless, certain and perfectly competitive world in which a firm faces the production function

$$Y = F(K, L). \tag{13.1}$$

Defining net investment, I, by

$$I = \frac{dK}{dt}, \tag{13.2}$$

the purpose of the model is to derive an expression for I as a function of exogenous parameters in the economy (prices of inputs and outputs).

The firm maximizes its present value as given

$$V = \int_0^\infty e^{-rt} (p_t Y_t - w_t L_t - q_t GI_t)\, dt \tag{13.3}$$

where Y = output;
p = price;
L = the flow of labour services;
GI = gross investment;
q = the price of capital goods;
r = the discount rate/rate of interest.
The maximization of (13.3) is constrained by (13.1) and

$$I_t = \frac{dK}{dt} = GI_t - \delta K_t \tag{13.4}$$

where δK_t is replacement investment (on the assumption that a constant proportion of the capital stock depreciates in each time period).

The maximization conditions (first-order) yield that

$$\frac{\partial Y_t}{\partial L_t} = \frac{w_t}{p_t} \tag{13.5}$$

$$\frac{\partial Y_t}{\partial K_t} = \frac{c_t}{p_t} \tag{13.6}$$

where $c_t \equiv q_t (r + \delta) - \dot{q}_t$ is called the user cost of capital. This user cost equals interest forgone plus depreciation minus capital gains. Given that prices are assumed known for all t, then $\dot{q}$ is known as well, and the firm chooses its inputs of K and L in time t to satisfy (13.5) and (13.6). Given the time-path of K to satisfy (13.6), call it K^*, we can get investment as dK^*/dt.

To obtain an aggregate investment equation, one just simply 'blows-up' the micro-result, assuming away aggregation problems.

As this model stands there is no real technological change included. Of course it may be that the time-path of p_t, q_t, or w_t reflects technological changes taking place somewhere, but this is really not considered. For example, one could argue that, because of advances in technology in capital goods production, q_t is falling over time. This will affect c_t and thus the choice of K^* and thus investment, (note that, given $\dot{q} < 0$, we cannot say whether c_t will rise or fall over time), but such a treatment does not really come to grips with the problem of technological change.

Schramm (1972) has taken the Jorgenson model and introduced disembodied technological change into the production function. He considers his production function as

$$Y_t = A e^{gt} K^\alpha L^\beta. \tag{13.7}$$

Given the Cobb–Douglas function, technological change is neutral under the Hicks, Harrod and Solow definitions. Using (13.5), (13.6), and (13.7), one may then derive

$$\log K_t^* = a_0 + a_1 \log \frac{w}{p} + a_2 \log \frac{c}{p} + a_3 t \qquad (13.8)$$

where

$$a_0 = \frac{\beta - 1}{\gamma} \log \alpha - \frac{\beta}{\gamma} \log \beta - \frac{1}{\gamma} \log A;$$

$$a_1 = \frac{\beta}{\gamma};$$

$$a_2 = \frac{1 - \beta}{\gamma};$$

$$a_3 = \frac{-g}{\gamma};$$

$$\gamma = \alpha + \beta - 1.$$

It is still the case that $c \equiv q\,(r + \delta) - \dot{q}$.

From (13.8) we may see that the term reflecting technological change enters as a time trend. The coefficient is $-g/(\alpha + \beta - 1)$ and thus is negative if there are increasing returns to scale and positive with decreasing returns to scale. (If $\alpha + \beta = 1$ the model cannot be solved.) In his empirical estimates Schramm finds that the time trend works well, with in most cases a positive, significant coefficient.

One could extend this analysis following the lines detailed in Chapter 11 on the determination of factor input demands in a world with technological change, where we showed that elasticities, biases, and scale economies were important. However, it should be clear that this approach cannot tell the whole story. Specifically, disembodiment is a useful but deceptive assumption. Embodied technological change is much to be preferred. Moreover, related to this, the apparent independence of depreciation (δ) and technological change is counterintuitive.

Feldstein and Foot (1971) have raised the issue of how 'the other half of gross investment' (δK) is determined, given that the proportionality hypothesis is at best a long-run property. They argue that the theory of the optimal durability of capital indicates that optimal-equipment life depends, *inter alia*, on the rate of technical progress. To quote,

because technical progress is to a certain extent embodied in capital goods, replacement investment reduces unit production costs. The rate of replacement investment will therefore be related to the potential opportunities to reduce production costs. (Feldstein and Foot, 1971)

In their empirical work, however, a proxy for this effect – the age of existing equipment – was not significant. If depreciation is greater the faster the rate of technological change, then what effect does this induce in the neoclassical investment model above? From (13.8) one may see that $\partial K^*/\partial \delta < 0$ because a higher δ raises the user cost of capital; thus one expects a rise in δ to imply less net investment. However, as gross investment has two parts, δK and I, which move in opposite directions, the overall effect on gross investment is uncertain.

The result, that the optimal scrapping date is earlier the faster the rate of technical progress, is confirmed by Nickell (1978, pp. 126–30) in a world of imperfect competition with irreversible investment. The model being used there is a vintage model, and given our preference for a treatment of embodied technological change it is to such a model that we now turn.

13.3 Vintage models

Both King (1972) and Koizumi (1969) have investigated investment in a putty-clay vintage model. If we follow the Koizumi model we assume that expected future output is known; that there is, *ex ante*, substitution possibilities but *ex post* no substitution; that the production function has constant returns to scale, with Hicks' neutral technical change embodied in capital; that prices are expected to remain constant; and that the capital good has an infinite physical life but an *expected* economic life T.

Let the *ex ante* production function for vintage v be written as

$$Y_v = Ae^{gv} K_v{}^{\alpha} L_v{}^{1-\alpha}. \tag{13.9}$$

Then the present value of cost per unit of output produced on this vintage is given by

$$q_t \frac{K_v}{Y_v} + \sum_{j=1}^{T} \frac{1}{(1+r_t)^j} w_t \frac{L_v}{Y_v} \tag{13.10}$$

where q_t is the price of machines in time t, w_t the wage, and r_t the discount-interest rate. To maximize profits given output we minimize costs, which yields condition

$$\frac{K_v}{L_v} = \frac{w_t}{q_t/c_t} \frac{\alpha}{1-\alpha} \tag{13.11}$$

where

$$c_t \equiv \sum_{j=1}^{T} \frac{1}{(1+r_t)^j},$$

and q_t/c_t corresponds to Jorgenson's user cost of capital. Given that the capital-labour ratio for new machines is optimal, from (13.11) we may calculate the cost of production on new machines, the relevant cost being given by (13.10).

The relevant cost of production on old machines, however, does not include the user cost; thus, although these old machines are less productive than the new, they may produce at cheaper costs. Let us consider that an amount of output Y_L can be produced at costs less than or equal to those on new machines; then we must install sufficient new machines to produce an amount of output $Y - Y_L$ to meet the output target. The number of machines required to meet this target yields investment in machines of vintage v, according to

$$I_v = \frac{K_v/L_v}{Y_v/L_v} \ (Y - Y_L). \tag{13.12}$$

Thus

$$I_v = \frac{1}{A} e^{-gv} \left(\frac{w_t}{q_t/c_t} \frac{\alpha}{1 - \alpha} \right)^{1-\alpha} (Y - Y_L). \tag{13.13}$$

To fully specify the investment function, we need to know something of Y_L. We may argue that the life of the oldest machine in use, T, will give us some information on this. The oldest machine has variable costs equal to total costs on new machines. Koizumi shows that, if prices and the rate of technical progress remain constant over time, then

$$T = \log_e \left\{ \frac{1}{g \, (1 - \alpha)} \right\},$$

and thus, the higher is the rate of technical progress, the lower is T. We may therefore argue that, for a given output Y, an increase in the rate of technical progress, through (13.13), will tend to reduce investment in current machines, for each is more productive; but there will also be an increase in investment, for the age of death of machines will be reduced, thus reducing Y_L. The overall effect is therefore uncertain.

   Although these models are an improvement on the models of disembodied change, they are not completely satisfactory. First, they consider the level of output to be exogenous rather than endogenous. Second, the models are firm-orientated and say nothing of the number of firms. If we consider Salter's more valid approach to investment in vintage models, he argues that

each period brings forth a new set of best practice unit requirements for labour, investment and materials . . . If the prices . . . do not change the improvement in best practice techniques allows the possibility of super normal profits. These will induce entrepreneurs to build such plants . . . until output is expanded sufficiently in relation to demand conditions (so that) super normal profits are eliminated. (Salter, 1966)

This approach makes both the number of firms and output endogenous. We do not know, however, of any literature that has used this approach to generate a vintage model-based investment function.

   One might also state that there has been considerable discussion of whether vintage models are a very satisfactory way of representing technology; c.f.

Gomulka (1976). Such objections to these models lead us to consider further approaches to modelling investment behaviour, although it should be stated that King (1972) does have some success in fitting his vintage model to data on UK investment.

13.4 The Keynesian theory of investment

Consider a given capital good, with purchase price q, that yields expected returns for each of H periods, R_h. We define the internal rate of return ρ as that discount rate which makes

$$q = \int_0^H \frac{R_h}{(1 + \rho)^h} \, dh \tag{13.14}$$

if $\rho \geqslant r$, the rate of interest, the machines is invested in. Let N be the stock of machines in existence, and assume

$$R_h = R_h(N) \quad \text{for all } h \quad R_h' < 0.$$

Then optimal N, N^*, will be determined as that number that makes $\rho = r$. Thus for any r we may say that N^* is determined. To put it another way, we may say that N^* is determined such that

$$q = \int_0^H \frac{R_h(N^*)}{(1 + r)^h} \, dh \tag{13.15}$$

which can be written more generally as

$$N^* = G(r, q). \tag{13.16}$$

If we let N_t^* be the desired stock of the machine in time t and N_{t-1} be the stock of the machine in existence, then for any (q, r) we can say that investment will be given by

$$I = N_t^* - N_{t-1} = G(r, q) - N_{t-1}. \tag{13.17}$$

However, we still need to determine q. Allow that there is a capital goods-supplying industry with an upward sloping supply curve where

$$I = I(q) \qquad\qquad I' > 0. \tag{13.18}$$

Then, from (13.17) and (13.18), we can solve for I given r and N_{t-1}; i.e., investment will take place at the rate at which the supply price of the machine is such that ρ is equal to the rate of interest r. To obtain aggregate investment we sum over all machines.

To introduce technological change into this framework we could consider:

1   changes to the population of machines over which choices are made;
2   effects on the supply curves for capital goods;

3   new products, new markets, or new resources affecting R_h, the expectations of returns. For some capital goods this may be a positive and for others a negative effect.

As far as we know, however, no such explicit modelling has actually taken place. Moreover, the story we tell at the end of this chapter, suggesting an alternative approach, is close to this Keynesian story and so we shall not try any further developments here.

13.5 The accelerator approach

The general accelerator approach can be summarized in an aggregate form as that gross investment in time t, GI_t, is given by

$$GI_t = (1 - \lambda)(K_t^* - K_{t-1}) + \delta K_{t-1} \tag{13.19}$$

where $(1 - \lambda)$ tells us of the speed of adjustment, K_{t-1} is the existing capital stock, K_t^* the desired capital stock, and δK_{t-1} depreciation. The optimal capital stock is assumed determined by a fixed relation to output:

$$K_t^* = \nu Y_t \tag{13.20}$$

and the rate of adjustment parameter λ, is determined by the decision lag and delivery lags in the investment process (it is not endogenous).

To consider technological change in this framework one can first refer back to our earlier discussions of replacement investment, but this is not very useful in providing any extra insight. What is of more interest is to consider the determination of K_t^*. For equation (13.20) to hold at all times, we must really consider either that factor prices remain constant over time or that there is a fixed coefficient production function. Let us assume the latter, so that

$$Y = \frac{1}{\nu}K = \frac{1}{u}L. \tag{13.21}$$

Now consider technological change. As should be immediately obvious, if technological change is Harrod-neutral and disembodied, so that (13.21) can be written as

$$Y = \frac{1}{\nu}K = \frac{1}{u}A(t)L, \tag{13.22}$$

then the model above is not affected at all by technological change. With any other type of disembodied change however we must write (13.21) as

$$Y = \frac{1}{\nu}B(t)K = \frac{1}{u}A(t)L. \tag{13.23}$$

If we now consider a naive accelerator model (where $\lambda = 1$), then we may write that

$$I_t = K_t^* - K_{t-1}^* = K_t - K_{t-1} = \frac{v}{B(t)} Y_t - \frac{v}{B(t-1)} Y_{t-1}. \quad (13.24)$$

Let

$$B(t) = B(1 + g)^t. \quad (13.25)$$

Then

$$I_t = \frac{v}{B(1 + g)^t} \{Y_t - (1 + g) Y_{t-1}\} . \quad (13.26)$$

From this it is clear that, unless output is increasing over time, net investment will be negative. One may also see that $\partial I / \partial g < 0$; thus a faster rate of technological change, given the time-path of Y, will mean lower net investment. The overall effect on gross investment depends on what happens to obsolescence.

To make technological change embodied we would need to move to a clay–clay vintage model, and as we have already looked at a putty–clay model we shall not do this.

Later work by, for example, Eisner and Strotz (1963) provides an endogenous determination of the adjustment parameter in these models, but that does not take us a great deal further on the question of technological change and investment.

Mansfield (1968) has considered a flexible accelerator type of model to which is added a term to reflect technological change. Allow $n(s)$ to be the number of innovations first occurring in year s, and assume it takes x years for all an industry's old capacity to be equipped with an innovation. Allow that, during each year of this period, $(100/x)$ per cent of the capacity at the beginning of the year is so equipped and that the cost per unit of capacity is q_j dollars for the jth innovation. Then expenditures necessary to introduce innovations is given as

$$E_t = \frac{\bar{q}_t}{x} K_{t-1} \sum_{s=t-x-1}^{s=t-1} n(s) \quad (13.27)$$

where $\bar{q}$ is the average value of q_j for innovations first occurring between $t - x - 1$ and $t - 1$. We then generate an overall investment equation, assuming $K_t^* = vY_t$:

$$I_t = (1 - \lambda)(K_t^* - K_{t-1}) + E_t. \quad (13.28)$$

Mansfield estimates this equation on data for petroleum and iron and steel. His results are summarized as follows:

although the fits are not particularly good the results are encouraging. All of the regression coefficients have the expected signs, all are statistically significant at the 0.05% level . . . They suggest that, using even a crude model which includes the timing of innovation, one can do a significantly better job of explaining variation in investment than if one uses the flexible 'capacity accelerator' alone. (Mansfield, 1968)

This empirical work of Mansfield is one of the few studies that takes account of the effect that specific innovations may have on investment expenditure.

It does have its problems; for example, the diffusion process is treated rather unsatisfactorily. What it does provide however is a good basis to justify an attempt to look further at investment and technological change on the basis of individual innovations, which is what we now do in an alternative approach.

13.6 An alternative approach

The theories discussed above indicate that there is reason to expect that changes in technology will significantly influence the level of investment. However, we have suggested that none of these models really adequately represents the role of new technology. In this section we hope to go some way further towards meeting that objective.

The models above have aggregation structures that are either aggregations over firms' demands for a generally defined capital aggregate or a summation over individual capital good (projects) demand functions. In this section our approach is allied to this latter Keynesian alternative, but we consider aggregation over technologies. We consider a world in which new products and processes are being continually produced. These products and processes are diffused across the economy, generating demands for capital goods. Aggregate investment is derived by summing across products and processes (technologies). We start by considering the reaction of investment to a specific product innovation, move to process innovation, and then consider aggregation.

13.6.1 Product innovation

Our first attempt to investigate the impact of new technology on investment concerns how productive capacity will be created in reaction to the appearance of a new product. It is reasonable to argue that, if a new product is 'invented', then to build up the output of this product one needs to construct a capital stock for the purpose. Spence (1979) has considered the formal analysis of this problem. Abstracting from uncertainty, he considers investment strategy in a new market in which firms face financial and other constraints that limit their growth rates. Moreover, he allows that as time proceeds an initial firm may be faced by an increasing number of competitors because of new entry.

Formally, allow that at time t there are n_t firms in the industry, and the ith firm has capital stock k_{it}. Allow that for each t the industry is in a short-run equilibrium and that operating profit for firm i is s_t^i (k_{it}, k_{jt}) where k_j is rivals' capital stock, and the partial derivatives $s_{1t}^i > 0$, $s_{2t}^i < 0$. Define m_{it} as investment by firm i in time t, and assume zero depreciation (making net and gross investment equal), so that $\dot{k}_{it} = m_{it}$. The firm then acts to maximize its present value:

$$V^i = \int_{\hat{T}_i}^{\infty} (s^i - rk_i) \, e^{-rt} \, dt \tag{13.29}$$

where r is the interest rate, and $\hat{T}_i$ is the date of firm i's entry to the market.

This maximization is made subject to three constraints: first, that investment is irreversible ($m_{it} \geqslant 0$), second, that there are physical limits to growth associated with the need to train new personnel and build and install plant and equipment (represented by $m_{it} \leqslant \phi\,(k_{it})$, $\phi' > 0$), and third, that after obtaining an initial equity all investment must be financed from retained earnings.

Taking the investment of other firms as given, Spence shows that the individual firm simply invests as rapidly as possible (given the constraints) up to some target level of investment and then stops. The optimal target level ($\bar{k}_i$) and stopping time (T_i) satisfy

$$\frac{dV^i}{dT_i} = 0 \tag{13.30}$$

and

$$\int_{T_i}^{\infty} \left(\frac{\partial s_t^i}{\partial k_{it}} - r \right) e^{-rt}\, dt = 0. \tag{13.31}$$

The time profile of k_i will follow the type of path shown in Fig. 13.1.

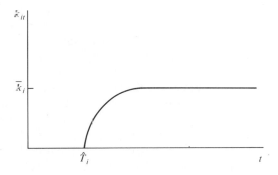

Fig. 13.1

Spence then proceeds to analyse the nature of the oligopoly game in the industry. It is argued that each firm i, on entry, will expand its capital stock as fast as possible, but the firm will attempt to pre-empt some of the market by having $\bar{k}_i$ higher than if such pre-emption were not possible. Thus we may consider that firm 1 will expand its capital stock as quickly as possible. When firm 2 enters it follows a similar time-path. Firm 1 will keep investing until $\bar{k}_1$ is reached and firm 2 will then continue to invest until it reaches its optimal $\bar{k}_2$ (which is dependent on $\bar{k}_1$).

The moral of this story is that the time-path of investment is

1 primarily dependent on the appearance of a potential new market;
2 conditioned by financial and physical limits to growth; and
3 heavily influenced by entry and the oligopoly game.

Now, as Spence himself states, there are a number of aspects of his model that limit its applicability — for example, no uncertainty. However, perhaps more important is the demand side. This is modelled by the $s(k)$ function. But from Part II above we have argued that new products follow a diffusion time-path that is often time-dependent, and this is not allowed for. Perhaps one way round this is to consider again Metcalfe's (1981) model discussed in Chapter 9. If one considers that model as relating to a new product, the predictions are that the growth of output of the product will be sigmoid (specifically logistic). This is a result dependent on similar financial constraints to those used by Spence. Now, as Metcalf assumes fixed coefficients in production, the net result implied is that the capital stock must also follow a sigmoid path. If we assume irreversible investment and no depreciation, this must mean that investment is positive until the point of inflexion on the growth curve of user ownership, with a peak prior to this, after which investment is zero (assuming non-negativity).

Now we have criticized this model of Metcalf because it does not really have a choice-theoretic base, but this link between investment and diffusion is an important one, and one worthy of further consideration by a look at process innovation.

13.6.2 *Process innovation*

If one is to look at the impact of a new process on investment, there are two sides of the problem to be investigated: (1) investment in the new process by new technology users, and (2) investment in new capacity in order to produce the new good embodying the new technology by capital goods producers. Thus process innovation has a two-pronged influence on investment.

Abstracting from changes in market shares of producing firms, we may argue that if Q_{it} is the output of capital goods of the new type i in time period t and the market is in short-run equilibrium for each t (supply and demand are equal), then investment associated with the new process will equal $Q_{it} + K(\dot{Q}_{it})$ where $K(\dot{Q}_{it})$ is the relationship between investment and increases in output in capital goods supply. In the simplest case, of fixed coefficients in capital goods supply, we may state that I_{it}, investment in time t associated with new process i, equals

$$I_{it} = Q_{it} + \alpha\dot{Q}_{it} \qquad (13.32)$$

where α is the capital–output ratio in the supplying industry.

In Chapter 9 above we have investigated a number of diffusion models that take account of both the supply of and demand for new technology. Our results suggested that the time-path of Q_i, and thus I_i, depended on

1   learning economies in production and the learning in use process;
2   further technological change;
3   the nature of the oligopoly game in use and production;
4   the characteristics of the new technology.

The models discussed in Chapter 9 have different predictions on the time-path of

Q_{it} depending upon the assumptions made. However, in a number of cases it is possible to argue that there are good reasons to expect that Q_{it} will follow some sigmoid path. We may thus argue that at least the first component of I_{it} will also follow this path. The second component of I_{it} is more problematical. If $\dot{Q}_{it} > 0$ there is no problem; there will be an extra contribution to I_{it}. However, if $\dot{Q}_{it} < 0$ we may not expect investment to have a negative component. Although the models above do not necessarily involve an explicit assumption that investment is irreversible, it does seem to be a reasonable assumption to make. We could thus write an expression for the investment associated with a new technology as

$$I_{it} = Q_{it} + \alpha\dot{Q}_{it} \qquad \dot{Q}_{it} \geqslant 0$$

$$= Q_{it} \qquad \dot{Q}_{it} < 0. \tag{13.33}$$

If we now plot in Fig. 13.2 a typical sigmoid path for the stock of new technology, x_{it}, we can generate a curve I_{it}, which has an initial spurt of activity as both users and producers build up their capital stock; then, as the investment of producers declines and as the growth of usage declines, investment falls away.

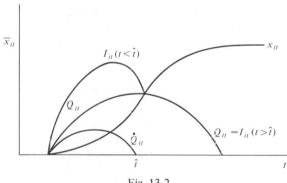

Fig. 13.2

We are thus arguing that with the appearance of new products or new processes there will be a spurt of investment that will die away as the industry matures.

13.6.3 The aggregate

The above discussion has centred on a particular new product or process. To discuss the aggregate we must sum over products and processes. One could simply write down an integral expression, but for present purposes it is informative to abstract somewhat. Consider that one can conceive of a representative new technology that is introduced (innovated) in time t, that will then follow a diffusion path as discussed above. Let the investment at all points on this diffusion path be written as $F(t, \tau)$ where τ is the number of years since innovation occurred and t is the year of innovation. Now let N_t be the number of new technologies introduced (innovated) in year t. We may then consider that investment in new processes in time T may be written as

$$\int_0^T N(T - t) F(T - t, t) \, dt.$$

To be somewhat more realistic, allow that a proportion $(1 - \delta)$ of all new technologies in each year are made obsolete by other new technologies entering the economy, whereupon we may state that investment due to the appearance of new technologies is

$$I_T = \int_0^T \delta N(T - t) F(T - t, t) \, dt. \tag{13.34}$$

This expression for investment has two main terms: that in the number of innovations, and that in $F(T - t, t)$. We discuss each separately.

The term in $F(T - t, t)$ represents the investment in time t in a technology first introduced in time $T - t$. This basically has two determinants, (1) the capital intensity of the new technology and (2) the speed of diffusion, which are not necessarily independent.

The term $N(T - t)$ represents the number of innovations in time $T - t$, which in turn is related to invention in prior years. In Part I we discussed forces influencing invention, innovation, and its direction. For example, one might argue that basic advances in science may yield higher $N(T - t)$; or, relevant to what we have to say below, entrepreneurial activity *à la* Schumpeter can be considered important.

If we now consider equation (13.34) we may concentrate on two basic scenarios.

1 Consider first a world where $N(t) = N$, a constant, so that a constant number of innovations are made each year. Consider second that these innovations are of such a type that $F(T - t, t)$ can be written as $\alpha F(t)$, i.e., where α represents capital requirements per addition to usage and $F(t)$ is the diffusion path of additions to use. We may now write that

$$I_T = \alpha \delta N \int_0^T F(t) \, dt. \tag{13.35}$$

If we assume logistic diffusion, then

$$I_T = \alpha \delta N \left\{ \frac{1}{1 + \exp(-a - bT)} - \frac{1}{1 + \exp(-a)} \right\}. \tag{13.36}$$

If T is large and a is small, we may approximate this by

$$I_T \simeq \frac{1}{2} \alpha \delta N \tag{13.37}$$

and investment for technological change is constant over time. One could extend this further by arguing that the final level of use of a new technology will be related to the general level of economic activity, but this will not add a great deal.

2   More interesting scenarios can be generated by allowing $N(t)$ or capital intensity to change over time. Thus, for example, if new technologies become more or less capital-intensive over time, or if capital intensity becomes cyclical over time, we will get specific time-paths of investment reflecting this. However cyclical activity in $N(t)$ is perhaps more interesting. The reason for this interest is that it is currently being argued that there is some empirical support for the Schumpeterian explanation of long waves in economic activity (see Freeman, Clark, and Soete, 1982), whereby bunching of innovations leads to investment and output cycles. Freeman *et al.* argue that the bunching occurs at different dates in different sectors, and that the capital intensity of innovations changes over time; however, their approach is basically to consider that the view of investment we are proposing has some empirical foundation.

If inventions are bunched, then this will obviously generate cyclical investment activity. The cycles of investment, however, will lag behind the innovation cycle because the maximum growth of capacity occurs some way into the diffusion process rather than at its beginnings.

One could consider further scenarios, but the outcome of this discussion is to argue that investment in aggregate will be related to

1   the rate at which new technologies are appearing;
2   the capital intensity of these new technologies;
3   the diffusion paths for new technologies.

13.7   Conclusions

We have argued that to a large extent technological change has been ignored in studies of investment behaviour, although there are good theoretical reasons why that should not be so. We have seen that, when technological change is introduced into the approaches standard in this literature, then both theoretically and empirically it has a role to play.

It was argued, however, that modifications to the mainline approaches may not really be sufficient to represent the manner in which technological change will influence investment expenditure; and thus an alternative approach, which essentially represents the consideration of the diffusion process and an aggregation across the numerous technologies being diffused at any one time, is suggested. It does seem only logical that, if we have a body of literature that attempts to explain the demand and supply of individual capital goods over time (the diffusion literature), we can supplement the literature on the determination of the capital aggregate by simply considering aggregating over the different technologies being diffused at any particular time.

References

Eisner, R. and Strotz, H. (1963), 'Determinants of Business Investment', in Commission on Money and Credit, *Impacts of Monetary Policy*, Prentice-Hall, Englewood Cliffs, NJ, pp. 60–138.

Feldstein, M.S. and Foot, D. (1971), 'The Other Half of Gross Investment: Replacement and Modernization Expenditures', *Review of Economics and Statistics*, 53, 49–58.

Freeman, C., Clark, J. and Soete, L. (1982), *Unemployment and Technical Innovation*, Frances Pinter, London.

Gomulka, S. (1976), 'Do New Factories Embody Best-practice Technology? New Evidence', *Economic Journal*, 86, 859–63.

Helliwell, J. (ed.) (1976), *Aggregate Investment*, Penguin, Harmonsworth.

Jorgenson, D. (1963), 'Capital Theory and Investment Behaviour', *American Economic Review*, 53, 247–59.

King, M.A. (1972), 'Taxation and Investment Incentives in a Vintage Model', *Journal of Public Economics*, 1, 121–47.

Koizumi, S. (1969), 'Technical Progress and Investment', *International Economic Review*, 10, 68–81.

Mansfield, E. (1968), *Industrial Research and Technological Innovation*, W.W. Norton, New York.

Metcalfe, J. (1981), 'Impulse and Diffusion in the Study of Technical Change', *Futures*, 5, 347–59.

Nickell, S. (1978), *The Investment Decision of Firms*, Cambridge University Press/Nisbet, Welwyn Garden City.

Salter, W.E.G. (1966), *Productivity and Technical Change* (2nd ed.), Cambridge University Press.

Schramm, R. (1972), 'Neoclassical Investment Models and French Private Manufacturing Investment', *American Economic Review*, 62, 553–63.

Spence, A.M. (1979), 'Investment Strategy and Growth in a New Market', *Bell Journal of Economics*, 10, 1–19.

Chapter 14

Technological Change and the Distribution of Income

14.1 Introduction

In perhaps no other area do the underlying political disputes inherent in the different approaches to economics surface so dramatically as in the area of income distribution. In this section we intend to be somewhat agnostic as to the choice of theory. Our purpose is to illustrate within the framework of several different theories of income distribution how changes in technology will impact on the distribution of the product between wages and profits.

For the majority of this piece we will discuss macroeconomic approaches to income distribution, although where relevant comments at a micro-level will be introduced. The majority of the literature, moreover, is concerned with the long run. We start as usual with the neoclassical approach, then move to two-sector, Kaleckian, and Schumpeterian approaches.

14.2 The neoclassical theory of income distribution

The neoclassical theory is essentially based on the premise that factors are rewarded according to their marginal products and the effect of technological change on the distribution of income will depend upon the bias of the change and the elasticity of substitution of the production function.

In Chapter 1 we proposed a number of definitions of neutrality, each definition being related to the impact on factor shares of technological change under certain assumptions about the time-path of factor inputs. Thus, if technology is Hicks-neutral, then factor shares would not change if the ratio of capital to labour remained constant. Harrod-neutral technological change guaranteed constant shares if the capital–output ratio was constant, and Solow-neutral change generated constant shares if the labour-output ratio was constant. However, these predictions on shares rely on the exogeneity of factor inputs, and we really expect these to be endogenous to the development of the economic system.

Ferguson (1968) sets out an exposition of this theory that is applicable to the economy as a whole or to any sector of it. We assume a production function,

$$Q = F(K, L, t) \quad F_K, F_L > 0, F_{KK}, F_{LL} < 0, \tag{14.1}$$

that is homogeneous of degree one in the homogeneous inputs, capital K, and labour L, with technology represented by the time term t. Competitive equilibrium is assumed to exist instantaneously, so that

$$r = F_K$$

$$w = F_L$$

where r and w denote the real rate of interest and the product wage. On equation (14.1) we may define

$$R = \frac{KF_{Kt} + FL_{Lt}}{KF_K + LF_L} \tag{14.2}$$

as the rate of technical change and

$$B = \frac{F_{Kt}}{F_K} - \frac{F_{Lt}}{F_L} \tag{14.3}$$

as the Hicks-bias of technical change (where F_{Kt} and F_{Lt} are the derivatives of the marginal products with respect to t). If technological change is Hicks-neutral, $B = 0$. If $B > 0$ technical change is Hicksian capital-using, and if $B < 0$ it is Hicksian labour-using. If we define

$$S = \frac{LF_L}{F} = \text{relative share of labour}$$

and

$$\sigma = \frac{F_K F_L}{FF_{KL}} = \text{elasticity of substitution,}$$

then one may show that, if technical change is Harrod-neutral, then

$$\left(1 - \frac{1}{\sigma}\right)R + SB = 0. \tag{14.4}$$

Ferguson (1968) shows that, if factors are paid their marginal products, then we may derive that the change in labour's relative share over time is given by

$$\frac{\dot{S}}{S} = -(1 - S)\left\{B + \left(1 - \frac{1}{\sigma}\right)\left(\frac{\dot{k}}{k}\right)\right\} \tag{14.5}$$

where a dot superscript refers to a time derivative and k is the capital–labour ratio. We may then argue as follows.

1 If technical progress is Hicks-neutral then $B = 0$ and we have

$$\frac{\dot{S}}{S} = -(1 - S)\left(1 - \frac{1}{\sigma}\right)\frac{\dot{k}}{k}. \tag{14.6}$$

From this Ferguson argues that 'neutral technical progress has no effect whatsoever on factor shares', for the rate of technical progress does not enter this expression and 'the relative share of labour increases or diminishes (for $\dot{k}/k > 0$) as the elasticity of substitution is less than or greater than unity'.

2 If $\dot{k}/k > 0$ and $\sigma < 1$, then the share of labour will be augmented if $B < 0$, i.e., if technological change is labour-using.

3 If technical progress is Harrod-neutral, then

$$B = -\frac{1}{S}\left(1 - \frac{1}{\sigma}\right)R \qquad (14.7)$$

and

$$\frac{\dot{S}}{S} = -(1 - S)\left(1 - \frac{1}{\sigma}\right)\left(\frac{\dot{k}}{k} - \frac{R}{S}\right). \qquad (14.8)$$

Thus with Harrod neutrality, if $\sigma = 1$ or $\dot{k}/k = R/S$, the relative shares will remain constant, this second condition being rather special.

Ferguson (1968) then extends to consider multi-sector models, but the above analysis is sufficient to give the flavour of the approach. What it is worth concentrating upon here is exactly what is being assumed exogenous. The story appears to be that $\dot{k}/k$ is exogenous, that is, that the rate of growth of the capital-labour ratio is determined independently of factor prices and technology. It would seem to be reasonable to argue that this can be considered a valid assumption only if:

1   resources are fully employed at each moment in time, so that for example there is full employment of labour at each moment in time. We have argued elsewhere (see Chapter 12) that this can be rationalized only by the use of Walrasian arguments, thus requiring the assumption of the existence of an auctioneer who changes prices in response to notional excess demands and supplies and the existence of a complete set of futures markets;
2   the supply of labour and capital are independent of factor prices and technology. Thus, if we assume a labour supply growing at an exogenous rate n, we would partially satisfy the required condition. However, to assume that the supply of capital and labour are independent of technology is stretching belief. Even if we consider only labour, the effects of technology on skill requirements and for example the availability of women to work is likely to affect labour supply.

However, if one is willing to accept that $\dot{k}/k$ can be taken as exogeneous, this theory states essentially that factor shares will move according to the bias in technological change. But in Chapter 4 we argued that the bias in technological change may be endogenous to the economic system through the effects of the Kennedy–Weizsacker innovation possibility frontier. It was argued in that chapter that technological change would tend to Harrod neutrality as long as the elasticity of substitution $\sigma < 1$. One might consider that, because in the above Harrod neutrality does not in general generate constant shares, the introduction of induced innovation would not generate constant shares. However this is not the case, for in the induced innovation literature $\dot{k}/k$ is an endogenous variable.

Specifically, Drandakis and Phelps (1966) allow labour to grow at a constant rate $n \geqslant 0$ and capital to grow such that

$$\dot{K} = \theta Q_t. \qquad (14.9)$$

The Kennedy–Weizsacker approach allows us to write that $B = B(S)$, $R = R(S)$. Now, given

$$\frac{\dot{Q}}{Q} = (1 - S)\frac{\dot{K}}{K} + S\frac{\dot{L}}{L} + R \qquad (14.10)$$

from the production function, we can write (14.11) from (14.10) and (14.9):

$$\frac{d(\dot{K}/K)}{dt} = \frac{\dot{K}}{K}\left\{R(S) - (1 - S)\left(\frac{\dot{K}}{K} - n\right)\right\}, \qquad (14.11)$$

and from the theory above we can rewrite (14.5):

$$\frac{\dot{S}}{S} = -(1 - S)\left\{B(S) + \left(1 - \frac{1}{\sigma}\right)\frac{\dot{k}}{k}\right\}. \qquad (14.12)$$

We thus have a pair of differential equations, (14.11) and (14.12), in S and K. If $\sigma < 1$, then there is an equilibrium solution to this pair that is globally stable and involves Harrod-neutral technical change and also constant factor shares. The shares will depend upon the shape of the invention possibility frontier.

Essentially, the story being told is that, given the current state of technology and the labour force, the condition for full employment of labour and capital yields a wage and capital rental rate and thus factor shares. These shares then influence the nature of the bias of the new technology invented and thus the new factor prices of the next period. As the factor shares and bias change over time the system will, if $\sigma < 1$, tend to an equilibrium with Harrod-neutral technological change.

The theory of induced bias and the resultant analysis of income distribution is perhaps one of the high spots of neoclassical theory. It is however open to a number of objections.

1   We criticized the induced bias hypothesis in Chapter 4 as somewhat unsatisfactory.
2   We have also criticized, elsewhere, an assumption of perfectly clearing markets.
3   The disembodiment hypothesis, and thus the comparative statics nature of the analysis, is somewhat unsatisfactory.
4   there is considerable doubt as to whether we can logically define the aggregate production function that underlies all this analysis.

Let us turn then to consider an alternative view.

14.3 The two-sector approach

In the Cambridge capital controversy debate (see Harcourt, 1972), a basic model used to illustrate the problems associated with the aggregate production function concept was Hicks's two-sector model. It seems to be useful therefore to analyse the impact of technological change on income distribution in this framework.

For this argument we use the following terminology:

π = price of the consumption good
α = input of machines per unit of consumption good output
β = input of labour per unit of consumption good output
C = output of consumption goods
p = price of a machine
a = input of machines per unit of machine output
b = input of labour per unit of machine output
τ = number of machines produced
T = total stock of machines
L = total workforce
w = wage rate
δ = depreciation coefficient
g = the growth rate
r = profit rate (interest rate)

All prices are expressed in terms of an arbitrary unit. When one chooses a commodity as a standard of value and sets its price equal to one, all prices are expressed in terms of this commodity. In long-period equilibrium the price relationships will be

$$\pi = \alpha p (r + \delta) + \beta w, \qquad (14.13)$$

$$p = a p (r + \delta) + b w, \qquad (14.14)$$

which, when π is set to unity, making the consumption good the standard of value, yields

$$w = \frac{1 - a(r + \delta)}{\beta \{1 + a(m - 1)(r + \delta)\}} \qquad (14.15)$$

where

$$m = \frac{\alpha b}{a \beta},$$

i.e., the ratio of sectoral factor intensities. Equation (14.15) is known as the factor price curve, and indicates the maximum wage that the given technology can support for any level of $(r + \delta)$. The curve is always downward-sloping but the curvature depends on m, i.e., on whether the consumption or capital good sector has the highest ratio of machine to labour input.

The quantity equations are the dual to the price equations. If the economy is under a regime of steady growth at a rate g, then

$$\tau = (g + \delta) T \qquad (14.16)$$

so that

$$L = b(g + \delta) T + \beta C \qquad (14.17)$$

$$T = a(g + \delta) T + \alpha C. \tag{14.18}$$

If one assumes that $L = 1$, so that all quantities are expressed in per head terms, then (14.17) and (14.18) yield

$$C = \frac{1 - a(g + \delta)}{\beta \{1 + a(m - 1)(g + \delta)\}}. \tag{14.19}$$

This relationship between C and g (consumption per head and the rate of growth) is mathematically identical to that between w and r in the factor price curve (FPC). Equation (14.19) is termed the optimal transformation curve (OTC).

Let us now consider that the economy is in an initial equilibrium using technology 1 when technology 2 appears. Allow that technology 2 is superior to technology 1, so that for any wage it can yield a higher rate of profit. Allow that the initial equilibrium was (r^*, w^*) with its implied income distribution. We now ask, what will be the w and r in the economy with the new technology, and what will be the resulting share allocation?

The way Hicks (1965) solves this problem is

1   to compare only full employment steady-state growth paths; and
2   to 'close' the system through a savings function and an assumption that savings and investment are always equal.

Hicks shows that we may, defining f as the share of capital in national income, i.e.,

$$f \equiv \frac{rpT}{wL + rpT}, \tag{14.20}$$

generate, under conditions (1) and (2) from (14.13)–(14.19), assuming $\delta = 0$,

$$\frac{1}{f} - 1 = \frac{1 - ra}{ra} \left(\frac{1 + (m - 1) an}{m} \right) \tag{14.21}$$

and r is to be determined by the savings function. In the simplest case, where all profits are saved and invested and all wages consumed, then $r^* = n$ and

$$\frac{1}{f} - 1 = \frac{1 - an}{an} \left(\frac{1 + (m - 1) an}{m} \right), \tag{14.22}$$

from which it is clear that the effect of technological change on factor shares will depend upon how a and m change. In the simplest case, where $m = 1$ and remains at unity, we may see that

$$f = \frac{1}{an} \tag{14.23}$$

and $\partial f/\partial a < 0$; in other words, an increase in capital intensity of the technology will reduce capital's share. In general however we may derive that

$$d \left(\frac{1}{f} - 1 \right) = \frac{da}{a} \left(\frac{an}{m} - \frac{1}{man} - na \right) + \frac{dm}{m} \left(\frac{2}{m} - \frac{1}{man} - \frac{an}{m} \right), \quad (14.24)$$

which unfortunately does not generate any neat general properties.

One should note however that we are again considering only full-employment positions, and also we are undertaking comparative statics, which is still rather unsatisfactory.

Dougherty (1974) considers the impact of technical change in a two-sector model under a number of different assumptions about the type of change. He does however avoid the issue to some extent by leaving the profit rate undefined. With a constant profit rate, it is shown that the profit share will rise if technical change in the consumption goods sector saves labour relative to machines, but will fall, first, if the change in the consumption goods sector saves machines relative to labour or, second, if technical change in the machine sector saves machines either totally or relatively.

14.4 The Kalecki approach

The third approach to the distribution of income that we consider is attributed to Kalecki. In the simplest version of the model (for a more detailed approach see Cowling, 1982), we let p be the price of output, which is assumed to be a mark-up on average variable cost (AVC) at rate k; thus,

$$p = kAVC \qquad (14.25)$$

and k is a function of monopoly power. Letting Q be total output, define

$$R = pQ. \qquad (14.26)$$

Let W = total wage bill and M be expenditure on raw material; then

$$R = k(W + M). \qquad (14.27)$$

We may then state that overheads (F) plus net profits (π) are given by

$$\pi + F = R - W - M$$

$$= (k - 1)(W + M). \qquad (14.28)$$

For the economy as a whole, national income Y is given as

$$Y = W + \pi + F. \qquad (14.29)$$

Thus

$$\frac{W}{Y} = \frac{W}{W + \pi + F} = \frac{1}{1 + (k - 1)(j + 1)} \qquad (14.30)$$

where $j \equiv M/W$.

From (14.30) it is clear that the wage share is inversely related to k, the degree of monopoly, and j, the ratio of material costs to wage costs. Now the role that technological change may play in affecting factor shares must be reasonably clear. The effect must come through either j or k. Consider the latter first. In essence, k will be related to monopoly power. We discuss at some length in Chapter 16 how technological change may affect market structure and thus monopoly power. A change that increases monopoly power may well increase the profit share, but it is not clear that new technology will always increase monopoly power.

If we consider j, we may argue from (14.30) that material using biases in technology (changes raising j) will lead to a falling wage share, and material-saving biases (changes reducing j) will lead to a rising wage share.

If we consider the two elements j and k together, we have argued above (Chapters 2 and 3) that greater monopoly power may through the Schumpeter hypothesis mean faster technological change (although the supporting evidence is not strong), and thus for any given biases in technological change a higher k may be associated with a faster changing j.

One might also consider a sort of induced-bias hypothesis of technological change. The level of j will influence whether it is 'better' (in the sense of generating greater cost reductions) to save wages or material costs; and thus the level of j may influence the changes in j.

These lines of enquiry have not, however, as far as we know, been pursued to any extensive degree.

14.5 Schumpeterian approaches

Several times in this book we have made reference to the Schumpeter view of technological change. It thus seems appropriate to discuss Schumpeter's vision of the share of wages and profits in national income. The starting point is the 'circular flow', a state of the world that repeats itself period after period. In this circular flow 'all consumption goods on hand will go to the services of labour and land employed in this period; hence all incomes are absorbed under the title of wages or rent of material agents.' Land and labour are paid their marginal products (Schumpeter, 1934). Profit appears in the economy only as entrepreneural profit, which is the surplus the entrepreneur receives for his path-breaking efforts. The entrepreneur sees and puts into effect new technological opportunities, which at the ruling prices yield a surplus or profit.

The story then unfolds, with the profit attracting imitators who compete away this entrepreneural profit. During the process labour and land continue to be paid their marginal product, but as the new technology spreads, so the average economy-wide marginal product rises and the returns to labour and land increase. As imitation proceeds, the entrepreneural profit is competed away and after a boom and a recession the economy settles in a new circular flow.

Now this essentially is a macroeconomic story, and must thus really refer to major innovations. However, there is no reason to limit one's insight so severely.

If entrepreneurial profit can arise from any innovation, then at any moment in time the share of profit in the economy will be related to the state of diffusion of the several technologies being diffused at that time and the number of technologies being diffused.

14.6 Empirics

Given that the work of Nelson and Winter is so closely allied to the Schumpeterian view of technological change, we can at this point see what results their simulation exercises yield on profit shares. We have discussed the character of their model elsewhere (Chapter 12), with its satisficing, behavioural base and its process of firm life and death and firm growth. In Nelson, Winter, and Schuette (1976) the output of their simulation exercise is recorded, and it is stated that 'the share of capital displays considerably more volatility than the Solow (observed) data. This behaviour may plausibly be attributed to the unrealistically effective functioning of our simulated labour market.' However, in a footnote it is also stated that 'It may also be the case that the Solow capital share data are unrealistically smooth.' In Nelson and Winter (1974) it is argued that the problem is a reflection of assuming too low a dividend rate in the simulation exercises; however, it is then stated that 'it is still not clear whether, without assuming labour-saving biases in local search, one can generate both a plausible share series and plausible behaviour of the capital-labour ratio in the same run.'

Of course, the Nelson and Winter exercises are not a real test of Schumpeter's theory of distribution, and thus not too much should be read into their results. Perhaps the interpretation that the model can generate a reasonable time-series for factor shares is as generous as one ought to be.

What, then, of the empirical validity of other theories of distribution, and what can empirical work tell us of the impact of technological change on factor shares? King and Regan, after a survey of the data on factor shares, argue that

the labour share fluctuates over the trade cycle, in the opposite direction to that taken by the level of economic activity. It shows more sign of increasing than of remaining constant in the long run, and it has risen significantly . . . in Britain and the United States over the last decade. The weight of the evidence clearly refutes the claim that labour's share is a constant. (King and Regan, 1976)

This last conclusion is an important one, for a supposed relative constancy of factor shares in the long run has been the basis of two controversies, the first as to whether shares have in fact been constant, and the second as to how it can be explained. The weight of the evidence now seems to be that shares have not remained constant: the evidence suggests that labour's share has increased in the long run. If one is to explain this through technological change, then

1   one must reject the hypothesis that the induced-bias story, at least in its equilibrium form, is appropriate, for that predicts constant shares (if the elasticity of substitution is less than unity);

2   the data are consistent with the neoclassical hypothesis if (see equation 14.5) $B + \{1 - (1/\sigma)\}$ $(\dot{k}/k) < 0$, which assuming, $\sigma < 1$ and $\dot{k}/k > 0$, implies that $B < 0$ and technological change is Hicksian labour-using; or, again assuming $\sigma < 1$, that Hicks-neutral technological change $(B = 0)$ is allied with $\dot{k}/k < 0$. However, a declining capital–labour ratio does not really fit the data (see Solow, 1957, for example). We must thus argue that $\dot{k}/k > 0$. If $\dot{k}/k > 0$ and $\sigma < 1$, then a rising wage share can come only from $B < 0$, i.e., labour-using technological change. A neoclassical explanation of the time-path of factor shares must thus reduce to either a labour-using (in a Hicksian sense) bias in technological change or to $\sigma > 1$.

The feasibility of explaining movements in factor shares using Kalecki-type arguments has been investigated extensively by Cowling (1982) for the UK; however, as far as we know technological change has not been incorporated in most empirical studies using this approach — Cowling does not even have one entry for 'technological change' in his index.

Although it is possible to use the two-sector model to see what changes in coefficients are consistent with the data on profit shares, it does not seem much more worthwhile than the neoclassical exercises. We can say what changes are consistent with the factor share data, but we do not really have any external information on the validity of assuming that such changes took place. Dougherty (1974) does however find that for the USA the data are consistent with labour-saving technical progress in the non-manufacturing sector.

Let us turn to the observation that in the short term the labour share moves contra-cyclically; i.e., in the upward phase of the trade cycle labour's share falls and the profit share increases and in the downward phase the profit share declines. One might think that this represents some support for a Schumpeterian view of factor shares, where profit is associated with innovation and the innovations occur in the upswing. Unfortunately, the cycles discussed by Schumpeter are long-period cycles with approximately a forty-year periodicity, whereas the empirical observations refer to much shorter cycles. It is not generally argued that these short cycles are the result of innovation waves, and thus the apparent support for a Schumpeterian view is rather illusory.

14.7 Conclusions

We have looked at the role that changes in technology can play in a number of different theories of income distribution, essentially illustrating that income distribution cannot be independent of the time-path of technological change. We have also argued that, over time, the share of labour in national income has increased. The theories suggest that:

1   if Nelson and Winter type approaches are used, then new technology must be labour-saving to be consistent with the data;

2   in a neoclassical framework, technological change must be labour-using in

order to explain the increase in factor shares (given an aggregate production function with an elasticity of substitution less than unity);

3 Dougherty argues that the data are consistent with a two-sector approach if technological change is labour-saving in the non-manufacturing sector;

4 to be consistent with the Kalecki approach, either there must have been a fall in the degree of monopoly (which is very difficult to substantiate), or the ratio of wage costs to material costs must have risen, or, as Cowling (1982) argues, the share of overheads (F) must have risen.

Obviously, technical change can not have been both labour-saving and labour-using; thus our conclusions cannot be neatly summarized. However, perhaps we are asking too much. Technological change is not the only factor that is going to influence income distribution (we have for example ignored union power), and thus to expect it to explain all the change in income distribution is too optimistic. If we simply argue that we have shown that one cannot ignore technological change when discussing income distribution, we seem to be on the most secure ground, but exactly how it operates to do so is another matter. Reflecting a personal view first expressed at the end of Chapter 12, the interaction of firms in a world of Schumpeterian competition as reflected in the work of Nelson and Winter is an approach that I feel has enormous appeal.

References

Cowling, K. (1982), *Monopoly Capitalism*, Macmillan, London.

Dougherty, C.R.S. (1974), 'On the Secular Macro-Economic Consequences of Technological Progress', *Economic Journal*, 84, 543–65.

Drandakis, E. and Phelps, E. (1966), 'A Model of Induced Invention, Growth and Distribution', *Economic Journal*, 76, 823–40.

Ferguson, C.E. (1968), 'Neoclassical Theory of Technical Progress and Relative Factor Shares', *Southern Economic Journal*, 34, 490–504.

Harcourt, G.C. (1972), *Some Cambridge Controversies in the Theory of Capital*, Cambridge University Press.

Hicks, Sir J.R. (1965), *Capital and Growth*, Oxford University Press.

King, J. and Regan, P. (1976), *Relative Income Shares*, Macmillan, London.

Nelson, R. and Winter, S.G. (1974), 'Neoclassical vs. Evolutionary Theories of Economic Growth: Critique and Prospectus', *Economic Journal*, 84, 886–905.

Nelson, R., Winter, S.G., and Schuette, H. (1976). 'Technical Change in an Evolutionary Model', *Quarterly Journal of Economics*, 90, 90–118.

Schumpeter, J.A. (1934), *The Theory of Economic Development*, Harvard University Press, Cambridge, Mass.

Solow, R.M. (1957), 'Technical Change and the Aggregate Production Function', *Review of Economics and Statistics*, 39, 312–20.

Chapter 15

Social Savings and Rates of Return
to Technological Change

15.1 Introduction

In the other chapters in this part of the book our discussion of the effects of technological change cover the impact on a number of leading economic indicators that may proxy social welfare, but we do not deal directly with the contribution that technological changes have made to social welfare. In this chapter we intend to fill this gap. We proceed by initially considering an exact approach to relating welfare gains to technological change, and then move to Mansfield's estimates of rates of return to R & D based essentially on this approach. We then consider the historians' social savings approach and the literature on the social returns to investment in new technology.

15.2 The theoretical framework

Consider an industry producing a homogeneous product facing an isoelastic demand curve

$$q = Ap^{-\eta} \tag{15.1}$$

where q is demand, p is price, and η is the elasticity of demand. For this product the total welfare contribution (W) is defined as social benefit minus cost (in the absence of externalities); i.e.,

$$W = TR + S + R - TC \tag{15.2}$$

where TR and TC are total revenue and total cost respectively, S is consumer's surplus, and R is the intra-marginal rents accruing to producers. Under constant returns to scale $R \equiv 0$, so that

$$W = \Pi + S \tag{15.3}$$

where $\Pi = TR - TC$; i.e., welfare equals the sum of consumer and producer surpluses. If we are to look at the impact of a technological change on welfare, we need to know how both producer and consumer surpluses are affected.

To pursue the argument a little further, define c as unit cost (assumed constant); then $\Pi = (p - c)q$. Following a suggestion by Dixit in Cowling *et al.*, (1980) (from which source this approach is developed), we can write

$$S = \int_{p}^{\bar{p}} Ap^{-\eta} \, dp \tag{15.4}$$

where $\bar{p}$ is some arbitrarily high level of price. Thus

$$W = (p - c)q + \int_{p}^{\bar{p}} A p^{-\eta} \, dp. \tag{15.5}$$

Therefore

$$W = \frac{A^2 (\bar{p})^{1-\eta}}{1 - \eta} + A p^{-\eta} \left(p \frac{\eta}{\eta - 1} - c \right). \tag{15.6}$$

We may think of a technological change as potentially affecting A and η, i.e., shifting the demand curve, and/or changing c, production costs, and/or affecting p, the price of the product. For any particular change one can calculate the change in W from (15.6) and thus the welfare contribution of the technological change. As an example, consider the simple case of a technological change that reduces c, with p, A, and η constant. Then from (15.6) we get

$$\Delta W = A p^{-\eta} \, \Delta c \tag{15.7}$$

which yields

$$\Delta W = - q \, \Delta c, \tag{15.8}$$

which of course is the increase in producers' profits.

However, one would not in general expect p, η, and c to be independent of one another; for example, in Cowling and Waterson (1976) it is argued that the p, c, and η in a homogeneous product industry are related to one another according to

$$\frac{p - c}{p} = \frac{H(1 + \mu)}{\eta} \tag{15.9}$$

where H is the Herfindahl index of concentration, and μ is a measure of conjectural variations. The work of Dasgupta and Stiglitz (see Chapter 3) further argues that H is endogenous to the technological change process.

If, however, we ignore this last complication and allow that price is related to unit cost according to

$$p = kc, \tag{15.10}$$

and that k, the mark-up, is not affected by the technological change, then from (15.6) we get

$$W = \frac{A^2 \bar{p}^{1-\eta}}{1 - \eta} + A c^{1-\eta} k^{-\eta} \left(k \frac{\eta}{\eta - 1} - 1 \right), \tag{15.11}$$

and after a change in c the change in W is given by

$$\frac{dW}{dc} = (1 - \eta)(A k^{-\eta}) \left(k \frac{\eta}{\eta - 1} - 1 \right) c^{-\eta}. \tag{15.12}$$

From (15.12) it is clear that the impact of a technological change on W will depend *inter alia* on the elasticity of demand and the mark-up. The simplest application of this to consider is one with perfect competition before and after the technological change, so that $p = c$ and $k = 1$, whereupon we get

$$\frac{dW}{dc} = (1 - \eta)(A) \left(\frac{\eta}{\eta - 1} - 1\right) c^{-\eta} \qquad (15.13)$$
$$= -Ac^{-\eta}$$

and the gain in welfare is simply equal to the increase in consumer's surplus, which in turn is related to the elasticity of demand.

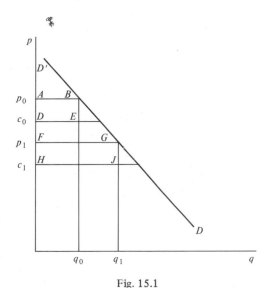

Fig. 15.1

To show these concepts diagramatically (for the linear as opposed to isoelastic demand curve), consider Fig. 15.1. Let DD' be the industry demand curve, and c_0, c_1 and p_0, p_1 be the costs and prices before and after innovation. The welfare gain derived from the innovation is equal to $\Delta\Pi + \Delta S = (FGHJ - ABDE) + ABFG$. So far this whole discussion has been a static one. If the gain is made for a number of years then the true welfare gain must be the present value of the flow of returns discounted at some rate r.

15.3 The returns to research and development

Mansfield *et al.* (1977) have generated some estimates of the social and private rates of return to innovations using essentially the framework in section 15.2. They consider seventeen innovations which they consider to be of three basic types:

1   a product innovation used by firms: in this case the social return is equal to the increase in consumers' surplus generated by a lower product price plus the increase in producer's surplus defined as the innovator's increase in profits;

2   a product innovation used by households: in this case the demand curve for the product is assumed not to move but the product is supplied at a lower price. The lower price generates an increase in consumer surplus. The gain in profits by innovators is used to measure the increase in producer's surplus;

3   a process innovation reducing producer costs: the analysis here follows that detailed above exactly.

In each case consumer surplus is constructed by use of pre- and post-innovation prices and a rough estimate of the elasticity of demand using

$$\Delta S = (p_0 - p_1)(q_1)\left(1 - \tfrac{1}{2}\eta\,\frac{p_0 - p_1}{p_1}\right). \qquad (15.14)$$

Producer's surplus is taken from innovator's profit records.

The time-profile over which the gains last is defined as essentially for the period from the date of innovation to the date when innovation would have occurred if the innovator had not already innovated. There is thus a comparison of welfare gains with a counter-factual path on which innovation occurs but at a later date.

The costs involved in generating the innovation are essentially the R & D costs, capitalized to the date of innovation. Rates of return are calculated relative to these costs.

Mansfield's private rates of return are calculated as the increase in innovator's profits relative to innovator's R & D expenditure. The rates of return here vary from 214 per cent to less than zero per cent, with a median 25 per cent return. In particular, in 30 per cent of the cases 'the private rate of return was so low that no firm, with the advantage of hindsight, would have invested in the innovation, but the social rate . . . was so high that from society's point of view, the investment was well worthwhile.' The social rate is defined as the welfare gain relative to R & D expenditure. Social rates of return are calculated to be in the range from 209 per cent to less than zero per cent with a median 56 per cent rate of return. Mansfield's analysis of the differences between social and private rates of return suggests that more important and more cheaply imitated innovations had the greatest difference. Mansfield's later work (Mansfield *et al.*, 1981) suggests that 60 per cent of *patented* innovations are imitated within four years.

Mansfield does not have much to say on the determinants of the level of social and private rates of return except to allude to riskiness as a cause of the generally high rates. However, if we refer back to Chapters 2 and 3 we might well argue that market structure will influence the rates of return quite strongly (see for example, Grabowski and Mueller, 1978).

This work of Mansfield follows on from work of Griliches, Fellner, and others (for a summary see Mansfield, 1972), but rather than detail this work we will now turn to the major users of this welfare-based approach, the economic historians.

15.4 Social savings and social rates of return to innovation

Using the 'social savings' approach, several authors have attempted to analyse the impact of major technological innovations on an economy's welfare. Thus studies that use this approach exist of, for example, the rise of the railways or the spread of steampower. The largest effort has centred around the application of the approach to the impact of railways in the US economy.

The original concept of 'social savings' may be defined as the loss that would be imposed on an economy if a most efficient technology were no longer available. In the railway case this would comprise the extra cost of transporting commodities by, say, ship and waggon. For freight transport only, Fishlow estimates these costs for the USA in 1859 as $134 million; Hawke's estimates for England and Wales in 1865 are around £28 million; and Metzer's for Russia in 1907 at 890 million roubles (see O'Brien, 1977). These are the extra costs that are estimated as would have arisen if the commodities actually shipped by the railways in the specified year had to be shipped by other means.

These social savings mean little in themselves, but come into focus as a percentage of GNP. Then the figures above are Fishlow, 3.3 per cent Hawke, 4 per cent, and Metzer, 4.6 per cent. Fogel's (1964) independent estimates for the USA in 1890 come to 4.7 per cent of GNP. These percentages seem small, but they do refer solely to the 'loss in GNP' that the absence of railways for one year would effect. O'Brien (1977) argues that, if no railways existed for the whole of 1859–90, then in 1890 GNP would have been some 10 per cent lower. Another way to consider the percentages is that for the UK the trend growth rate in 1865 was less than 4 per cent, and thus the cost of compensating for the loss of railways for one year would be more than a whole year's growth.

Fogel's estimates differ somewhat from the other three. Keeping everything else constant, Fogel (1964) estimates a 9 per cent social saving, but he argues that in the absence of railways the different structure of the economy would mean that it would have been more suited to the alternative modes of transport and losses would have amounted to only 4.7 per cent. However, taking this argument further, O'Brien states

If economic historians hope to measure the effect of an innovation . . . the associated counterfactual would *not* be an economy compelled to operate without railways for a single year, but rather an hypothetical economy which derived absolutely no benefits at all from railways because they were never constructed. (O'Brien, 1977, pp. 33–4)

The costs of not having railways would thus certaintly be greater than the costs of closure for one year, but how much greater depends so much on the posited counterfactual.

The social savings we have been discussing so far refer to railways. In Chapter 10 we discussed the spread of steam engines in the UK and it thus seemed appropriate to give some idea of the sorts of numbers being calculated for savings on these machines. Von Tunzelman's (1978) first results refer to the social savings of James Watt. The social savings as calculated take the same pattern of output for the counterfactual world as the real world, with, in the counterfactual world, all Watt engines replaced by atmospheric engines. It is then calculated that, if in 1800 all Watt engines had been replaced by atmospheric engines, increased costs would have amounted to 0.11 per cent of GNP. Von Tunzelman feels that even this figure may be biased upwards. When extending from a consideration of the Watt engine to a consideration of all steam engines relative to non-steam power in 1800, social savings in the order of 0.2 per cent of GNP are considered relevant. However, in 1800 the diffusion of steam was very limited, and in later years social savings may well have been much greater.

To throw further light on the social savings concept, it is worth turning now to consider an approach to similar issues that is more comparable with that of Mansfield — the historian's search for an estimate of the social rate of return to investment in new technology. McLelland (1972), in a most informative discussion of the issues involved, argues that the average social rate of return to a project in a single year equals the increases in consumer and producer surplus plus any externalities *as a ratio to gross investment in the project*. If, as is correct, one considers that the project has a life longer than one year, then ratios of present values or the calculations of internal rates of return are appropriate. Note that the externalities introduced here summarize all effects of the investment not captured by consumer and producer surplus.

If price equalled marginal costs before and after the innovation and marginal costs were constant, then in the absence of externalities the gain in social welfare in a given year can be measured solely by the increase in consumers' surplus as shown in equation (15.13) above. If p_0 is the price the product would have without innovation and p_1 its price with, then with a linear demand curve one can measure this gain as (ref. Fig. 15.1) $(p_0 - p_1)q_1$ where q_1 is the output of the product with price p_1.

In measuring social savings, Hawke (1970) (among others) calculates social savings (ref. Fig. 15.1) as $(c_0 - p_1)q_1$. McLelland argues that there has been a tendency in the past to measure additions to consumer surplus by the social savings thus defined. As should be clear, however, this is not generally acceptable. If we have perfect competition with and without railways, then p_0 and c_0 will be the same, and so with perfect competition and constant costs this measure of social savings and additions to consumer surplus coincide, — without these conditions they do not. However, McLelland further argues that Fogel and others calculate social savings not as $(c_0 - p_1)q_1$ but as $(p_a - p_1)q_1$ where p_a is not a proxy for c_0 but is in fact the price of alternative transport services in a world with railways (not the price in a world without railways). In this case social savings and additions to consumer surplus do not coincide.

There is therefore, considerable doubt as to

1   what welfare implications can be read into the estimates of social savings; and
2   to what extent the estimates of social savings can be used to generate estimates of the social rate of return.

On this latter point it is worth referring to work that has attempted to remove the perfect competition assumption and to consider more than one time period.

Paul David (1975) extends Fogel's earlier analysis to include producer's surplus but excludes externalities. Nerlove (1966) generates rates of return, and Fishlow (1975) explicitly considers the multi-period nature of the problem. However, McLelland argues, all three use discredited proxies for consumers' surplus (social savings), thus weakening the results. An alternative approach, used by Mercer (1970), is to consider that the increase in consumer surplus can be proxied by changes in land rents, for a discussion of the problems of which one is referred to McLelland (1972).

Despite McLelland's reservations of the use of social savings estimates for estimates of gains in consumer surplus, we need to give some idea of the social rates of return being calculated by the historians. Fishlow put the average rate on US railroads before the Civil War at about 15 per cent per year. Hawke calculates a return for England and Wales of 15–20 per cent per year. However, as McLelland states, it is difficult to evaluate these figures (e.g., is 15 per cent high or low?) without some standard of comparison, especially as the economist's comparisons would involve marginal rather than average rates.

One should note at this point that the question being asked by the economic historians is somewhat different from that of Mansfield. Mansfield is looking at rates of return on investment in developing new technology, i.e., at returns to R & D. The historians are looking at returns to investment embodying the new technology. Both approaches basically consider average rates of return, but if one could get at marginal rates Mansfield's approach would be directed towards the question, should we invest more in R & D?, whereas the historians' question is, should there have been greater use of the new technology? In both cases comparisons of marginal social rates of return with marginal social rates on other alternatives would be the correct one. The estimate of social rates is very difficult to obtain, as is a transformation of an average rate for any innovation into a marginal rate.

One should also state that marginal rates of return to investment should fall as the new technology spreads. In particular, if entrepreneurs are rational, diffusion of the new technology should proceed to the point where marginal *private* rates of return to the innovation equal the cost of funds (risk-adjusted), although social marginal rates may be higher than this.

In case one should give the wrong impression, it is fair to say that not all work on the impact of railways or other new technologies has concentrated on the social savings approach. There is a considerable literature on the backward and forward linkages of railways into other industries and its effects there (see O'Brien, 1977). This, however, only leads one to the realization that considering the impact of a large new technology such as railways is really a general and not a partial-equilibrium problem. In fact, the basic problem with all these approaches to calculating the return to the investment in new technology centres around the largely partial equilibrium nature of the approaches discussed above. There are three implications of this inherent in the above.

1 The approach discussed above assumes that any resources released in the economy by the technological innovation (or any extra resources used) can be sold at existing prices (purchased at existing prices). If released resources remain unemployed the calculations no longer hold. The basic assumption therefore is that the economy is in a full-employment equilibrium and any resources released can be sold without changes in the prices of the resources.

2 When calculating rates of return a comparison of actual social welfare with that on some counterfactual path (that without the technological change) must be employed. Thus for example in the railroad case the calculation requires knowledge of what prices would have been had railroads not appeared. In Mansfield's study the counterfactual path involved delayed innovation as opposed to no innovation at all. The construction of counterfactual paths is notoriously difficult, and the greater the technological shock the greater the difficulty. Partial equilibrium analysis allows one to limit the problem of constructing the counterfactual, but if one considers an innovation such as railways the widespread repercussions must make a partial-equilibrium approach to calculating the counterfactual rather inappropriate. This problem becomes all the greater if one considers that the counterfactual will essentially represent the 'opportunity cost' path. In its construction one must therefore consider what would have been developed with the R & D funds and where the investment used in installing the new technology would have gone (allowing of course that R & D and investment may have been somewhat lower because of the lower expected returns). This takes one into the realms of impossibility.

Moreover, technological change may be considered cumulative. One technological advance lays the basis for future advances; if one changes the basic advance it will have repercussions into the distant future. As McLelland states, to consider the question of whether an economy would have been better off without its entire railway system leads one to consider that

such a large reallocation of investment might well have seriously altered the returns on investment alternatives and thereby the willingness of the community to save and invest. The task of determining what those dollars could have earned if diverted to uses other than railroads becomes extremely complex. (McLelland, 1972)

He in fact argues that the reasonable question to ask is, would the American economy of the nineteenth century have been better off with a little more, or a little less, investment in the railroad sector? This standard *marginal* question is much more suited to partial analysis.

3   For the partial equilibrium analysis to be acceptable, the demand curve for the goods in question must be independent of the investment in the new technology. For example, in the railway case, if income increases through lower transport costs and this causes an increase in the demand for transport services (shifts the demand curve), then the partial-equilibrium analysis breaks down (see White, 1976).

The outcome of the above discussion is that, if we are considering a technological change of any other than local significance, then a general equilibrium rather than a partial equilibrium analysis is desirable. Moreover, once one turns to general equilibrium analysis it is soon realized that the above approaches suffer from one further problem: they are essentially restricted to closed-economy analysis. In an open economy some of the gain from the new technology may 'leak' overseas. In Chapter 17 the effects of technological change on welfare in a general equilibrium for an open economy are explored. The approach there is essentially restricted to comparisons of Walrasian equilibria, but the theory indicates that technological advances do not always yield a domestic welfare gain. No pretence is made to even attempt a practical application of the theory to any particular example.

Finally, before moving on we should say something of externalities. In a sense, what comprises an externality depends upon how partial is one's approach. Fishlow however 'guesses' that the inclusion of externalities would almost double his calculated value for the social rate of return to 20 or 30 per cent.

15.5 Conclusions

In this chapter we have attempted to discuss the efforts that have been made to obtain direct estimates of either gains in social welfare or social rates of return being derived from technological change. The basic framework used for these analyses is a partial-equilibrium one and essentially two questions are being asked: (1) what is the rate of return to R & D? (2) what is the rate of return to investment embodied in new technology? On the way to answering (2) the concept of 'social savings' has been discussed.

The work above suggests that the social rate of return to R & D is high (compared with, say, average profit rates) although private rates of return are somewhat lower. The returns to be achieved from investment in installing new technology in the historical literature are calculated in a range from 10 to 30 per cent, although each estimate is much disputed. Of particular relevance is the view that, the further into the diffusion is the calculation of return made, the lower should one expect to find the marginal rate of return.

Calculations of social savings as a very rough-and-ready proxy for the impact of a new technology on social welfare indicate orders of magnitude that are smaller than one might expect. Figures for a major innovation such as railways in the order of 3-5 per cent of GNP may suggest that technological change is not quite as important as first thought. Although the figures are much disputed, and division of social savings by GNP is a division by a large number, one should not be too surprised by the result. The figures are to be read as the increased cost of doing what is actually done but using the second most efficient method. If this second most efficient method is not too inefficient, the social savings calculated will be small.

Overall, this literature suggests that there are significant returns to be achieved by society from technological innovation. However, any quantification of rates of return must at least be considered with a pinch of salt.

References

Cowling, K. and Waterson, M. (1976), 'Price-Cost Margins and Market Structure', *Economica*, 43, 267–74.

Cowling, K. *et al.* (1980), *Mergers and Economic Performance*, Cambridge University Press.

David, P. (1975), *Technical Choice, Innovation and Economic Growth*, Cambridge University Press.

Fishlow, A. (1965), *American Railroads and the Transformation of the Ante-Bellum Economy*, Harvard University Press, Cambridge, Mass.

Fogel, R.W. (1964), *Railroads and American Economic Growth, Essays in Econometric History*, Johns Hopkins Press, Baltimore.

Grabowski, H.G., and Mueller, D.C. (1978), 'Industrial Research and Development, Intangible Capital Stocks and Firm Profit Rates', *Bell Journal of Economics*, 9, 328–43.

Hawke, G.R. (1970), *Railways and Economic Growth in England and Wales, 1840-1870*, Clarendon Press, Oxford.

Mansfield, E. (1972), 'Contribution of R & D to Economic Growth in the United States', *Science*, 175, 477–86.

Mansfield, E. *et al.* (1977), *The Production and Application of New Industrial Technology*, Norton, New York.

Mansfield, E. *et al.* (1981), 'Imitation Costs and Patents: An Empirical Study', *Economic Journal*, 91, 907–18.

McLelland, P. (1972), 'Social Rates of Return on American Railroads in the Nineteenth Century', *Economic History Review*, XXV, 471–88.

Mercer, L.J. (1970), 'Rates of Return for Land-Grant Railroads: The Central Pacific System', *Journal of Economic History*, XXX, 602–26.

Nerlove, M. (1966), 'Railroads and American Economic Growth', *Journal of Economic History*, XXCI, 112–15.

O'Brien, P. (1977), *The New Economic History of the Railways*, Croom Helm, London.

Von Tunzleman, G. (1978), *Steam Power and British Industrialisation to 1860*, Clarendon Press, Oxford.

White, C.M. (1976), 'The Concept of Social Saving in Theory and Practice', *The Economic History Review*, XXIX, 82–100.

Chapter 16

The Impact of Technological Change on Market Structure

16.1 Introduction

In Chapters 2 and 3 we investigated the relationship between R & D and market structure and patenting activity and market structure mainly in terms of how the number and size distribution of firms will affect the rate of technological advance. This direction of causality is the classic one, market structure affecting market performance. It has long been recognized, however, that the direction of causality may also be reversed, so that there is a feedback from market performance to market structure. In this chapter we investigate one aspect of this feedback mechanism — the effect of technological change on market structure, a feedback especially associated with Phillips (1971).

We can start by referring back to Chapter 2. In that chapter we discussed the work of Dasgupta and Stiglitz in some detail. One of the predictions of their analysis was that, in a market with free entry, the rate of technological advance and market structure were determined simultaneously, the parameters influencing both essentially being consumer preferences and technological opportunities. However, in that work there is assumed to be instantaneous diffusion of new technological discoveries and, moreover, patent protection preventing imitation. Mansfield's paper (Mansfield *et al.*, 1981) belies the validity of the second assumption, and Part II of this book is concerned with lags in diffusion. Thus, although the underlying moral of the Dasgupta and Stiglitz analysis is an important one, it does not give us the whole story.

16.2 Economies of scale

In the literature on the determinants of the number and size distribution of firms it is usually considered that a major influence is the extent of scale economies or minimum efficient firm size (MES) relative to the size of the market. The prediction is that, the larger is MES relative to the size of market, the fewer firms there will be. To take the story further, one might consider scale economies relative to plant size. One can define MES for plants rather than firms, and then the combination of plant and firm MES will influence the number of firms and the degree of multi-plant operations.

If we consider first the question of plant level scale economies, Scherer, after a literature review, concludes that

While some developments in technology lead to smaller optimal plant scales relative to market volume, others are pushing in the opposite direction. It is too early to assess confidently how these forces balance out, and how they will affect scale economies and market concentration in decades to come. The problem is an important one, worthy of much more attention. (Scherer, 1970, p. 89)

This we do not find to be a very surprising conclusion. Perhaps an initial supposition that patterns of technological change would increase plant MES is related to a belief that most technological changes are capital-using, and higher capital intensities might imply higher plant MES. It is not difficult however to think of new technologies that are not capital using; for example, computer hardware advances may well have been capital-saving.

If plant-level MES has shown no consistent pattern, what of firm-level MES? Here I believe there are two points of relevance to consider.

1   Williamson (1975) argues that a basic constraint on the size of the firm is that as it becomes large the problems of control and co-ordination start to outweigh the benefits to be derived from large physical scale. A major innovation, the multi-divisional form of managerial organization, enables firms to overcome these problems of control and thus to bypass this constraint on firm size. If this is combined with recent advances in the information industry, i.e., computers, it would seem that over the last three decades technology has led to a possible increase in firm MES. This could lead to increasing concentration if markets have not grown apace.
2   The MES for the firm will be related to the degree of scale economies in the process of generating and implementing new technology. Support for the Schumpeterian hypothesis that there are scale economies in the R & D process would support a view that the technological change process leads to increasing concentration. However, we have shown in Chapters 2 and 3 that the support for the hypothesis is weak.

One can however argue that one does not need such scale economies for R & D expenditures to affect market structures. We are thinking here not of entry barriers or the success or failure of R & D, which we consider below, but more in terms of the imperfections of the capital market. If one has a technologically advancing industry it is quite possible that the need to undertake R & D to maintain competitiveness gives larger firms an advantage. The better credit rating of larger firms can make their R & D financing cheaper. Moreover, in certain industries in which government funding of R & D is of great importance (e.g., aircraft in the UK), the reluctance of government to fund competing projects leads to increased concentration. Thus, for example, in the UK, government pressures led to mergers in both the aerospace and computer industries, both of which relied extensively on government funding for R & D.

One final point in this discussion of firm-level economies and technological change. One can argue that diversified firms may get greater returns for R & D if they are able to internalize the profit from unexpected discoveries in their R & D expenditures. If this is the case, then the process of technological change will encourage diversification.

16.3 Entry barriers

The effect of technological change on scale economies is only one part of the argument. The influence of such changes on market structure has been developed much further. The second line we consider is the concept of technological change as an entry barrier. A barrier to entry may be considered as any factor that makes costs higher for the potential entrant relative to the costs of the extant producers, although barriers may also exist through absolute capital requirements — the higher are the requirements for entry, the fewer potential entrants there are.

Kamien and Schwartz (1982, pp. 70-5) argue that the necessity to carry on R & D activities in order to remain a viable competitor in certain industries constitutes an entry barrier, for it increases the absolute capital requirements for entry and moreover may raise the MES of operation of a new firm because of the need to recoup R & D expenditures. It is further argued that effective entry may require some technical advance so the cost and risk of research constitutes an entry barrier. However, by the same token one might argue that in a world without complete property rights in new technology certain advances can lead to reductions in the absolute capital requirements for entry. New technology that is capital-saving and can be copied may make entry easier. We have already argued that there is no evidence that technological advance has meant any systematic trends in scale economies, so barriers may not have been raised through this mechanism.

It is commonly argued that the ownership of patents by existing producers represents an entry barrier. Mansfield's work, however, on the ease with which patents can be 'designed around' (Mansfield *et al.*, 1981) makes one doubt this. However as we argued in Chapter 2, patents are not the only way to protect knowledge. Of particular importance may be firm-specific knowledge. Here we are thinking of learning-by-doing effects. An existing company may build up knowledge through use of a technology that a potential entrant would not have on entry and the entrant may thus have higher costs. This is a barrier considered by Spence (1977). The importance of such economies as an entry barrier depends upon

1   the extent of the economies to be derived from learning: the greater they are, the more disadvantaged may be the entrant; and
2   how quickly the learning economies can be realized. If they are realized after only a small volume of gross output has been produced they will be relatively unimportant; if it takes a large production run then they may represent a considerable barrier to be overcome by the entrant.

Even given this, one should note that, if such learning economies are embodied in a firm's labour force, then that labour force could well set up a new firm to enter the industry with no cost disadvantage. Such breakaways seem fairly common in, for example, computers and semi-conductors.

If the learning economies exist on the consumer side, so that in a world of rapidly changing technology the cheapest form of information for a consumer is brand identification, then the entry barrier may be more difficult to overcome. The role of brand loyalty as a barrier to entry has, however, been somewhat disputed (e.g. Schmalansee, 1974, and Cubbin, 1981).

Finally, we may consider technological change being used as an entry barrier through the deliberate actions of extant firms. We discussed protective R & D in Chapter 3, where it was argued that such R & D, used to generate advances that were not marketed except when a firm's dominant position was challenged, was equivalent as a concept to Spence's (1977) 'excess capacity' barrier to entry.

Mueller and Tilton (1969) and Pavitt and Wald (1971) see the maturity of the industry as important in defining the role of technological change as an entry barrier. It is essentially argued that in the early stages of the life of an industry small firms have most opportunities because scale economies are unimportant, market shares are volatile, and brand loyalty is weak. However, as the technology matures, scale, efficiency, and brand loyalty build up, making opportunities for small firms fewer and entry more difficult.

This cyclical approach to the problem has most recently been investigated by Gort and Klepper (1982). They suggest five stages in the evolution of the market with respect to the number of producers within it, starting with the first commercial introduction of a new product:

1   from commercial introduction by the first producer through to a sharp increase in the rate of entry of new competitors: the length of this stage depends on market size, the ease of entry, and the number of potential entrants;
2   a period of sharp increases in the number of producers;
3   a period during which net entry is zero, with entrants and exiting firms cancelling each other out;
4   a period of negative net entry;
5   a second period of approximately zero net entry.

The basic hypothesis is that entry is related to advances in product technology and these arise from both internal and external sources. The more important are external sources the more likely is entry. It is hypothesized that in the early stages of the cycle most advances are external, whereas in the late stages they are internal. The story told is then that in the early stages of the product life-cycle the high profitability of existing producers and a high rate of external innovation generates high levels of entry. As technology matures and the potential rate of product improvement declines, the rate of innovation declines and so does entry. This is reinforced by the accumulation of experience by existing producers, the reducing profit levels of these producers owing to earlier entry, and a decline in the number of potential entrants because of earlier entry. As the industry starts to approach normal profit rates, the pressure for technological competition internal to the industry increases, forcing out the less successful and raising entry barriers further. The exit rate then rises sharply.

Gort and Klepper (1982) apply this analysis to the growth of forty-six products in the USA. They argue that their empirical work supports their basic hypothesis, and moreover their evidence lends little support to other explanations of new entry into a market. Among their findings they argue as follows.

1   The markets for most new products pass through at least five distinguishable phases in their evolution.
2   There appears to be an association between rises and declines in the rate of innovation and the rate of new entry into new markets.
3   The character, importance, and sources of innovations appear to change over the product cycle.
4   The structure of markets is shaped to an important degree by discrete events such as technical change and the flow of information among existing and potential producers.

The least one can draw from these results is that technological change is an important influence on market structure. One must have some doubts about the analysis; for example, the *ad hoc* nature of the hypothesis is worrying (why, for example, does the potential for product improvement decline over time?), and the concentration on product as opposed to process innovation is of some concern; but even so the strength of the results gives support to the basic hypotheses presented.

16.4 Success breeds success

It might appear that the obvious starting point for an analysis of the impact of technological change on market structure is an hypothesis that states that a technological advance will enable a firm to oust rivals from the market and that this will lead to monopoly power in the market. This presumes however that only one firm is making technological advances, that there is no imitation, and that there is no entry. Our discussion above suggests that all these conditions are not generally found. However, one need not take such an extreme form of the hypothesis; a much weaker but equally valid form is that one technological success yields advantages that allow the firm, over time, to build up market power — the success-breeds-success hypothesis.

To make the hypothesis operational, one can argue that successful innovation generates profits that may be reinvested in further technological change, which enables the firm to capitalize on its success. If success does breed success, then it will mean a tendency towards higher levels of concentration. We discussed in Chapter 2 how successful innovation can influence the growth of firms by this route, referring to Mansfield (1968, pp. 126–9), who shows that successful innovators grow faster than other firms (although their advantage declines over time). Grabowski and Mueller (1978) also show that one can support a link from

R & D intensity to profitability, making the first part of the hypothesis support-able. Grabowski's (1968) study of R & D in chemicals, drugs, and petroleum lends some support to the hypothesis, past R & D success leading to greater current R & D effort. Also in the general literature on the growth of firms there is some evidence to support the proposition that growth promotes growth. The work of Singh and Whittington (1968) on 2000 UK firms found that above-average growth rates in one period were associated with above-average growth rates in a later period. Downie (1958) suggests combining with the advantages of success a 'stimulant to change' argument. The less successful need to innovate to survive, while the more successful have the resources but not the same incen-tive. These counteracting forces generate expansion paths.

However, the success-breeds-success hypothesis is not necessary to generate increases in concentration over time. Gibrat's Law of Proportionate Effect argues that, even if the growth rate of the firm is independent of its size, and in fact even if growth rates are randomly distributed across firms, the outcome of the growth process will be that the concentration of the industry will increase over time. If success does breed success, then concentration will increase faster over time. Essentially, the concentration increases because firms grow at different rates.

We cannot consider this discussion of stochastic growth processes complete until we have discussed another piece of work closely allied with the success-breeds-success hypothesis, the work of Nelson and Winter (1978). The model is essentially the same as that described in Chapter 12, where firms satisfice, undertake search and invest, and generally act in a behavioural manner. The core of the work is a variation on Gibrat's Law, including a serial correlation of growth rates (success breeds success) and restraints on growth as firms grow large. The principle is that it is an attempt to see the implications for industrial structure of Schumpeterian competition. The key relationships of the model that is simulated are that

1   firms always produce at capacity levels, capacity being a constant times the capital stock;
2   profits in excess of some target rate of return (which increases with firm size) lead to investment, which is however constrained by credit opportunities;
3   R & D spending of the firm is proportional to firm size, and R & D leads to the discovery of better technology;
4   the demand curve is a constant, unit-elastic, curve, and price is determined to clear the market given capacity production;
5   there is no entry.

The advantages of size through the R & D and investment assumptions generate the success-breeds-success phenomena. The outcome of the simulation is that:
some firms track emerging technological opportunities with greater success than others; the former tend to prosper and grow, the latter to suffer losses and

decline. Growth confers advantages that make further success more likely, while decline breeds technological obsolescence and further decline. As these processes operate through time there is a tendency for concentration to develop even in an industry initially composed of many equal sized firms. (Nelson and Winter, 1978, p. 541)

Of particular interest is the finding, confirming Phillips's (1971) result, that an environment with abundant technological opportunities and difficult imitation tended to produce more rapid increases in concentration.

Essentially, Nelson and Winter's work is an attempt to formalize the success-breeds-success hypothesis, but basing it on R & D and investment behaviour and then investigating the impact on firm growth; i.e., the process of Schumpeterian competition is being modelled. Nelson and Winter argue that the model 'does a respectable job of generating firm size distributions that are at least superficially realistic'.

16.5 Empirics

The outcome of all the discussion above is that there are a number of reasons as to why technological change should have an effect on market structure. The reasons fall into essentially three classes: (1) the effect on minimum efficient scale for firms and plants; (2) the effect on entry; (3) the working of stochastic growth process. The obvious question to ask now is, given the increases in concentration that have been observed in most developed economies over the last thirty years, to what extent can these increases be attributed to technological change, or to what extent has technological change prevented further increases in concentration? There does not seem to be any clear answer to these questions. Consider, however, the case of the UK. Prais argues that in the UK, although firm concentration has increased in the twentieth century, plant concentration has not. This tends to suggest, perhaps, that optimum plant size has not increased faster than market size. We have already argued that we cannot observe any systematic tendency for technology to lead to higher plant MES, so perhaps this is as it should be. The increase in firm level concentration is another matter.

Work on explaining changes in firm concentration has centred upon the relative contribution of mergers and the differential internal growth rates of firms. A survey of the literature (see Cowling *et al*., 1980) suggests that anything from 50 to 100 per cent of increases in concentration are due to merger activity. Hannah and Kay (1977) suggest that nearly all the increase in concentration is due to merger activity, leaving no role for differential internal growth. Now, unless merger activity is just the institutional realization of the stochastic growth processes, this suggests that the success-breeds-success story has little role to play in increasing concentration. Hart (1979) especially disputes the Hannah and Kay result, giving greater stress to differential internal growth. Our own feeling is that about 50 per cent of the increase being due to merger and 50 per cent due to differential growth is a fair reading of the evidence, which may leave some role for the success-breeds-success hypothesis.

To the extent that merger is a reaction to the possibility of achieving firm-level scale economies, one could argue that technological change has affected concentration levels. However, the work on the effectiveness of mergers (see for example Cowling *et al.*, 1980, or Meeks, 1977) suggests that mergers have not led to the realization of scale economies, which throws some doubt on the argument. Moreover, Prais (1976) has shown that the increase in concentration in the UK, at least to 1950, can be explained by the Gibrat hypothesis without the success-breeds-success component, so again little role is left for technological change. The only conclusion one can really draw from this evidence is that there are very good reasons for expecting the whole process of technological change to affect market structure, but this evidence does not yield strong support for the hypothesis.

At the more micro-level, Temin (1979) argues that to some degree new technology led to the rise to dominance of the American Tobacco Company, Eastman Kodak, and IBM. However, in his study of the pharmaceutical industry he concludes that it is difficult to know whether technological change increased market power. He argues that changes in technology and the regulating environment can explain the dramatic increase in the size of drug firms, with associated decreases in price competition; but he finds that the apparent failure of industry profits to rise relative to average manufacturing profits leads to some doubt as to whether market power did in fact increase.

16.6 Conclusions

There seems to be several good reasons to expect technological change to affect the size distribution of firms, either through scale economies, entry barriers, or the success-breeds-success hypothesis. The empirical evidence however does not allow one to quantify the effects of technological change, or even to say there is overwhelming evidence to support the basic hypothesis. There is however some supporting circumstantial evidence.

References

Cowling, K. *et al.* (1980), *Mergers and Economic Performance*, Cambridge University Press.

Cubbin, J. (1981), 'Advertising and the Theory of Entry Barriers', *Economica*, 48, 289–98.

Downie, J. (1958), *The Competitive Process*, Duckworth, London.

Gort, M. and Klepper, S. (1982), 'Time Paths in the Diffusion of Product Innovations', *Economic Journal*, 92, 630–53.

Grabowski, H.G. (1968), 'The Determinants of Industrial Research and Development: A Study of the Chemical Drug and Petroleum Industries', *Journal of Political Economy*, 76, 292–306.

Grabowski, H.G. and Mueller, D.C. (1978), 'Industrial Research and Development, Intangible Capital Stocks, and Firm Profit Rates', *Bell Journal of*

Economics, 9, 328–43.

Hannah, L. and Kay, J. (1977), *Concentration in Modern Industry*, Macmillan, London.

Hart, P.E. (1979), 'On Bias and Concentration', *Journal of Industrial Economics*, XXVII, 211–26.

Kamien, M. and Schwarz, N. (1982), *Market Structure and Innovation*, Cambridge University Press.

Mansfield, E. (1968), *Industrial Research and Technological Innovation*, W.W. Norton, New York.

Mansfield, E. *et al*. (1981), 'Imitation Costs and Patents: An Empirical Study', *Economic Journal*, 91, 907–81.

Meeks, G. (1977), *Disappointing Marriage – A Study of the Gains from Merger*, DAE Occasional Paper no. 51, Cambridge University Press.

Mueller, D.C. and Tilton, J.E. (1969), 'Research and Development Costs as a Barrier to Entry', *Canadian Journal of Economics*, 2, 570–9.

Nelson, R. and Winter, S.G. (1978), 'Forces Generating and Limiting Concentration under Schumpeterian Competition', *Bell Journal of Economics*, 9, 524–48.

Pavitt, K. and Wald, S. (1971), *The Conditions for Success in Technological Innovation*, OECD, Paris.

Phillips, A. (1971), *Technology and Market Structure: A Study of the Aircraft Industry*, Heath, Lexington Books, Lexington, Mass.

Prais, S.J. (1976), *The Evolution of Giant Firms in Britain*, Cambridge University Press.

Scherer, F.M. (1970), *Industrial Market Structure and Economic Performance*, Rand McNally, Chicago.

Schmalensee, R. (1974), 'Brand Loyalty and Barriers to Entry', *Southern Economic Journal*, 40, 579–88.

Singh, A. and Whittington, G. (1968), *Growth Profitablility and Valuation*, DAE Occasional Paper no. 7, Cambridge University Press.

Spence, A.M. (1977), 'Entry Capacity Investment and Oligopolistic Pricing', *Bell Journal of Economics*, 8, 534–44.

Temin, P. (1979), 'Technology, Regulation, and Market Structure in the Modern Pharmaceutical Industry', *Bell Journal of Economics*, 10, 429–46.

Williamson, O.E. (1975), *Markets and Hierarchies: Analysis and Antitrust Implications*, Free Press, New York.

Chapter 17

Technological Change and the Open Economy

17.1 Introduction

The theory of international trade falls essentially into two distinct parts: the pure theory of international trade and the theory of the balance of payments. The former is micro-orientated and concerned with a barter economy; the latter is directed towards the analysis of macroeconomic phenomena in a world with money. In the former the exchange rate is irrelevant; in the latter it is a key variable for analysis. The theory of international trade is one of the show pieces of modern economic theory, with a wide burgeoning literature, many disputes, and many propositions. It is not our intention to try and survey this literature in anything like its entirety. Instead, we limit ourselves to essentially two basic questions:

1   What can this theory tell us of the role of technology differences between countries in the determination of the pattern of trade?
2   What does the theory tell us of the impact of technological change on an open economy and on the trading equilibrium?

This first question is a new one. The second question feeds directly into our analysis in other parts of this book, particularly our analysis of the impact of new technology on output and employment (see Chapter 12).

The way we proceed is similar to many trade books. We first consider the pure theory and then consider the balance of payments literature. The majority of our work, especially in the pure theory, is centred on what is now called the standard model of international trade (sometimes the Heckscher–Ohlin–Samuelson (HOS) model), although we do refer to other models where relevant. The HOS model is essentially a general-equilibrium model, and our results are therefore general-equilibrium results. However, one must be careful in distinguishing different types of general-equilibrium models. The pure theory is primarily concerned with Walrasian models, in which all markets clear by price adjustments in response to notional excess demands. Such models reflect changes in parameters in the equilibrium price vector, all markets clearing. In general, therefore, changes in parameters do not generate economic disequilibria. However, one can construct, and we do refer to, non-Walrasian general-equilibrium models with quantity adjustments in reaction to effective disequilibria. In such models changes in underlying parameters do yield (temporary) equilibria in which there can be markets that do not clear. The effects of parameter changes will then be seen not in changes in the price vector but in terms of underutilized resources. The pure theory of international trade is nearly totally concerned with Walrasian equilibria. Only when we discuss balance of payments theories do we meet non-Walrasian equilibria.

The literature in this field, as we have already stated, is large. Much of it can be found in a number of different textbooks. We have found Sodersten (1980), Chacholiades (1978), Kemp (1969), and Dixit and Norman (1980) to be particularly relevant. Much of the work on the standard model in this chapter is related to the approach by Dixit and Norman, which although at times rather terse and difficult does have the advantage of being in many cases more general than most literature. We turn then to consider the pure theory of international trade.

17.2 The pure theory of international trade

17.2.1 The standard model

Following Dixit and Norman (1980), we consider that consumer behaviour can be represented by one consumer who maximizes a utility function $f(c)$ by choice of his consumption vector c, the elements of which are his consumption of each good. His consumption vector is chosen subject to the constraint that, at the given vector of prices p, his total expenditure must be less than or equal to his total income; i.e., the consumer solves the programme

$$\max_{c} \{f(c)\} \ st \ pc \leqslant y. \tag{17.1}$$

The result of this maximization is a demand for each good. We write the vector of demands for the goods as $d(p, y)$. The problem thus stated can however be considered from the point of view of its dual, where the consumer problem is to choose c to minimize the expenditure necessary to achieve a given level of utility. This yields an expenditure function $e(p, u)$, i.e.,

$$e(p, u) = \min_{c} \{pc\} \ st f(c) \geqslant u \tag{17.2}$$

where u is that u which satisfies

$$y = e(p, u). \tag{17.3}$$

The expenditure function thus generated has the property that its partial derivative with respect to p equals the Hicksian compensated demand function $c(p, u)$. Thus,[1]

$$e_p(p, u) = c(p, u). \tag{17.4}$$

The supply of goods is determined by firms maximizing profits facing given prices. Let v be the vector of net inputs of primary factors, the quantities of which are given and are not traded between countries, and x be the vector of net output of goods. Given v and prices p, x is chosen so that the fixed resources

---

[1] We use subscripts to represent partial derivatives in this section.

are used to maximize the value of output to yield a revenue function $r(p, v)$; i.e.,

$$r(\mathbf{p}, \mathbf{v}) = \max_{\mathbf{x}} \{\mathbf{px}\} \quad \text{subject to } (\mathbf{x}, \mathbf{v}) \text{ feasible} \quad (17.5)$$

The partial derivatives $r_\mathbf{p}(\mathbf{p}, \mathbf{v})$ yield the vector of goods supplied.

Let us now consider the international equilibrium assuming only two countries. Following Dixit and Norman (1980), this equilibrium requires that the sum (across both countries) of demands and supplies must be equal for each good, and that a national income identity holds for each country. Using lower case variables for the home country and upper case for the foreign, we may then state that the general equilibrium requires that

$$e(\mathbf{p}, u) = r(\mathbf{p}, \mathbf{v}) \quad (17.6)$$

$$E(\mathbf{P}, U) = R(\mathbf{P}, \mathbf{V}) \quad (17.7)$$

and

$$e_\mathbf{p}(\mathbf{p}, u) + E_\mathbf{P}(\mathbf{P}, U) = r_\mathbf{p}(\mathbf{p}, \mathbf{v}) + R_\mathbf{P}(\mathbf{P}, \mathbf{V}) \quad (17.8)$$

with factor prices in each country given by $\mathbf{w} = r_\mathbf{v}(\mathbf{p}, \mathbf{v})$, $\mathbf{W} = R_\mathbf{V}(\mathbf{P}, \mathbf{V})$. Let the number of goods be n. The solution to the general equilibrium will then involve vectors $\mathbf{p}$ and $\mathbf{P}$ (of dimension $n - 1$) of relative prices for goods and two utility levels. However, under free trade the relative prices of goods will be the same in each country; thus $\mathbf{p}$ and $\mathbf{P}$ will differ only by a scale factor. Given the equilibrium price vector, net imports (for the home country), $\mathbf{m}$, will be given as $r_\mathbf{p}(\mathbf{p}, u) - e_\mathbf{p}(\mathbf{p}, u)$, and we may thus determine which goods are imported and which are exported.

It can be shown that trade will occur if the two countries' autarky prices differ. A country will tend to import goods that are relatively more expensive there than in the rest of the world in autarky and export goods that are relatively cheap in a no-trade equilibrium. Differences in autarky prices can be attributed to differences in factor endowments, technologies, or tastes. Thus we can say that trade results from these basic differences. The Ricardian explanation of trade is based on technology differences; the Heckscher–Ohlin explanation is based on differences in factor endowments. In the present circumstances it does not seem of particular interest to try to select one of these factors in preference to any other. It is our view that, in general, all will be relevant. What is of interest, however, is not to what extent technology differences under autarky explain the pattern of international trade, but the more relevant question, given that international trading is taking place, of what effect a change in technology will have on the pattern of trade or the nature of the general equilibrium.

17.2.2 The impact of technological change

The results in this area are concerned primarily with the comparative statics properties of the standard model. We assume an initial general equilibrium and then allow technology to change in one (the home) country. We then investigate the impact on trade, the terms of trade, welfare, output, etc., by comparing the new equilibrium with the old. One is thus considering a once-for-all change in technology only, and following the literature this change is disembodied and costless.

As results are really forthcoming only in the special case of two goods, consider a model with only two goods in it. Allow one of these to be the numeraire, with a price of unity in each country, so that the two country price vectors are $(1, p)$ and $(1, P)$ and in the absence of trade taxes $p = P$. By Walras's Law one market-clearing condition is redundant; allow it to be for the numeraire good. One can then write the equilibrium conditions as

$$e(1, p, u) = r(1, p, v, \theta) \tag{17.9}$$

$$E(1, P, U) = R(1, P, V) \tag{17.10}$$

$$e_p + E_P = r_p + R_P \tag{17.11}$$

$$p = P \tag{17.12}$$

where θ is a shift parameter in the home country revenue function that allows for technological change in the non-numeraire good.

To find the impact of a change in θ, totally differentiate (17.9)–(17.12); then, realizing that

$$m \equiv e_p - r_p = - M = - (E_p - R_p),$$

$$e_p \equiv c(p, u)$$

(the Hicksian compensated demand function),

$$e_{pu} = c_u = c_y y_u = c_y e_u,$$

and

$$E_{pu} = C_Y E_U,$$

where c_y, and C_Y are the derivatives of compensated demand with respect to money income, we may generate that

$$e_u \frac{du}{d\theta} = r_\theta - m \frac{dp}{d\theta} \tag{17.13}$$

$$E_U \frac{dU}{d\theta} = m \frac{dp}{d\theta} \tag{17.14}$$

$$\frac{dr_p}{d\theta} = \frac{r_{pp} dp}{d\theta} + r_{p\theta} \tag{17.15}$$

$$\frac{dp}{d\theta} = \frac{r_{p\theta} - c_y r_\theta}{S - m(c_y - C_Y)} \tag{17.16}$$

where $S \equiv e_{pp} + E_{pp} - r_{pp} - R_{pp} < 0$ and $S - m(c_y - C_Y) < 0$ if the Walrasian equilibrium is to be stable.

Equations (17.13) and (17.14) tell us of what happens to home and foreign welfare; (17.15) tells us of the effect of technological changes on the production of the non-numeraire good. These three all depend on changes in relative prices, which is given by (17.16). Given that p is the price of the non-numeraire good in terms of the numeraire, the change in p also tells us of what happens to the terms of trade. If $dp/d\theta > 0$ and the non-numeraire good is exported ($m < 0$), then the terms of trade improve. If $dp/d\theta < 0$ and $m < 0$, the terms of trade worsen. As one can see, the foreign country derives benefits only by changes in the terms of trade. Under a small-country assumption when technological changes in the home country do not affect the terms of trade only the home country will benefit.

From equations (17.13)–(17.16) it is clear that the effect of technological change depends on r_θ and $r_{p\theta}$, i.e., on the way in which technological changes affect the revenue function. Consider the following example. Dixit and Norman show that, if technological change is product augmenting, so that the production function for the non-numeraire good can be written as $x = \theta F(v)$, then

$$r_\theta = r_p = x$$

and

$$r_p = x + \frac{\partial x}{\partial p}$$

where x, as previously defined, is the notional supply of the non-numeraire good. Then we see that from (17.16) that sign $(dp/d\theta) = $ sign $(c_y r_\theta - r_{p\theta}) = $ sign $(c_y x - x - \partial x/\partial p)$

$$= \text{sign } \{x(c_y - 1) - \partial x/\partial p\}, \tag{17.17}$$

which, given $0 < c_y < 1$ and $\partial x/\partial p > 0$, implies $dp/d\theta < 0$. Thus the good experiencing the technological change gets relatively cheaper, and from (17.14) the foreign country has higher utility if it is a net importer of the good and lower utility if it is an exporter. From (17.13) it is clear that a sufficient condition for domestic welfare to increase is that $m > 0$, i.e., if it is a net importer of the non-numeraire good; in other words, product-augmenting technological change will always yield a welfare gain, if it occurs in a net-importing industry. From (17.15) one can see that the supply of the commodity at a given price will increase with θ for a given p, the supply curve shifting by $x + \partial x/\partial p$, but as p falls the actual increase in output will be reduced. A sufficient condition for the output of the non-numeraire industry to be higher in the new equilibrium is that $|\partial p/\partial \theta| < 1$.

The product-augmenting case is in fact the simple case. Dixit and Norman investigate two other cases, the first being general factor augmentation and the second, product-specific factor augmentation. In such situations it is possible for the home country to become worse off because of a change in technology. If, for example, a country experiences technical progress that augments its abundant factor, it could find its terms of trade deteriorating to such an extent that it is worse off with the new technology than with the old. If one has product-specific factor augmentation, then the effect on welfare depends on the elasticity of substitution in production and one cannot derive general propositions regarding the effects of factor-augmenting technical change in export or import industries. In one particular case, however, where production functions are Cobb–Douglas and thus the elasticity of substitution is unity, one can show that the rest of the world is bound to benefit from factor-augmenting technical progress in the home country's export industry.

This analysis depends upon assumptions of constant returns to scale and perfect competition. If one introduces non-constant (increasing) returns to scale and imperfect competition, it is much more difficult to generate any specific results — one cannot even guarantee that there will be gains from trade. However, one can show that trade is likely to provide greater product variety and a reduction in monopolistic distortions, relative to autarky. On the question of technological change in such an environment, it has been argued that, if technological change is higher under monopolistic competition (a supposition without complete support — see Chapters 2 and 3), then the growth rate of supply will be higher under monopolistic competition than under free market conditions. This implies that the terms of trade will move against (exporting) monopolized countries. Thus the more monopolized is the exporting sector, the worse are the prospects for the terms of trade. As we have seen above, this could lead to a deterioration in a country's welfare.

In the above analysis, also, we have investigated only the comparative statics question of what happens when the home country introduces new technology and the foreign country does not. Moreover, the technology being introduced did not involve any completely new goods — it could be represented by shifts in the production function. What if new technology means new products, and what if as time goes on a technical change is copied overseas? Both these aspects are at the centre of what are now labelled the 'neotechnology theories of trade'.

The basis of these theories is attributed to Posner (1961) and Vernon (1966). Essentially, a product innovation occurs in an innovating country, and for a while, at least, that country has a monopoly in the product (in the model above no monopoly power results from technological change). While the monopoly exists other countries have to import the product. (This does not necessarily make the exporting country better off. If the new product replaced an old monopolized product it may be worse off. If it replaces a non-monopolized product it is more likely to yield increases in welfare. Even then, the relative resource intensity of the new product, as in the discussion above, will be important.)

The new product will be a net export. However, as time proceeds, other countries are likely to begin to imitate the new technology, and as they do so the innovator's monopoly power will be eroded away. As the imitation occurs, so net exports will decline. Moreover, as other countries acquire the new technology comparative advantage may operate against the innovator, with the goods eventually becoming a net import. Thus an expanding industry based on a new technology may turn into an importing industry as international diffusion proceeds. Moreover, as Vernon (1966) argues, and as we have argued in other chapters (e.g., Chapter 9), new technology tends to change over time, so that the input requirements to produce a new good may change over time. Thus in the early life of a new good production may require skilled labour, which makes international imitation difficult. However, further technological change may 'routinize' production, making imitation easier. In fact, one may feed in here all the discussions of diffusion in Part II. The overall story is that an innovation leads to exports through differences in knowledge and/or production capabilities, but over time the innovator's advantages are eroded and exports fall back. To extend further, the innovating country needs another innovation to keep up its exports. Krugman (1979) shows that, if we consider two countries having the same technology and factor endowments (i.e., labour) but one is an innovator of new products and the other an imitator, then if the innovator can keep up a flow of new products it can maintain a higher standard of living (wages) than the imitator, the size of the differential being dependent on the speed at which the imitator reacts. Any slow-down of innovation or speeding up of diffusion will narrow the differential and may even lead to a fall in the innovator's living standards.

To tell the complete story with these neotechnology theories of trade one needs an explanation of the international diffusion of technology. In Chapter 8 we considered this issue and specifically discussed the work of Teece (1977) on the transfer of technology. The data sample in that study largely covered multinational companies, and it seems a fair generalization to state that the transfer of technology, and through these neotechnology theories the international division of labour, will depend to a large extent on the behaviour of such companies. Moreover, multinational companies will by their size tend to have extensive market power in each of the markets in which they operate. In such a world of multinational companies the basic assumptions of Walrasian analysis are unlikely to hold. The neotechnology theories may be more applicable. But as we showed in Chapter 8, we still have only limited insight into the transfer of technologies by these companies, and thus we are not yet in a position to draw any firm conclusions as to how technology will affect the pattern of trade.

However, we can try to summarize up to this point. The pure theory of international trade, by comparing Walrasian equilibria, tells us that technological change is going to impact on the nature of the trading equilibrium. A process innovation at home with the rest of the world's technology remaining unchanged will shift the notional supply curve of the innovating industry outwards for given factor prices. This suggests some improvement in trade performance. The innovation will lead to changes in factor prices and goods prices (the terms of trade) dependent upon the nature of the technology embodied. These price changes may negate any welfare gains that the new technology directly generated. This approach ignored any market power that the innovator might obtain through technological advance. The neotechnology theories concentrate on this aspect. A product innovation generates improved trading performance until the competitive advantage is removed through emulation. We have argued however that in such theories the role of multinational companies is of paramount importance, and further work is needed on this. We have also argued that the existence of multinational companies throws considerable doubt on whether the standard model of trade is really applicable to the real world.

17.2.3 Empirics

The role of technological change in trade has been the subject of a certain amount of empirical work over the years.[2] Early studies by Posner (1961), Freeman (1963), and Hufbauer (1966) are good examples. The sort of results obtained can be illustrated by Gruber *et al.* (1967), who show that in the USA industries that are heavy R & D spenders are successful exporters. The basic underlying philosophy for these studies is that technological change increases 'competitiveness' and thus net exports. If this is the case, then the use of what is considered an 'input' measure of technological change (R & D) may not be an appropriate indicator of technological superiority to relate to export performance.

Pavitt and Soete (1980) and Soete (1981) have tried to overcome this problem by using patenting activity as a measure of technological superiority. We discussed in Chapter 2 the shortcomings of this indicator, some of which however have been overcome in these studies by using patenting activity in the USA, i.e., foreign (to the USA) patents issued. In Pavitt and Soete forty industrial sectors are considered and a regression equation is run on each sector across ten OECD countries of the form

$$\frac{X_{ij}}{h_i} = a_i + b_i \frac{N_{ij}}{h_i} \tag{17.18}$$

where i = industry (sector);
j = country;
X_{ij} = exports in industry i of country j;
N_{ij} = number of US patents in industry i issued to country j;
h = population.

A statistically significant estimate of b is generated in twenty-three of the forty sectors, suggesting a positive relationship between export success and innovative activity. In Soete (1981) a more sophisticated exercise is performed.

---

[2] Soete (1982) is a useful reference for this aspect of the literature.

A regression of the form

$$\log XSHA_{ij} = \beta_{0i} + \beta_{1i} \log NSHA_{ij} + \beta_{2i} \log KL_j + \beta_{3i} \log POP_j + \beta_{4i} DIST_j$$
$$(17.19)$$

is run where

$XSHA_{ij}$	=	share of country j's exports of industry i in total OECD exports of industry i for 1977;
$NSHA_{ij}$	=	share of country j's 1963–77 US patents in industry i in total OECD patents in industry i;
KL_j	=	gross fixed capital formation divided by total employment for each country j for 1977;
POP_j	=	population of country j in 1977;
$DIST_j$	=	a proxy measure of physical distance from 'world centre' for country j.

The equation is estimated for forty industrial sectors again across the ten OECD countries. Significant F statistics were obtained for all but three sectors with β_{1i} significantly different from zero in twenty-eight sectors. The elasticities of export shares relative to patent shares differed considerably across industries, being positive in all but two of the industries with significant estimates for β_{1i}. Values varied from 1.26 for aircraft to 0.26 for non-ferrous metal products.

These results suggest that technological differences as reflected in patents are significant influences on trade flows. The work, as Soete states, ignores technology imports and is somewhat static, but the indications are there. One might also consider that the work does not really adequately separate out price effects from technology effects, nor does it consider those other factors that have been hypothesized to influence trade. It would also be somewhat illuminating to separate out product and process innovations, for one might feel that the failure to use new processes could be overcome by lower factor prices or lower exchange rates whereas a failure to product-innovate may have much greater effect on a country's trade flows. Moreover, product innovation is much more in tune with the original Vernon and Posner hypotheses which Soete claims to be testing. Finally, we might note that these tests are essentially testing partial-equilibrium hypotheses. In the general-equilibrium analysis we might expect to find gains in export shares having knock-on effects that remove competitive advantage.

17.3 The theory of the balance of payments

Thus far we have ignored any way in which technological change may impact upon the trading balance of the economy, and how the economy may react to a trading imbalance. The theory of the balance of payments considers such issues. Moreover, it introduces money into the model and its domain is that of macro rather than microeconomics.

One of the obvious differences between trade within and across national boundaries is that the latter involves different currencies. The relative price of two currencies is the exchange rate. One may construct for any one country a demand and supply schedule for foreign currency that will determine the equilibrium exchange rate (that rate equating demand and supply) on the basis that a lower exchange rate stimulates exports and reduces imports whereas a higher exchange rate reduces exports and stimulates imports. However, whether the economy should at any time have this equilibrium exchange rate will depend, *inter alia*, on whether it is operating under a fixed or floating exchange rate regime.

Trading equilibrium requires that the demand and supply of foreign currency must be equal. The demand and supply functions will have exchange rates, prices, and the levels of economic activity as arguments. If a disequilibrium exists, then adjustment must involve changes in at least one of these three arguments. There exist in the literature a number of different approaches to the adjustment question. It is most appropriate to start with the monetary approach.

The monetary approach to the balance of payments is essentially a Walrasian equilibrium theory of international trade involving financial assets. Basically, it is an extension of the standard model of trade with claims on future production held in the form of financial assets. It assumes perfectly flexible prices of goods and factors and therefore instantaneous attainment of equilibria on all these markets. It cannot therefore, for example, show unemployment. The monetary theory essentially argues (for more detail see Sodersten, 1980) that, under fixed exchange rates, a balance of payments disequilibrium will lead to changes in domestic money stocks, and thus prices, that will feed back on to the balance of payments until there is balance, and money stocks no longer change. Thus, any surplus generated through technological advances will be eroded away by changes in money stocks until balance is re-established. The new equilibrium will involve a different price vector. If one has flexible rates of exchange, then the exchange rate will change to preserve balance in each period and there will be no consequent changes in money supplies. Thus an improvement in technology will no longer generate a trade surplus but will generate a revaluation with the exchange rate increasing. Any improvements in trading performance will be reflected not in surpluses but in the exchange rate. Whether fixed or flexible rates are in operation can be particularly relevant as to what industries will be affected as the adjustment to a new equilibrium takes place. If rates are fixed, then the domestic surplus yields an increased money stock and increased demands in the economy for all (non-inferior) goods, this driving up domestic prices to clear the foreign balance. If the exchange rate rises, then those industries with high price elasticities of demand may be most affected. In this case the argument is similar to that used by Forsyth and Kay (1980) to discuss the impact of North Sea oil. The discovery of oil leads to a revaluation of sterling, manufacturing can no longer compete at the high exchange rate, and thus jobs are lost in manufacturing. Being a Walrasian theory, this approach cannot really predict unemployment — prices adjust to remove it. But this argument does suggest that whether exchange rates are fixed or flexible will influence the nature of the impact on the economy of a technological change.

We have argued in Chapter 12 that the set of conditions required to justify Walrasian analysis are not really acceptable. We thus move to consider non-Walrasian modes of analysis by looking at the possibility of technological unemployment in an open economy.

17.4 Technological unemployment in an open economy

Given the equilibrium nature of the monetary theory of the balance of payments, we do not find in such models that new technology affects employment. To pursue this line of enquiry we must move away from such general equilibrium models.

Neary (1981) analyses the impact of technical change (represented by an increase in productivity) in a temporary equilibrium model of the open economy. (Dixit and Norman, 1980, also construct such a temporary equilibrium model but do not investigate technical change *per se*.) Neary shows that in a temporary equilibrium, when both output and labour markets are characterized by excess supply, technological progress must reduce employment.

Flatters (1979) considers a model of a small open economy in which full Walrasian equilibrating forces are prevented from working through a fixed real-wage assumption. With traded and non-traded goods sectors, perfect competition, and constant returns to scale, he shows that productivity growth might decrease employment if it has an overall labour-saving bias or if it occurs in the non-traded goods sector.

To amplify the argument further, we will look at a development of the Keynesian model discussed in Chapter 12. Although this model does not include all the general-equilibrium links inherent in the monetary approach to the balance of payments discussed above, its predictions are informative. In the three-sector diagram, Fig. 17.1, we have three markets represented:

1 In part C of the diagram we have the home labour market with notional demand and supply curves DD' and SS' respectively, both labour supply and demand being considered functions of the real wage (w/p).
2 In part B we represent the home production function relating domestic output Y to labour input L. Only one good is produced which may be traded or consumed or invested at home.
3 In part A we have
   (a) a domestic aggregate supply curve HH' which is derived from DD' and the production function, showing the notional supply of goods of domestic producers for any domestic price level p, assuming a given money wage w;
   (b) a curve WW', at price P_w/e (world price divided by the exchange rate), indicating that the rest of the world is willing to meet any demands for goods at this price; and

(c) an aggregate demand curve, showing the demands for goods (including exports) at any price p. Implicit within this curve is the assumption of money market equilibrium.

Let us assume that

consumption	C	$= C(D) \quad C_D > 0$
investment	I	$= I(r)$
exports	X	$= X(P_w/pe)$
total demand	D	$= C + I + X$
money demand	M^D	$= L\,(pD, r)$
money supply	M^S	$= \bar{M}$
money market equilibrium	M^S	$= M^D$

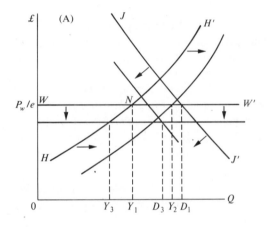

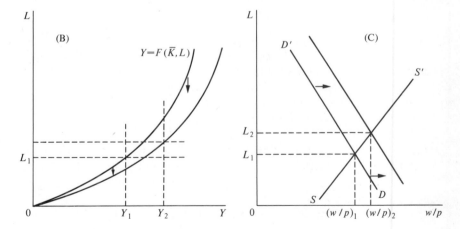

Fig. 17.1

From this set of equations we may solve for D as a function of p, P_W, e, and $\bar{M}$, which is the curve plotted as JJ'. If the rest of the world offers a perfectly elastic supply of goods at price P_W/e, then price at home cannot rise above this level. Thus the aggregate supply curve of goods to the home economy (including domestic production and imports) will be HNW'. Equilibrium of aggregate demand and supply occurs at the intersection of JJ' and HNW', yielding total demand in the economy, in Fig. 17.1(A), represented as D_1. Assume that, if demand is greater than supply, domestic producers will supply and sell all they wish at the current price, and then domestic production will be Y_1 and imports will be $D_1 - Y_1$. At output Y_1 domestic employment is L_1, which we have allowed to be full employment.

We now undertake two comparative statics exercises:

1   we allow technological change at home (but not overseas) to shift the production function, keeping P_W/e constant; and
2   we then allow technological change overseas (with no change at home) to reduce P_W/e.

In the first case we allow the change to shift the production function downwards and the demand for labour curve outwards, also implying a shift of HH' to the right.

Assuming that the new curve HH' does not cross JJ' to the right of D_1, price will remain at the world price P_W/e. At P_W/e domestic production must now be higher, domestic production replacing imports. Whether this leads to increased or reduced employment will depend on the elasticity of demand, the degree of scale economies in production, and the bias in the technical change (as detailed in Chapter 11). It is possible that at the new output level Y_2 there will be unemployment. If no further changes occur, the technological change will have led to increased domestic production, an improved balance of payments, and uncertain effects on employment and the real wage. Now, in reaction to the new disequilibrium position,

1   if there is unemployment then, given that labour is paid its marginal product, the real wage will be greater than $(w/p)_2$ and the money wage if flexible will fall, shifting HH' further to the right until the labour market clears at $(w/p)_2$ and L_2; if the money wage is not flexible this does not happen;
2   if the improved balance of payments leads to an increased money stock, then JJ' will shift to the right until the balance of payments balances.

With such flexibility, the new equilibrium will involve higher domestic output, higher real wages, and higher employment. (If there is not wage flexibility there may be unemployment resulting from the technological change.)

Let us now consider an innovation overseas, not imitated at home, implying a fall in P_W/e. We make comparisons with the initial full-employment equilibrium position. The fall in P_W/e will lead to a lower WW' curve and a shift of JJ' to the left (through reductions in export demand), yielding, say, home production

Y_3 and demand D_3. (Depending on the price elasticity of exports and home demand, D_3 may be greater or less than D_1.) The initial impact however is that home production, Y_1, will fall, generating lower employment. At this lower employment level the marginal product of labour and thus the real wage must be higher (given price is lower, this may or may not imply changes in the money wage, shifting HH'). Whether the unemployment is removed will depend upon whether wages respond to the resulting unemployment. Wage reductions will shift HH' to the right, leading to increases in domestic production and employment. If wages are perfectly flexible, full employment can be re-established at Y_1 with the same real wage but lower wages and prices. In the absence of wage change the technological change overseas will create lower output and unemployment at home. Of course, one could look for relief of this situation through exchange rate changes, but the moral is clear: flexibility in prices is necessary for full employment.

In Chapter 12 we have expanded on these arguments and criticized the model for a closed economy so we do not really need to do that here. There are two points worth considering here, however. The first is whether product as opposed to process innovation affects our results. If product innovation occurs overseas and the innovating product replaces, directly or indirectly, home-produced products, then it is possible for unemployment to be a likely outcome. For example, if the introduction of video recorders, produced overseas, leads to expenditure being switched out of domestic products, in the absence of price changes, such a technological change may reduce labour demand. Only if reductions in domestic wages result and lead to import substitution for other goods can full employment be re-established.

The second point is that in Chapter 12 we emphasized how technology multiplier effects (and other compensation effects) may offset direct reductions in employment from new technology. In an open economy these effects may leak overseas. If this is the case, then compensation effects discussed in that chapter may be even weaker than suggested — although, of course, in a full general equilibrium framework with all prices flexible and feedback from the rest of the world, full employment could always be re-established.

Given the theory, do we have any empirical evidence that is of relevance? The role of foreign trade in determining the impact of microelectronics on employment has been much stressed. In a number of studies (see Stoneman *et al.*, 1982) it is argued that a failure to innovate where other countries do will drastically affect employment opportunities, but innovation faster than others will act effectively to compensate for any direct job losses. If one looks at Whitley and Wilson (1982), for example (see Chapter 12), a major compensating element arises through increased international competitive advantage. However, much of this work depends upon 'belief' rather than strict empirical testing, of which we have no examples.

17.5 Conclusions

In this chapter we have considered (1) the pure theory of trade and (2) the macroeconomics of an open economy. We have argued that technological differences between countries are one of the factors influencing trade flows, and that changes in technology may have transitional and permanent effects on trade flows. We argued that exporting success can be related to technological progressiveness, but in an open economy technological advance does not necessarily lead to welfare gains.

At the macroeconomic level we investigated Walrasian and non-Walrasian models of the open economy, arguing that the economy will, at least temporarily, yield higher output and lower imports from technological advance at home, and lower output and higher imports from advance overseas, but that the impact of technological change on the level and pattern of employment will depend on the degree of wage flexibility in the economy and the flexibility of exchange rates. Even if employment is not affected, prices and thus the distribution of income will be.

In this analysis we again see the conflict, which has been apparent elsewhere, between Walrasian or equilibrium modes of analysis and non-Walrasian approaches. The Walrasian view is one involving perfect markets that function efficiently. There seem to be good reasons for considering that the real world is not like that, in which case one takes a much less optimistic view of how the economy functions. In a world with multinational companies and generally imperfect markets, Walrasian predictions may be irrelevant. However, the analysis of such imperfect worlds still has a number of stages to pass through before it can match the sophistication of the Walrasian analysis. A major advance in this area would involve greater understanding of the role of multinational companies in the transfer of technological knowledge, and given that, we may then have greater insight into the international division of labour and the impact of new technology on this division.

References

Chacholiades, M. (1978), *International Trade Theory and Policy*, McGraw-Hill, New York.
Dixit, A. and Norman, V. (1980), *Theory of International Trade*, Cambridge University Press/Nisbet, Welwyn Garden City.
Flatters, F. (1979), 'Will Improvements in Productivity Necessarily Increase Employment in a Small Open Economy? Queens University, Canada, mimeo.
Forsyth, P. and Kay, J. (1980), 'The Economic Implications of North Sea Oil Revenues', *Fiscal Studies*, 1, 1–28.
Freeman, C. (1963), 'The Plastics Industry: A Comparative Study of Research and Innovation', *National Institute Economic Review*, 26, 22–62.
Gruber, W.H. *et al.* (1967), 'The R & D Factor in International Trade and International Investment of United States Industries', *Journal of Political Economy*, 75, 20–37.
Hufbauer, G.C. (1966), *Synthetic Materials and the Theory of International*

Trade, Harvard University Press, Cambridge, Mass.

Kemp, M.C. (1969), *The Pure Theory of International Trade*, Prentice-Hall, Englewood Cliffs, NJ.

Krugman, P. (1979), 'A Model of Innovation, Technology Transfer, and the World Distribution of Income', *Journal of Political Economy*, 87, 253–66.

Neary, J.P. (1981), 'On the Short-run effects of Technological Progress', *Oxford Economic Papers*, 33, 224–33.

Pavitt, K. and Soete, L. (1980), 'Innovative Activities and Export Shares: Some Comparisons between Industries and Countries', in K. Pavitt (ed.), *Technical Innovation and British Economic Performance*, Macmillan, London.

Posner, M. (1961), 'International Trade and Technical Change', *Oxford Economic Papers*, 13, 323–41.

Sodersten, B. (1980), *International Economics*, (2nd ed.), Macmillan, London.

Soete, L. (1981), 'A General Test of Technological Gap Trade Theory', *Weltwirtschaftliches Archiv*, 117, 638–66.

Soete, L. (1982), 'Innovation and International Trade: What We Know and What We Do Not Know', paper prepared for the Symposium on Technical Change, Technical Change Centre, London.

Stoneman, P. *et al.* (1982), 'Information Technology, Productivity and Employment', in *Micro-electronics, Robotics and Jobs*, ICCP, no. 7, OECD, Paris.

Teece, D.J. (1977), 'Technology Transfer by Multinational Firms: The Resource Cost of Transferring Technological Know-how', *Economic Journal*, 87, 242–61.

Vernon, R. (1966), 'International Investment and International Trade in the Product Cycle', *Quarterly Journal of Economics*, 80, 190–207.

Whitley, J. and Wilson, R. (1982), 'Quantifying the Employment Effects of Micro-Electronics', *Futures*, 14, 486–96.

Chapter 18

By Way of a Conclusion

It is tempting, when writing a conclusion to a book of this length and scope, to step sideways from the material presented and to start deriving some policy implications based on the material. I have resisted this temptation, first because the whole question of technology policy could fill a volume on its own, second because the material has not been ordered to facilitate such a discussion, and third because there does seem to be a more relevant matter to discuss. The presentation of the material in this volume has been categorized into separate parts and chapters, and to some degree this has the disadvantage that at times the overall picture is clouded — one loses sight of the wood for the trees. It seemed relevant, therefore, to use this chapter to provide the outline of the technological change process that may be missing from the detailed presentation above.

We start by assuming that new technological opportunities are coming forth at each and every moment in time. These opportunities may arise from changes in basic knowledge or changes in the economic environment. They may be exogenous or endogenous to the economic system, but they represent the starting point of our argument. As these opportunities arise and become known, some will be selected and used for the generation of new products and processes. Those selected will be either more widely known about and/or more relevant, and/or will be expected to be more profitable than those not selected. As the selected opportunities are pursued and yield additions to the 'state of the art', it becomes possible for these advances to be patented. The patent is there to grant the 'inventor' some property rights in his invention in order that he may generate some return from that invention and thus encourage further advances. However, the property rights may not be completely inviolable, the advance may be copied, and the patent system may not be necessary to guarantee a return to the inventor. Those advances that seem to represent the greatest potential and are most cheaply copied will be those most imitated.

From the large number of additions to the state of the art a further selection process will yield a series of innovations of new products and processes. From the numerous sources of advances the commercial sphere will pick up potential innovations, and by expenditure on R & D will develop marketable products. We suggest that a key element in the selection process will again be expected profitability.

The R & D process is a competitive activity, with firms competing with each other to place technologically new products on the market or to gain market share by developing new cost-reducing processes. This competitive process will have winners and losers; some firms will introduce new products and processes before others. New firms will enter on the basis of new technology and other firms will exit as the competitive struggle leaves them behind. The need to succeed in order to survive will lead technology into directions where the chances of success are greatest. For given costs of development, the new processes developed are likely to be those yielding greatest cost reductions, and the new products are likely to be those with the greatest potential market. In a capitalist system this leads one to argue that the bias in technological change will tend towards saving a factor more when it has a high share in costs than when it has a low share. It also leads one to argue that new products are more likely to be introduced where the market is large and potential competition is small.

As the developers of new processes introduce these into their production activities, so will their costs and prices change, affecting market shares and yielding increases in producer and consumer surplus. Moreover, as they introduce these processes their input demands may also change.

As innovation proceeds over time a large number of new products will be appearing on the market. There is now a further selection process, whereby some of the advances are widely accepted and others are not accepted or have only limited acceptance. Some of the new advances will be opening up completely new markets; others will be displacing older products or even other recent advances.

Let us discuss first of all those new products that represent other industries' new processes. Users of new processes will tend to select those processes that have the highest expected rates of return and about which they are informed. The rate of return will be dependent on expected revenue changes and expected cost changes, which will depend on the prices being charged for the new technology. As new processes need to be manufactured, there may well develop new industries producing new capital goods, all developing, changing, and maturing over time. Dependent on entry, further technological advances, scale economies, and the degree of protection provided by the patent system, the new industry may develop in more or less concentrated ways. Its degree of concentration relative to that of the potential users of its products will influence the price of its products and thus the growth of demand for those products. As the supplying industry grows in response to the growth of demand for its products, so will investment take place and its employment level expand.

Given the price of the new technology, the competitive game, information flows, further technological advances and similar factors will impinge on the speed of take-up of the new technology. As the new technology is being taken up the using firms may well be changing in size; the competitive structure of the industry may change; its output and prices will start to change in response to the new technical conditions, and its labour and capital requirements may begin to change. Moreover, as more firms take up the new technology so the macro-environment will start to change.

As the new technology is used more widely, so the average cost of production should fall, and the potential arises for increases in consumer welfare. At the same time, the use of new technology can yield profit to its users. As these profits arise, so the attractiveness of this new technology is illustrated and thus the desirability of imitation. However, as the new processes are being used so the old are being replaced. This may for a while stimulate their further development, but will in general imply the decline of old industries. As new firms arise and old firms die, as new industries rise and fall, so the demand for labour will be subject to a continual recomposition in terms of level, skill and geographic location. As these impacts occur, so technological unemployment may arise.

Consider now the case of those new products that are being sold to consumers. Again, rates of take-up will depend upon information, price levels, further technological changes, etc. The new products may lead to the creation of new supplying industries which will develop over time. As the development proceeds prices will change and ownership may extend. If the new product is preferred to the old, or is an addition to choice, welfare will increase. Moreover, as the supplying industry expands so will employment and capital demands, although maybe at the expense of an old industry. Also, if the new product is exported improvements in trading performance may result, although such improvements may be eroded as development proceeds overseas.

As the new technologies are being diffused and prices, incomes, employment, trade, etc., are being affected, further new technologies are coming forward. Those that are developed may be conditional upon previous technological advances. Those that are accepted and used may be dependent on the price regime, which in turn is conditional on previous technological changes. The system is becoming a closed circle. However, as the technological change system proceeds so the economy develops. Its output, employment, investment, income distribution, market structure, and welfare levels are all being affected by, and in turn are affecting, the development of new technology. In other economies similar processes are occurring with overlaps and informational exchanges between economies. The international trading market provides a meeting-place whereby the technologies of different economies compete. Again, some will be losers and some winners. If the same countries always win then they are likely to prosper at the expense of others. If some countries always lose they are unlikely to prosper.

The nature of the development of the economy that is likely to result from this complicated and involved process of technological change is not necessarily going to be ideal. Reductions in prices and increases in consumer welfare may result, but these may only be generated at a cost. New technology may de-skill labour and routinize the labour process. It may generate environmental pollution. It may dehumanize the work process or create undesirable social implications. But nobody pretends that, in a world where decisions are made on the basis of private costs and benefits, that world will normally behave in a socially optimal way. To achieve social optimality requires decisions to be made with respect to social costs and benefits. But this takes us back to where we came in and refused to be tempted. The social control of technology, or technology policy more widely, must wait until another occasion.

Author Index

Subject Index

3 1994 01554 7711

"*Troopers* finds SF's first grand master in excellent command of the language, mixing technical terms, 'modern' slang, and plain old words in the unstylish style that is his hallmark. It is a cross between eloquent and workmanlike, and it does not just get the job done, it propels the story. Heinlein brings worlds alive with his prose, whether they are boot camps or bug holes." —*Science Fiction Weekly*

"The single most influential book that I have ever read. A lot of the ideas and values that Heinlein offers in *Starship Troopers* have had a profound effect on me and the way that I have molded my life . . . [It] is a story about social ills we are faced with today. Unlike many authors, Heinlein offers ideas and options as to possible reforms." —*SF Site*

"What makes *Starship Troopers* such an important book is in its pioneering approach to dramatizing military themes in an SF context. Unless I miss my guess, *Troopers* was the first SF novel in which military life was depicted in a manner believable to readers who had actually served.

"As a fast-paced piece of action storytelling, *Starship Troopers* mostly races along . . . The humanity Heinlein bestowed upon characters, the gritty realism of their conflicts, in what had largely been *Flash Gordon* territory up to that point, was a significant step in science fiction maturation.

"Love 'em or hate 'em, the novel's 'controversial' politics [are] another feather in its cap. A novel in a genre [once] dedicated to escapist juvenilia challenges adult readers to question their assumptions and consider such ideas as duty, altruism, and patriotism under the harsh light of scrutiny . . . [*Troopers*] is doing you an intellectual favor." —SF Reviews.net

P9-ELG-519

continued . . .

"A serious moral tract about the obligations of citizenship and the nobility of the individual willing to sacrifice himself for the greater good . . . *Troopers* has to be seen as partly a celebration of victory in World War II with its unsung citizen-heroes, partly a reflection of the Cold War and its attendant anxiety, and partly a reaction to growing popular discontent which originated with the inconclusive Korean War and culminated in the anti-war movement of the 1960s."
        —*SF Crowsnest*

"[An] incredible classic of science fiction . . . Heinlein grabs the attention of the reader from the very beginning . . . with 'I always get the shakes before a drop.' That simple line illustrates the beauty of this work; it's not just about action, though there is certainly plenty of that. Instead it's about what goes through the mind of a trooper.

"Heinlein not only combines futuristic action with psychological insight here, but also manages to throw in some social commentary as well. Whether or not the reader would agree with Heinlein's ideas, the concepts are still intriguing . . . It brilliantly blends action and intellect to provide an entertaining, thought-provoking experience for readers of all ages. It's one of my personal favorite books, and I highly recommend it to everyone. [It] gets better every time one reads it, for one is always discovering some new idea hidden within the pages. I have never even remotely tired of reading it, and I'm sure I never will."
        —*The 11th Hour*

Books by Robert A. Heinlein

Starship Troopers

Robert A. Heinlein

ACE BOOKS, NEW YORK

THE BERKLEY PUBLISHING GROUP
Published by the Penguin Group
Penguin Group (USA) LLC
375 Hudson Street, New York, New York 10014

USA • Canada • UK • Ireland • Australia • New Zealand • India • South Africa • China

penguin.com

A Penguin Random House Company

STARSHIP TROOPERS

An Ace Book / published by arrangement with The Robert A. & Virginia Heinlein Prize Trust

Copyright © 1959 by The Robert A. & Virginia Heinlein Prize Trust.
Penguin supports copyright. Copyright fuels creativity, encourages diverse voices,
promotes free speech, and creates a vibrant culture. Thank you for buying an authorized
edition of this book and for complying with copyright laws by not reproducing, scanning,
or distributing any part of it in any form without permission. You are supporting writers
and allowing Penguin to continue to publish books for every reader.

Ace Books are published by The Berkley Publishing Group.
ACE and the "A" design are trademarks of Penguin Group (USA) LLC.

For information, address: The Berkley Publishing Group,
a division of Penguin Group (USA) LLC,
375 Hudson Street, New York, New York 10014.

ISBN: 978-0-441-78358-8

PUBLISHING HISTORY
G. P. Putnam's Sons edition / 1959
Berkley edition / May 1968
Ace mass-market edition / May 1987
Ace trade paperback edition / July 2006
Ace premium edition / May 2010

PRINTED IN THE UNITED STATES OF AMERICA

72   71   70   69   68   67

Cover art by Steve Stone.
Cover design by Annette Fiore DeFex.

A much abridged version of this book was published in *Fantasy & Science Fiction*
magazine under the title "Starship Soldier."

This is a work of fiction. Names, characters, places, and incidents either are the product
of the author's imagination or are used fictitiously, and any resemblance to actual persons,
living or dead, business establishments, events, or locales is entirely coincidental.

If you purchased this book without a cover, you should be aware that this book is
stolen property. It was reported as "unsold and destroyed" to the publisher, and neither
the author nor the publisher has received any payment for this "stripped book."

Acknowledgments

The stanza from "The 'Eathen" by Rudyard Kipling, at the head of Chapter VII, is used by permission of Mr. Kipling's estate. Quotations from the lyrics of the ballad "Rodger Young" are used by permission of the author, Frank Loesser.

TO "SARGE" ARTHUR GEORGE SMITH—
SOLDIER, CITIZEN, SCIENTIST—AND TO ALL
SERGEANTS ANYWHEN WHO HAVE LABORED
TO MAKE MEN OUT OF BOYS.

R.A.H.

/

CH:01

Come on, you apes! You wanta live forever?

—Unknown platoon sergeant, 1918

I always get the shakes before a drop. I've had the injections, of course, and hypnotic preparation, and it stands to reason that I can't really be afraid. The ship's psychiatrist has checked my brain waves and asked me silly questions while I was asleep and he tells me that it isn't fear, it isn't anything important—it's just like the trembling of an eager race horse in the starting gate.

I couldn't say about that; I've never been a race horse. But the fact is: I'm scared silly, every time.

At D-minus-thirty, after we had mustered in the drop room of the *Rodger Young*, our platoon leader inspected us. He wasn't our regular platoon leader, because Lieutenant Rasczak had bought it on our last drop; he was really the platoon sergeant, Career Ship's Sergeant Jelal. Jelly was a Finno-Turk from Iskander around Proxima—a swarthy little man who looked like a clerk, but I've seen him tackle two berserk privates so big he had to reach up to grab them, crack their heads together like coconuts, step back out of the way while they fell.

Off duty he wasn't bad—for a sergeant. You could even call him "Jelly" to his face. Not recruits, of course, but anybody who had made at least one combat drop.

But right now he was on duty. We had all each inspected our combat equipment (look, it's your own neck—see?), the acting platoon sergeant had gone over us carefully after he mustered us, and now Jelly went over us again, his face mean, his eyes missing nothing. He stopped by the man in front of me, pressed the button on his belt that gave readings on his physicals. "Fall out!"

"But, Sarge, it's just a cold. The Surgeon said—"

Jelly interrupted. "'But Sarge!'" he snapped. "The Surgeon ain't making no drop—and neither are you, with a degree and a half of fever. You think I got time to chat with you, just before a drop? *Fall out!*"

Jenkins left us, looking sad and mad—and I felt bad, too. Because of the Lieutenant buying it, last drop, and people moving up, I was assistant section leader, second section, this drop, and now I was going to have a hole in my section and no way to fill it. That's not good; it means a man can run into something sticky, call for help and have nobody to help him.

Jelly didn't downcheck anybody else. Presently he stepped out in front of us, looked us over and shook his head sadly. "What a gang of apes!" he growled. "Maybe if you'd all buy it this drop, they could start over and build the kind of outfit the Lieutenant expected you to be. But probably not—with the sort of recruits we get these days." He suddenly straightened up, shouted, "I just want to remind you apes that each and every one of you has cost the gov'ment, counting weapons, armor, ammo, instrumentation, and training, everything, including the way you overeat—has cost, on the hoof, better'n half a million. Add in the thirty cents you are actually worth and that runs to quite a sum." He glared at us. "So bring it back! We can spare you, but we can't spare that fancy suit you're wearing. I don't want any heroes in this outfit; the Lieutenant wouldn't like it. You got a job to do, you go down, you do it, you keep your ears open for recall, you show up for retrieval on the bounce and by the numbers. Get me?"

He glared again. "You're supposed to know the plan. But some of you ain't got any minds to hypnotize so I'll sketch it out. You'll be dropped in two skirmish lines, calculated two-thousand-yard intervals. Get your bearing on me as soon as you hit, get your bearing and distance on your squad mates, both sides, while you take cover. You've wasted ten seconds already, so you smash-and-destroy whatever's at hand until the flankers hit dirt." (He was talking about me—as assistant section leader I was going to be left flanker, with nobody at my elbow. I began to tremble.)

"Once they hit—straighten out those lines!—equalize those intervals! Drop what you're doing and do it! Twelve seconds. Then advance by leapfrog, odd and even, assistant section leaders minding the count and guiding the envelopment." He looked at me. "If you've done this properly—which I doubt—the flanks will make contact as recall sounds . . . at which time, home you go. Any questions?"

There weren't any; there never were. He went on, "One more word—This is just a raid, not a battle. It's a demonstration of firepower and frightfulness. Our mission is to let the enemy know that we could have destroyed their city—but didn't—but that they aren't safe even though we refrain from total bombing. You'll take no prisoners. You'll kill only when you can't help it. But the entire area we hit is to be smashed. I don't want to see any of you loafers back aboard here with unexpended bombs. Get me?" He glanced at the time. "Rasczak's Roughnecks have got a reputation to uphold. The Lieutenant told me before he bought it to tell *you* that he will always have his eye on you every minute . . . and that he expects your names to *shine*!"

Jelly glanced over at Sergeant Migliaccio, first section leader. "Five minutes for the Padre," he stated. Some of the boys dropped out of ranks, went over and knelt in front of Migliaccio, and not necessarily those of his creed, either—Moslems, Christians, Gnostics, Jews, whoever wanted a word with him before a drop, he was there. I've heard tell that there used to be military outfits whose chaplains did not fight alongside the others, but I've never been able to see how that could work. I mean, how can a chaplain bless anything he's not willing to do himself? In any case, in the Mobile Infantry, *everybody* drops and *everybody* fights—chaplain and cook and the Old Man's writer. Once we went down the tube there wouldn't be a Roughneck left aboard—except Jenkins, of course, and that not his fault.

I didn't go over. I was always afraid somebody would see me shake if I did, and, anyhow, the Padre could bless me just as handily from where he was. But he came over to me as the last stragglers stood up and pressed his helmet against mine to speak privately. "Johnnie," he said quietly, "this is your first drop as a non-com."

"Yeah." I wasn't really a non-com, any more than Jelly was really an officer.

"Just this, Johnnie. Don't buy a farm. You know your job; do it. Just do it. Don't try to win a medal."

"Uh, thanks, Padre. I shan't."

He added something gently in a language I don't know, patted me on the shoulder, and hurried back to his section. Jelly called out, "Tenn . . . *shut*!" and we all snapped to.

"Pla*toon*!"

"Sec*tion*!" Migliaccio and Johnson echoed.

"By sections—port and starboard—prepare for drop!"

"Sec*tion*! Man your capsules! *Move!*"

"Squad!"—I had to wait while squads four and five manned their capsules and moved on down the firing tube before my capsule showed up on the port track and I could climb into it. I wondered if those old-timers got the shakes as they climbed into the Trojan Horse? Or was it just me? Jelly checked each man as he was sealed in and he sealed me in himself. As he did so, he leaned toward me and said, "Don't goof off, Johnnie. This is just like a drill."

The top closed on me and I was alone. "Just like a drill," he says! I began to shake uncontrollably.

Then, in my earphones, I heard Jelly from the centerline tube: "Bridge! Rasczak's Roughnecks . . . ready for drop!"

"Seventeen seconds, Lieutenant!" I heard the ship captain's cheerful contralto replying—and resented her calling Jelly "Lieutenant." To be sure, our lieutenant was dead and maybe Jelly would get his commission . . . but we were still "Rasczak's Roughnecks."

She added, "Good luck, boys!"

"Thanks, Captain."

"Brace yourselves! Five seconds."

I was strapped all over—belly, forehead, shins. But I shook worse than ever.

It's better after you unload. Until you do, you sit there in total darkness, wrapped like a mummy against the acceleration, barely able to breathe—and knowing that there is just nitrogen around you in the capsule even if you could get your helmet open, which you can't—and knowing that the capsule is surrounded by the firing tube anyhow and if the ship gets hit before they fire you, you haven't got a prayer, you'll just die there, unable to move, helpless. It's that endless wait in the dark that causes the shakes—thinking that they've forgotten you . . . the ship has been hulled and stayed in orbit, dead, and soon you'll buy it, too, unable to move, choking. Or it's a crash orbit and you'll buy it that way, if you don't roast on the way down.

Then the ship's braking program hit us and I stopped shaking. Eight gees, I would say, or maybe ten. When a female pilot handles a ship there is nothing comfortable about it; you're going to have bruises every place you're strapped. Yes, yes, I know they make better pilots than men do; their reactions are faster, and they can tolerate more gee. They can get in faster, get out faster, and thereby improve everybody's chances, yours as well as theirs. But that still doesn't make it fun to be slammed against your spine at ten times your proper weight.

But I must admit that Captain Deladrier knows her trade. There was no fiddling around once the *Rodger Young* stopped braking. At once I heard her snap, "Centerline tube . . . *fire!*" and there were two recoil bumps as Jelly and his acting platoon sergeant unloaded—and immediately: "Port and starboard tubes—*automatic fire!*" and the rest of us started to unload.

Bump! and your capsule jerks ahead one place—*bump!* and it jerks again, precisely like cartridges feeding into the chamber of an old-style automatic weapon. Well, that's just what we were . . . only the barrels of the gun were twin launching tubes built into a spaceship troop carrier and each cartridge was a capsule big enough (just barely) to hold an infantryman with all field equipment.

Bump!—I was used to number three spot, out early;

now I was Tail-End Charlie, last out after three squads. It makes a tedious wait, even with a capsule being fired every second; I tried to count the bumps—*bump!* (twelve) *bump!* (thirteen) *bump!* (fourteen—with an odd sound to it, the empty one Jenkins should have been in) *bump!*—

And *clang!*—it's my turn as my capsule slams into the firing chamber—then WHAMBO! the explosion hits with a force that makes the Captain's braking maneuver feel like a love tap.

Then suddenly nothing.

Nothing at all. No sound, no pressure, no weight. Floating in darkness . . . free fall, maybe thirty miles up, above the effective atmosphere, falling weightlessly toward the surface of a planet you've never seen. But I'm not shaking now; it's the wait beforehand that wears. Once you unload, you can't get hurt—because if anything goes wrong it will happen so fast that you'll buy it without noticing that you're dead, hardly.

Almost at once I felt the capsule twist and sway, then steady down so that my weight was on my back . . . weight that built up quickly until I was at my full weight (0.87 gee, we had been told) for that planet as the capsule reached terminal velocity for the thin upper atmosphere. A pilot who is a real artist (and the Captain was) will approach and brake so that your launching speed as you shoot out of the tube places you just dead in space relative to the rotational speed of the planet at that latitude. The loaded capsules are heavy; they punch through the high, thin winds of the upper atmosphere without being blown too far out of position—but just the same a platoon is bound to disperse on the way down, lose some of the perfect formation in which it unloads. A sloppy pilot can make this still worse, scatter a strike group over so much terrain that it can't make rendezvous for retrieval, much less carry out its mission. An infantryman can fight only if somebody else delivers him to his zone; in a way I suppose pilots are just as essential as we are.

I could tell from the gentle way my capsule entered the atmosphere that the Captain had laid us down with as near zero lateral vector as you could ask for. I felt happy—not only a tight formation when we hit and no time wasted, but also a pilot who puts you down properly is a pilot who is smart and precise on retrieval.

The outer shell burned away and sloughed off—unevenly, for I tumbled. Then the rest of it went and I straightened out. The turbulence brakes of the second shell bit in and the ride got rough . . . and still rougher as they burned off one at a time and the second shell began to go to pieces. One of the things that helps a capsule trooper to live long enough to draw a pension is that the skins peeling off his capsule not only slow him down, they also fill the sky over the target area with so much junk that radar picks up reflections from dozens of targets for each man in the drop, any one of which could be a man, or a bomb, or anything. It's enough to give a ballistic computer nervous breakdowns—and does.

To add to the fun your ship lays a series of dummy eggs in the seconds immediately following your drop, dummies that will fall faster because they don't slough. They get under you, explode, throw out "window," even operate as transponders, rocket sideways, and do other things to add to the confusion of your reception committee on the ground.

In the meantime your ship is locked firmly on the directional beacon of your platoon leader, ignoring the radar "noise" it has created and following you in, computing your impact for future use.

When the second shell was gone, the third shell automatically opened my first ribbon chute. It didn't last long but it wasn't expected to; one good, hard jerk at several gee and it went its way and I went mine. The second chute lasted a little bit longer and the third chute lasted quite a while; it began to be rather too warm inside the capsule and I started thinking about landing.

The third shell peeled off when its last chute was gone and now I had nothing around me but my suit armor and a plastic egg. I was still strapped inside it, unable to move; it was time to decide how and where I was going to ground. Without moving my arms (I couldn't) I thumbed the switch for a proximity reading and read it when it flashed on in the instrument reflector inside my helmet in front of my forehead.

A mile and eight-tenths— A little closer than I liked, especially without company. The inner egg had reached steady speed, no more help to be gained by staying inside it, and its skin temperature indicated that it would not open automatically for a while yet—so I flipped a switch with my other thumb and got rid of it.

The first charge cut all the straps; the second charge exploded the plastic egg away from me in eight separate pieces—and I was outdoors, sitting on air, and could *see*! Better still, the eight discarded pieces were metal-coated (except for the small bit I had taken proximity reading through) and would give back the same reflection as an armored man. Any radar viewer, alive or cybernetic, would now have a sad time sorting me out from the junk nearest me, not to mention the thousands of other bits and pieces for miles on each side, above, and below me. Part of a mobile infantryman's training is to let him see, from the ground and both by eye and by radar, just how confusing a drop is to the forces on the ground— because you feel awful naked up there. It is easy to panic and either open a chute too soon and become a sitting duck (do ducks really sit?—if so, why?) or fail to open it and break your ankles, likewise backbone and skull.

So I stretched, getting the kinks out, and looked around . . . then doubled up again and straightened out in a swan dive face down and took a good look. It was night down there, as planned, but infrared snoopers let you size up terrain quite well after you are used to them. The river that cut diagonally through the city was almost below me and coming up fast, shining out clearly with a higher temperature than the land. I didn't care which side of it I landed on but I didn't want to land in it; it would slow me down.

I noticed a flash off to the right at about my altitude; some unfriendly native down below had burned what was probably a piece of my egg. So I fired my first chute at once, intending if possible to jerk myself right off his screen as he followed the targets down in closing range. I braced for the shock, rode it, then floated down for about twenty seconds before unloading the chute—not wishing to call attention to myself in still another way by not falling at the speed of the other stuff around me.

It must have worked; I wasn't burned.

About six hundred feet up I shot the second chute . . . saw very quickly that I was being carried over into the river, found that I was going to pass about a hundred feet up over a flat-roofed warehouse or some such by the river . . . blew the chute free and came in for a good enough if rather bouncy landing on the roof by means of the suit's jump jets. I was scanning for Sergeant Jelal's beacon as I hit.

And found that I was on the wrong side of the river; Jelly's star showed up on the compass ring inside my helmet far south of where it should have been—I was too far north. I trotted toward the river side of the roof as I took a range and bearing on the squad leader next to me, found that he was over a mile out of position, called, "Ace! Dress your line," tossed a bomb behind me as I stepped off the building and across the river. Ace answered as I could have expected—Ace should have had my spot but he didn't want to give up his squad; nevertheless he didn't fancy taking orders from me.

The warehouse went up behind me and the blast hit me while I was still over the river, instead of being shielded by the buildings on the far side as I should have been. It darn near tumbled my gyros and I came close to tumbling myself. I had set that bomb for fifteen seconds . . . or had I? I suddenly realized that I had let myself get excited, the worst thing you can do once you're on the ground. "Just like a drill," that was the way, just as Jelly had warned me. Take your time and do it right, even if it takes another half second.

As I hit I took another reading on Ace and told him again to realign his squad. He didn't answer but he was already doing it. I let it ride. As long as Ace did his job, I could afford to swallow his surliness—for now. But back aboard ship (if Jelly kept me on as assistant section leader) we would eventually have to pick a quiet spot and find out who was boss. He was a career corporal and I was just a term lance acting as corporal, but he was under me and you can't afford to take any lip under those circumstances. Not permanently.

But I didn't have time then to think about it; while I was jumping the river I had spotted a juicy target and I wanted to get it before somebody else noticed it—a lovely big group of what looked like public buildings on a hill. Temples, maybe . . . or a palace. They were miles outside the area we were sweeping, but one rule of a smash & run is to expend at least half your ammo outside your sweep area; that way the enemy is kept confused as to where you actually are—that and keep moving, do everything fast. You're always heavily outnumbered; surprise and speed are what saves you.

I was already loading my rocket launcher while I was checking on Ace and telling him for the second time to straighten up. Jelly's voice reached me right on top of that on the all-hands circuit: "Pla*toon*! By leapfrog! *Forward!*"

My boss, Sergeant Johnson, echoed, "By leapfrog! Odd numbers! *Advance!*"

That left me with nothing to worry about for twenty seconds, so I jumped up on the building nearest me, raised the launcher to my shoulder, found the target and pulled the first trigger to let the rocket have a look at its target—pulled the second trigger and kissed it on its way, jumped back to the ground. "Second section, even numbers!" I called out . . . waited for the count in my mind and ordered, "*Advance!*"

And did so myself, hopping over the next row of buildings, and, while I was in the air, fanning the first row by the river front with a hand flamer. They seemed to be wood construction and it looked like time to start a good fire—with luck, some of those warehouses would house oil products, or even explosives. As I hit, the Y-rack on my shoulders launched two small H.E. bombs a couple of hundred yards each way to my right and left flanks but I never saw what they did as just then my first rocket hit—that unmistakable (if you've ever seen one) brilliance of an atomic explosion. It was just a peewee, of course, less than two kilotons nominal yield, with tamper and implosion squeeze to produce results from a less-than-critical mass—but then who wants to be bunk mates with a cosmic catastrophe? It was enough to clean off that hilltop and make everybody in the city take shelter against fallout. Better still, any of the local yokels who happened to be outdoors and looking that way wouldn't be seeing anything else for a couple of hours—meaning *me*. The flash hadn't dazzled me, nor would it dazzle any of us; our face bowls are heavily leaded, we wear snoopers over our eyes—and we're trained to duck and take it on the armor if we do happen to be looking the wrong way.

So I merely blinked hard—opened my eyes and stared straight at a local citizen just coming out of an opening in the building ahead of me. He looked at me, I looked at him, and he started to raise something—a weapon, I suppose—as Jelly called out, "Odd numbers! *Advance!*"

I didn't have time to fool with him: I was a good five hundred yards short of where I should have been by then. I still had the hand flamer in my left hand; I toasted him and jumped over the building he had been coming out of, as I started to count. A hand flamer is primarily for incendiary work but it is a good defensive anti-personnel weapon in tight quarters; you don't have to aim it much.

Between excitement and anxiety to catch up I jumped too high and too wide. It's always a temptation to get the most out of your jump gear—but *don't do it!* It leaves you hanging in the air for seconds, a big fat target. The way to advance is to skim over each building as you come to it, barely clearing it, and taking full advantage of cover while you're down—and never stay in one place more than a second or two, never give them time to target in on you. Be somewhere else, anywhere. Keep moving.

This one I goofed—too much for one row of buildings, too little for the row beyond it; I found myself coming down on a roof. But not a nice flat one where I might have tarried three seconds to launch another pee-wee A-rocket; this roof was a jungle of pipes and stanchions and assorted ironmongery—a factory maybe, or some sort of chemical works. No place to land. Worse still, half a dozen natives were up there. These geezers are humanoid, eight or nine feet tall, much skinnier than we are and with a higher body temperature; they don't wear any clothes and they stand out in a set of snoopers like a neon sign. They look still funnier in daylight with your bare eyes but I would rather fight them than the arachnids—those Bugs make me queasy.

If these laddies were up there thirty seconds earlier when my rocket hit, then they couldn't see me, or anything. But I couldn't be certain and didn't want to tangle with them in any case; it wasn't that kind of a raid. So I jumped again while I was still in the air, scattering a handful of ten-second fire pills to keep them busy, grounded, jumped again at once, and called out, "Second section! Even numbers! . . . Advance!" and kept right on going to close the gap, while trying to spot, every time I jumped, something worth expending a rocket on. I had three more of the little A-rockets and I certainly didn't intend to take any back with me. But I had had pounded into me that you *must* get your money's worth with atomic weapons—it was only the second time that I had been allowed to carry them.

Right now I was trying to spot their waterworks; a direct hit on it could make the whole city uninhabitable, force them to evacuate it without directly killing anyone—just the sort of nuisance we had been sent down to commit. It should—according to the map we had studied under hypnosis—be about three miles upstream from where I was.

But I couldn't see it; my jumps didn't take me high enough, maybe. I was tempted to go higher but I remembered what Migliaccio had said about not trying for a medal, and stuck to doctrine. I set the Y-rack launcher on automatic and let it lob a couple of little bombs every time I hit. I set fire to things more or less at random in between, and tried to find the waterworks, or some other worth-while target.

Well, there was *something* up there at the proper range—waterworks or whatever, it was big. So I hopped on top of the tallest building near me, took a bead on it, and let fly. As I bounced down I heard Jelly: "Johnnie! Red! Start bending in the flanks."

I acknowledged and heard Red acknowledge and switched my beacon to blinker so that Red could pick me out for certain, took a range and bearing on his blinker while I called out, "Second Section! Curve in and envelop! Squad leaders acknowledge!"

Fourth and fifth squads answered, "Wilco"; Ace said, "We're already doin' it—pick up your feet."

Red's beacon showed the right flank to be almost ahead of me and a good fifteen miles away. Golly! Ace was right; I would have to pick up my feet or I would never close the gap in time—and me with a couple of hundred-weight of ammo and sundry nastiness still on me that

I just had to find time to use up. We had landed in a V formation, with Jelly at the bottom of the V and Red and myself at the ends of the two arms; now we had to close it into a circle around the retrieval rendezvous . . . which meant that Red and I each had to cover more ground than the others and still do our full share of damage.

At least the leapfrog advance was over with once we started to encircle; I could quit counting and concentrate on speed. It was getting to be less healthy to be anywhere, even moving fast. We had started with the enormous advantage of surprise, reached the ground without being hit (at least I hoped nobody had been hit coming in), and had been rampaging in among them in a fashion that let us fire at will without fear of hitting each other while they stood a big chance of hitting their own people in shooting at us—if they could find us to shoot at, at all. (I'm no games-theory expert but I doubt if any computer could have analyzed what we were doing in time to predict where we would be next.)

Nevertheless the home defenses were beginning to fight back, co-ordinated or not. I took a couple of near misses with explosives, close enough to rattle my teeth even inside armor, and once I was brushed by some sort of beam that made my hair stand on end and half paralyzed me for a moment—as if I had hit my funny bone, but all over. If the suit hadn't already been told to jump, I guess I wouldn't have got out of there.

Things like that make you pause to wonder why you

ever took up soldiering—only I was too busy to pause for anything. Twice, jumping blind over buildings, I landed right in the middle of a group of them—jumped at once while fanning wildly around me with the hand flamer.

Spurred on this way, I closed about half of my share of the gap, maybe four miles, in minimum time but without doing much more than casual damage. My Y-rack had gone empty two jumps back; finding myself alone in sort of a courtyard I stopped to put my reserve H.E. bombs into it while I took a bearing on Ace—found that I was far enough out in front of the flank squad to think about expending my last two A-rockets. I jumped to the top of the tallest building in the neighborhood.

It was getting light enough to see; I flipped the snoopers up onto my forehead and made a fast scan with bare eyes, looking for anything behind us worth shooting at, anything at all; I had no time to be choosy.

There was something on the horizon in the direction of their spaceport—administration & control, maybe, or possibly even a starship. Almost in line and about half as far away was an enormous structure which I couldn't identify even that loosely. The range to the spaceport was extreme but I let the rocket see it, said, "Go find it, baby!" and twisted its tail—slapped the last one in, sent it toward the nearer target, and jumped.

That building took a direct hit just as I left it. Either a skinny had judged (correctly) that it was worth one of their buildings to try for one of us, or one of my own mates was getting mighty careless with fireworks. Either way, I didn't want to jump from that spot, even a skimmer; I decided to go through the next couple of buildings instead of over. So I grabbed the heavy flamer off my back as I hit and flipped the snoopers down over my eyes, tackled a wall in front of me with a knife beam at full power. A section of wall fell away and I charged in.

And backed out even faster.

I didn't know what it was I had cracked open. A congregation in church—a skinny flophouse—maybe even their defense headquarters. All I knew was that it was a very big room filled with more skinnies than I wanted to see in my whole life.

Probably not a church, for somebody took a shot at me as I popped back out—just a slug that bounced off my armor, made my ears ring, and staggered me without hurting me. But it reminded me that I wasn't supposed to leave without giving them a souvenir of my visit. I grabbed the first thing on my belt and lobbed it in—and heard it start to squawk. As they keep telling you in Basic, doing something constructive at once is better than figuring out the best thing to do hours later.

By sheer chance I had done the right thing. This was a special bomb, one each issued to us for this mission with instructions to use them if we found ways to make them effective. The squawking I heard as I threw it was the bomb shouting in skinny talk (free translation): "I'm a thirty-second bomb! I'm a thirty-second bomb! Twenty-nine! . . . twenty-eight! . . . twenty-seven!—"

It was supposed to frazzle their nerves. Maybe it did; it

certainly frazzled mine. Kinder to shoot a man. I didn't wait for the countdown; I jumped, while I wondered whether they would find enough doors and windows to swarm out in time.

I got a bearing on Red's blinker at the top of the jump and one on Ace as I grounded. I was falling behind again—time to hurry.

But three minutes later we had closed the gap; I had Red on my left flank a half mile away. He reported it to Jelly. We heard Jelly's relaxed growl to the entire platoon: "Circle is closed, but the beacon is not down yet. Move forward slowly and mill around, make a little more trouble—but mind the lad on each side of you; don't make trouble for *him*. Good job, so far—don't spoil it. Pla*toon*! By sections . . . *Muster!*"

It looked like a good job to me, too; much of the city was burning and, although it was almost full light now, it was hard to tell whether bare eyes were better than snoopers, the smoke was so thick.

Johnson, our section leader, sounded off: "Second section, call off!"

I echoed, "Squads four, five, and six—call off and report!" The assortment of safe circuits we had available in the new model comm units certainly speeded things up; Jelly could talk to anybody or to his section leaders; a section leader could call his whole section, or his noncoms; and the platoon could muster twice as fast, when seconds matter. I listened to the fourth squad call off while I inventoried my remaining firepower and lobbed one bomb toward a skinny who poked his head around a corner. He left and so did I—"Mill around," the boss man had said.

The fourth squad bumbled the call off until the squad leader remembered to fill in with Jenkins' number; the fifth squad clicked off like an abacus and I began to feel good . . . when the call off stopped after number four in Ace's squad. I called out, "Ace, where's Dizzy?"

"Shut up," he said. "Number six! Call off!"

"Six!" Smith answered.

"Seven!"

"Sixth squad, Flores missing," Ace completed it. "Squad leader out for pickup."

"One man absent," I reported to Johnson. "Flores, squad six."

"Missing or dead?"

"I don't know. Squad leader and assistant section leader dropping out for pickup."

"Johnnie, you let Ace take it."

But I didn't hear him, so I didn't answer. I heard him report to Jelly and I heard Jelly cuss. Now look, I wasn't bucking for a medal—it's the assistant section leader's *business* to make pickup; he's the chaser, the last man in, expendable. The squad leaders have other work to do. As you've no doubt gathered by now the assistant section leader isn't necessary as long as the section leader is alive.

Right that moment I was feeling unusually expendable, almost expended, because I was hearing the sweetest sound in the universe, the beacon the retrieval boat would land on, sounding our recall. The beacon is a robot rocket, fired ahead of the retrieval boat, just a spike that buries itself in the ground and starts broadcasting that welcome, welcome music. The retrieval boat homes in on it automatically three minutes later and you had better be on hand, because the bus can't wait and there won't be another one along.

But you don't walk away on another cap trooper, not while there's a chance he's still alive—not in Rasczak's Roughnecks. Not in any outfit of the Mobile Infantry. You try to make pickup.

I heard Jelly order: "Heads up, lads! Close to retrieval circle and interdict! On the bounce!"

And I heard the beacon's sweet voice: "—*to the everlasting glory of the infantry, shines the name, shines the name of Rodger Young!*" and I wanted to head for it so bad I could taste it.

Instead I was headed the other way, closing on Ace's beacon and expending what I had left of bombs and fire pills and anything else that would weigh me down. "Ace! You got his beacon?"

"Yes. Go back, Useless!"

"I've got you by eye now. Where is he?"

"Right ahead of me, maybe quarter mile. Scram! He's *my* man."

I didn't answer; I simply cut left oblique to reach Ace about where he said Dizzy was.

And found Ace standing over him, a couple of skinnies flamed down and more running away. I lit beside him. "Let's get him out of his armor—the boat'll be down any second!"

"He's too bad hurt!"

I looked and saw that it was true—there was actually a *hole* in his armor and blood coming out. And I was stumped. To make a wounded pickup you get him out of his armor . . . then you simply pick him up in your arms—no trouble in a powered suit—and bounce away from there. A bare man weighs less than the ammo and stuff you've expended. "What'll we *do?*"

"We carry him," Ace said grimly. "Grab ahold the left side of his belt." He grabbed the right side, we manhandled Flores to his feet. "Lock on! Now . . . by the numbers, stand by to jump—one—*two!*"

We jumped. Not far, not well. One man alone couldn't have gotten him off the ground; an armored suit is too heavy. But split it between two men and it can be done.

We jumped—and we jumped—and again, and again, with Ace calling it and both of us steadying and catching Dizzy on each grounding. His gyros seemed to be out.

We heard the beacon cut off as the retrieval boat landed on it—I saw it land . . . and it was too far away. We heard the acting platoon sergeant call out: "In succession, prepare to embark!"

And Jelly called out, "Belay that order!"

We broke at last into the open and saw the boat standing on its tail, heard the ululation of its take-off warning—saw the platoon still on the ground around it, in interdiction circle, crouching behind the shield they had formed.

Heard Jelly shout, "In succession, man the boat—*move!*"

And we were *still* too far away! I could see them peel off from the first squad, swarm into the boat as the interdiction circle tightened.

And a single figure broke out of the circle, came toward us at a speed possible only to a command suit.

Jelly caught us while we were in the air, grabbed Flores by his Y-rack and helped us lift.

Three jumps got us to the boat. Everybody else was inside but the door was still open. We got him in and closed it while the boat pilot screamed that we had made her miss rendezvous and now we had *all* bought it! Jelly paid no attention to her; we laid Flores down and lay down beside him. As the blast hit us Jelly was saying to himself, "All present, Lieutenant. Three men hurt—but all present!"

I'll say this for Captain Deladrier: they don't make any better pilots. A rendezvous, boat to ship in orbit, is precisely calculated. I don't know how, but it is, and you don't change it. You *can't.*

Only she did. She saw in her scope that the boat had failed to blast on time; she braked back, picked up speed again—and matched and took us in, just by eye and touch, no time to compute it. If the Almighty ever needs an assistant to keep the stars in their courses, I know where he can look.

Flores died on the way up.

CH:02

It scared me so, I hooked it off,
Nor stopped as I remember,
Nor turned about till I got home,
Locked up in mother's chamber.
Yankee Doodle, keep it up,
Yankee Doodle dandy,
Mind the music and the step,
And with the girls be handy.

I never really intended to join up.

And certainly not the infantry! Why, I would rather have taken ten lashes in the public square and have my father tell me that I was a disgrace to a proud name.

Oh, I had mentioned to my father, late in my senior year in high school, that I was thinking over the idea of volunteering for Federal Service. I suppose every kid does, when his eighteenth birthday heaves into sight— and mine was due the week I graduated. Of course most of them just think about it, toy with the idea a little, then go do something else—go to college, or get a job, or something. I suppose it would have been that way with me . . . if my best chum had not, with dead seriousness, planned to join up.

Carl and I had done everything together in high school—eyed the girls together, double-dated together, been on the debate team together, pushed electrons together in his home lab. I wasn't much on electronic theory myself, but I'm a neat hand with a soldering gun; Carl supplied the skull sweat and I carried out his instructions. It was fun; anything we did together was fun. Carl's folks didn't have anything like the money that my father had, but it didn't matter between us. When my father bought me a Rolls copter for my fourteenth birthday, it was Carl's as much as it was mine; contrariwise, his basement lab was mine.

So when Carl told me that he was not going straight on with school, but would serve a term first, it gave me to pause. He really meant it; he seemed to think that it was natural and right and obvious.

So I told him I was joining up, too.

He gave me an odd look. "Your old man won't let you."

"Huh? How can he stop me?" And of course he couldn't, not legally. It's the first completely free choice anybody gets (and maybe his last); when a boy, or a girl, reaches his or her eighteenth birthday, he or she can volunteer and nobody else has any say in the matter.

"You'll find out." Carl changed the subject.

So I took it up with my father, tentatively, edging into it sideways.

He put down his newspaper and cigar and stared at me. "Son, are you out of your mind?"

I muttered that I didn't think so.

"Well, it certainly sounds like it." He sighed. "Still . . . I should have been expecting it; it's a predictable stage in a boy's growing up. I remember when you learned to walk and weren't a baby any longer—frankly you were a little hellion for quite a while. You broke one of your mother's Ming vases—on purpose, I'm quite sure . . . but you were too young to know that it was valuable, so all you got was having your hand spatted. I recall the day you swiped one of my cigars, and how sick it made you. Your mother and I carefully avoided noticing that you couldn't eat dinner that night and I've never mentioned it to you until now—boys have to try such things and discover for themselves that men's vices are not for them. We watched when you turned the corner on adolescence and started noticing that girls were different—and wonderful."

He sighed again. "All normal stages. And the last one, right at the end of adolescence, is when a boy decides to join up and wear a pretty uniform. Or decides that he is in love, love such as no man ever experienced before, and that he just has to get married right away. Or both." He smiled grimly. "With me it was both. But I got over each of them in time not to make a fool of myself and ruin my life."

"But, Father, I wouldn't ruin my life. Just a term of service—not career."

"Let's table that, shall we? Listen, and let *me* tell *you* what you are going to do—because you *want* to. In the first place this family has stayed out of politics and cultivated its own garden for over a hundred years—I see no reason for you to break that fine record. I suppose it's the influence of that fellow at your high school—what's his name? You know the one I mean."

He meant our instructor in History and Moral Philosophy—a veteran, naturally. "Mr. Dubois."

"Hmmph, a silly name—it suits him. Foreigner, no doubt. It ought to be against the law to use the schools as undercover recruiting stations. I think I'm going to write a pretty sharp letter about it—a taxpayer has *some* rights!"

"But, Father, he doesn't do that at all! He—" I stopped, not knowing how to describe it. Mr. Dubois had a snotty, superior manner; he acted as if none of us was really *good* enough to volunteer for service. I didn't like him. "Uh, if anything, he discourages it."

"Hmmph! Do you know how to lead a pig? Never mind. When you graduate, you're going to study business at Harvard; you know that. After that, you will go on to the Sorbonne and you'll travel a bit along with it, meet some of our distributors, find out how business is done elsewhere. Then you'll come home and go to work. You'll start with the usual menial job, stock clerk or something, just for form's sake—but you'll be an executive before you can catch your breath, because I'm not getting any younger and the quicker you can pick up the load, the better. As soon as you're able and willing, you'll be boss. There! How does that strike you as a program? As compared with wasting two years of your life?"

I didn't say anything. None of it was news to me; I'd thought about it. Father stood up and put a hand on my shoulder. "Son, don't think I don't sympathize with you; I do. But look at the real facts. If there were a war, I'd be the first to cheer you on—and to put the business on a war footing. But there isn't, and praise God there never will be again. We've outgrown wars. This planet is now peaceful and happy and we enjoy good enough relations with other planets. So what is this so-called 'Federal Service'? Parasitism, pure and simple. A functionless organ, utterly obsolete, living on the taxpayers. A decidedly expensive way for inferior people who otherwise would be unemployed to live at public expense for a term of years, then give themselves airs for the rest of their lives. Is that what *you* want to do?"

"Carl isn't inferior!"

"Sorry. No, he's a fine boy . . . but misguided." He frowned, and then smiled. "Son, I had intended to keep something as a surprise for you—a graduation present. But I'm going to tell you now so that you can put this nonsense out of your mind more easily. Not that I am afraid of what you might do; I have confidence in your basic good sense, even at your tender years. But you are troubled, I know—and this will clear it away. Can you guess what it is?"

"Uh, no."

He grinned. "A vacation trip to Mars."

I must have looked stunned. "Golly, Father, I had no idea—"

"I meant to surprise you and I see I did. I know how you kids feel about travel, though it beats me what anyone sees in it after the first time out. But this is a good time for you to do it—by yourself; did I mention that?—and get it out of your system . . . because you'll be hardpressed to get in even a week on Luna once you take up your responsibilities." He picked up his paper. "No, don't thank me. Just run along and let me finish my paper—I've got some gentlemen coming in this evening, shortly. Business."

I ran along. I guess he thought that settled it . . . and I suppose I did, too. Mars! And on my own! But I didn't tell Carl about it; I had a sneaking suspicion that he would regard it as a bribe. Well, maybe it was. Instead I simply told him that my father and I seemed to have different ideas about it.

"Yeah," he answered, "so does mine. But it's *my* life."

I thought about it during the last session of our class in History and Moral Philosophy. H. & M. P. was different from other courses in that everybody had to take it but nobody had to pass it—and Mr. Dubois never seemed to care whether he got through to us or not. He would just point at you with the stump of his left arm (he never bothered with names) and snap a question. Then the argument would start.

But on the last day he seemed to be trying to find out what we had learned. One girl told him bluntly: "My mother says that violence never settles anything."

"So?" Mr. Dubois looked at her bleakly. "I'm sure

the city fathers of Carthage would be glad to know that. Why doesn't your mother tell them so? Or why don't *you*?"

They had tangled before—since you couldn't flunk the course, it wasn't necessary to keep Mr. Dubois buttered up. She said shrilly, "You're making fun of me! Everybody knows that Carthage was destroyed!"

"You seemed to be unaware of it," he said grimly. "Since you do know it, wouldn't you say that violence had settled their destinies rather thoroughly? However, I was not making fun of you personally; I was heaping scorn on an inexcusably silly idea—a practice I shall always follow. Anyone who clings to the historically untrue—and thoroughly immoral—doctrine that 'violence never settles anything' I would advise to conjure up the ghosts of Napoleon Bonaparte and of the Duke of Wellington and let them debate it. The ghost of Hitler could referee, and the jury might well be the Dodo, the Great Auk, and the Passenger Pigeon. Violence, naked force, has settled more issues in history than has any other factor, and the contrary opinion is wishful thinking at its worst. Breeds that forget this basic truth have always paid for it with their lives and freedoms."

He sighed. "Another year, another class—and, for me, another failure. One can lead a child to knowledge but one *cannot* make him think." Suddenly he pointed his stump at me. "You. What is the moral difference, if any, between the soldier and the civilian?"

"The difference," I answered carefully, "lies in the field of civic virtue. A soldier accepts personal responsibility for the safety of the body politic of which he is a member, defending it, if need be, with his life. The civilian does not."

"The exact words of the book," he said scornfully. "But do you understand it? Do you *believe* it?"

"Uh, I don't know, sir."

"Of course you don't! I doubt if any of you here would recognize 'civic virtue' if it came up and barked in your face!" He glanced at his watch. "And that is all, a final all. Perhaps we shall meet again under happier circumstances. Dismissed."

Graduation right after that and three days later my birthday, followed in less than a week by Carl's birthday—and I still hadn't told Carl that I wasn't joining up. I'm sure he assumed that I would not, but we didn't discuss it out loud—embarrassing. I simply arranged to meet him the day after his birthday and we went down to the recruiting office together.

On the steps of the Federal Building we ran into Carmencita Ibañez, a classmate of ours and one of the nice things about being a member of a race with two sexes. Carmen wasn't my girl—she wasn't anybody's girl; she never made two dates in a row with the same boy and treated all of us with equal sweetness and rather impersonally. But I knew her pretty well, as she often came over and used our swimming pool, because it was Olympic length—sometimes with one boy, sometimes with another. Or alone, as Mother urged her to—Mother considered her "a good influence." For once she was right.

She saw us and waited, dimpling. "Hi, fellows!"

"Hello, *Ochee Chyornya*," I answered. "What brings you here?"

"Can't you guess? Today is my birthday."

"Huh? Happy returns!"

"So I'm joining up."

"Oh . . ." I think Carl was as surprised as I was. But Carmencita was like that. She never gossiped and she kept her own affairs to herself. "No foolin'?" I added, brilliantly.

"Why should I be fooling? I'm going to be a spaceship pilot—at least I'm going to try for it."

"No reason why you shouldn't make it," Carl said quickly. He was right—I know now just how right he was. Carmen was small and neat, perfect health and perfect reflexes—she could make competitive diving routine look easy and she was quick at mathematics. Me, I tapered off with a "C" in algebra and a "B" in business arithmetic; she took all the math our school offered and a tutored advance course on the side. But it had never occurred to me to wonder why. Fact was, little Carmen was so ornamental that you just never thought about her being useful.

"We—uh, I," said Carl, "am here to join up, too."

"And me," I agreed. "Both of us." No, I hadn't made any decision; my mouth was leading its own life.

"Oh, wonderful!"

"And I'm going to buck for space pilot, too," I added firmly.

She didn't laugh. She answered very seriously, "Oh, how grand! Perhaps in training we'll run into each other. I hope so."

"Collision courses?" asked Carl. "That's a no-good way to pilot."

"Don't be silly, Carl. On the ground, of course. Are you going to be a pilot, too?"

"*Me?*" Carl answered. "I'm no truck driver. You know me—Starside R&D, if they'll have me. Electronics."

"'Truck driver' indeed! I hope they stick you out on Pluto and let you freeze. No, I don't—good luck! Let's go in, shall we?"

The recruiting station was inside a railing in the rotunda. A fleet sergeant sat at a desk there, in dress uniform, gaudy as a circus. His chest was loaded with ribbons I couldn't read. But his right arm was off so short that his tunic had been tailored without any sleeve at all . . . and, when you came up to the rail, you could see that he had no legs.

It didn't seem to bother him. Carl said, "Good morning. I want to join up."

"Me, too," I added.

He ignored us. He managed to bow while sitting down and said, "Good morning, young lady. What can I do for you?"

"I want to join up, too."

He smiled. "Good girl! If you'll just scoot up to room

201 and ask for Major Rojas, she'll take care of you." He
looked her up and down. "Pilot?"

"If possible."

"You look like one. Well, see Miss Rojas."

She left, with thanks to him and a see-you-later to
us; he turned his attention to us, sized us up with a total
absence of the pleasure he had shown in little Carmen.
"So?" he said. "For what? Labor battalions?"

"Oh, no!" I said. "I'm going to be a pilot."

He stared at me and simply turned his eyes away.
"You?"

"I'm interested in the Research and Development
Corps," Carl said soberly, "especially electronics. I understand the chances are pretty good."

"They are if you can cut it," the Fleet Sergeant said
grimly, "and not if you don't have what it takes, both in
preparation and ability. Look, boys, have you any idea
why they have me out here in front?"

I didn't understand him. Carl said, "Why?"

"Because the government doesn't care one bucket of
swill whether you join or not! Because it has become
stylish, with some people—too many people—to serve a
term and earn a franchise and be able to wear a ribbon in
your lapel which says that you're a vet'ran . . . whether
you've ever seen combat or not. But if you *want* to serve
and I can't talk you out of it, then we have to take you,
because that's your constitutional right. It says that
everybody, male or female, shall have his born right to pay
his service and assume full citizenship—but the facts are
that we are getting hard pushed to find things for all the

volunteers to do that aren't just glorified K.P. You can't all be real military men; we don't need that many and most of the volunteers aren't number-one soldier material anyhow. Got any idea what it takes to make a soldier?"

"No," I admitted.

"Most people think that all it takes is two hands and two feet and a stupid mind. Maybe so, for cannon fodder. Possibly that was all that Julius Caesar required. But a private soldier today is a specialist so highly skilled that he would rate 'master' in any other trade; we can't afford stupid ones. So for those who insist on serving their term—but haven't got what we want and must have—we've had to think up a whole list of dirty, nasty, dangerous jobs that will either run 'em home with their tails between their legs and their terms uncompleted . . . or at the very least make them remember for the rest of their lives that their citizenship is valuable to them because they've paid a high price for it. Take that young lady who was here—wants to be a pilot. I hope she makes it; we always need good pilots, not enough of 'em. Maybe she will. But if she misses, she may wind up in Antarctica, her pretty eyes red from never seeing anything but artificial light and her knuckles callused from hard, dirty work."

I wanted to tell him that the least Carmencita could get was computer programmer for the sky watch; she really was a whiz at math. But he was talking.

"So they put me out here to discourage you boys. Look at this." He shoved his chair around to make sure that we could see that he was legless. "Let's assume that you don't wind up digging tunnels on Luna or playing human guinea pig for new diseases through sheer lack of talent; suppose we do make a fighting man out of you. Take a look at *me*—this is what you may buy . . . if you don't buy the whole farm and cause your folks to receive a 'deeply regret' telegram. Which is more likely, because these days, in training or in combat, there aren't many wounded. If you buy at all, they likely throw in a coffin— I'm the rare exception; I was lucky . . . though maybe you wouldn't call it luck."

He paused, then added, "So why don't you boys go home, go to college, and then go be chemists or insurance brokers or whatever? A term of service isn't a kiddie camp; it's either real military service, rough and dangerous even in peacetime . . . or a most unreasonable facsimile thereof. Not a vacation. Not a romantic adventure. Well?"

Carl said, "I'm here to join up."

"Me, too."

"You realize that you aren't allowed to pick your service?"

Carl said, "I thought we could state our preferences?"

"Certainly. And that's the last choice you'll make until the end of your term. The placement officer pays attention to your choice, too. First thing he does is to check whether there's any demand for left-handed glass blowers this week—that being what you think would make you happy. Having reluctantly conceded that there is a need for your choice—probably at the bottom of the Pacific— he then tests you for innate ability and preparation. About once in twenty times he is forced to admit that everything matches and you get the job . . . until some practical joker gives you dispatch orders to do something very different. But the other nineteen times he turns you down and decides that you are just what they have been needing to field-test survival equipment on Titan." He added meditatively, "It's chilly on Titan. And it's amazing how often experimental equipment fails to work. Have to have real field tests, though—laboratories just never get all the answers."

"I can qualify for electronics," Carl said firmly, "if there are jobs open in it."

"So? And how about you, bub?"

I hesitated—and suddenly realized that, if I didn't take a swing at it, I would wonder all my life whether I was anything but the boss's son. "I'm going to chance it."

"Well, you can't say I didn't try. Got your birth certificates with you? And let's see your IDs."

Ten minutes later, still not sworn in, we were on the top floor being prodded and poked and fluoroscoped. I decided that the idea of a physical examination is that, if you *aren't* ill, then they do their darnedest to make you ill. If the attempt fails, you're in.

I asked one of the doctors what percentage of the victims flunked the physical. He looked startled. "Why, we *never* fail anyone. The law doesn't permit us to."

"Huh? I mean, excuse me, Doctor? Then what's the point of this goose-flesh parade?"

"Why, the purpose is," he answered, hauling off and hitting me in the knee with a hammer (I kicked him, but not hard), "to find out what duties you are physically able to perform. But if you came in here in a wheel chair and blind in both eyes and were silly enough to insist on enrolling, they would find something silly enough to match. Counting the fuzz on a caterpillar by touch, maybe. The only way you can fail is by having the psychiatrists decide that you are not able to understand the oath."

"Oh. Uh . . . Doctor, were you already a doctor when you joined up? Or did they decide you ought to be a doctor and send you to school?"

"*Me?*" He seemed shocked. "Youngster, do I look that silly? I'm a civilian employee."

"Oh. Sorry, sir."

"No offense. But military service is for ants. Believe me. I see 'em go, I see 'em come back—when they do come back. I see what it's done to them. And for what? A purely nominal political privilege that pays not one centavo and that most of them aren't competent to use wisely anyhow. Now if they would let medical men run things—but never mind that; you might think I was talking treason, free speech or not. But, youngster, if you've got savvy enough to count ten, you'll back out while you still can. Here, take these papers back to the recruiting sergeant—and remember what I said."

I went back to the rotunda. Carl was already there. The Fleet Sergeant looked over my papers and said glumly, "Apparently you both are almost insufferably healthy—except for holes in the head. One moment, while I get some witnesses." He punched a button and two female clerks came out, one old battle-ax, one kind of cute.

He pointed to our physical examination forms, our birth certificates, and our IDs, said formally: "I invite and require you, each and severally, to examine these exhibits, determine what they are and to determine, each independently, what relation, if any, each document bears to these two men standing here in your presence."

They treated it as a dull routine, which I'm sure it was; nevertheless they scrutinized every document, they took our fingerprints—again!—and the cute one put a jeweler's loupe in her eye and compared prints from birth to now. She did the same with signatures. I began to doubt if I was myself.

The Fleet Sergeant added, "Did you find exhibits relating to their present competence to take the oath of enrollment? If so, what?"

"We found," the older one said, "appended to each record of physical examination a duly certified conclusion by an authorized and delegated board of psychiatrists stating that each of them is mentally competent to take the oath and that neither one is under the influence of alcohol, narcotics, other disabling drugs, nor of hypnosis."

"Very good." He turned to us. "Repeat after me—

"I, being of legal age, of my own free will—"

"'I,'" we each echoed, "'being of legal age, of my own free will—'"

"—without coercion, promise, or inducement of any

sort, after having been duly advised and warned of the meaning and consequences of this oath—

"—do now enroll in the Federal Service of the Terran Federation for a term of not less than two years and as much longer as may be required by the needs of the Service—"

(I gulped a little over that part. I had always thought of a "term" as two years, even though I knew better, because that's the way people talk about it. Why, we were signing up for *life*.)

"I swear to uphold and defend the Constitution of the Federation against all its enemies on or off Terra, to protect and defend the Constitutional liberties and privileges of all citizens and lawful residents of the Federation, its associated states and territories, to perform, on or off Terra, such duties of any lawful nature as may be assigned to me by lawful direct or delegated authority—

"—and to obey all lawful orders of the Commander-in-Chief of the Terran Service and of all officers or delegated persons placed over me—

"—and to require such obedience from all members of the Service or other persons or non-human beings lawfully placed under my orders—

"—and, on being honorably discharged at the completion of my full term of active service or upon being placed on inactive retired status after having completed such full term, to carry out all duties and obligations and to enjoy all privileges of Federation citizenship including but not limited to the duty, obligation and privilege of exercising sovereign franchise for the rest of my natural life unless stripped of honor by verdict, finally sustained, of court of my sovereign peers."

(*Whew!*) Mr. Dubois had analyzed the Service oath for us in History and Moral Philosophy and had made us study it phrase by phrase—but you don't really feel the *size* of the thing until it comes rolling over you, all in one ungainly piece, as heavy and unstoppable as Juggernaut's carriage.

At least it made me realize that I was no longer a civilian, with my shirttail out and nothing on my mind. I didn't know yet what I was, but I knew what I wasn't.

"So help me God!" we both ended and Carl crossed himself and so did the cute one.

After that there were more signatures and fingerprints, all five of us, and flat colorgraphs of Carl and me were snapped then and there and embossed into our papers. The Fleet Sergeant finally looked up. "Why, it's 'way past the break for lunch. Time for chow, lads."

I swallowed hard. "Uh . . . Sergeant?"

"Eh? Speak up."

"Could I flash my folks from here? Tell them what I—Tell them how it came out?"

"We can do better than that."

"Sir?"

"You go on forty-eight hours leave now." He grinned coldly. "Do you know what happens if you don't come back?"

"Uh . . . court-martial?"

"Not a thing. Not a blessed thing. Except that your papers get marked, *Term not completed satisfactorily*, and you never, never, never get a second chance. This is our cooling-off period, during which we shake out the overgrown babies who didn't really mean it and should never have taken the oath. It saves the government money and it saves a power of grief for such kids and their parents— the neighbors needn't guess. You don't even have to tell your parents." He shoved his chair away from his desk. "So I'll see you at noon day after tomorrow. If I see you. Fetch your personal effects."

It was a crumbly leave. Father stormed at me, then quit speaking to me; Mother took to her bed. When I finally left, an hour earlier than I had to, nobody saw me off but the morning cook and the houseboys.

I stopped in front of the recruiting sergeant's desk, thought about saluting and decided I didn't know how. He looked up. "Oh. Here are your papers. Take them up to room 201; they'll start you through the mill. Knock and walk in."

Two days later I knew I was not going to be a pilot. Some of the things the examiners wrote about me were:—*insufficient intuitive grasp of spatial relationships . . . insufficient mathematical talent . . . deficient mathematical preparation . . . reaction time adequate . . . eyesight good.* I'm glad they put in those last two; I was beginning to feel that counting on my fingers was my speed.

The placement officer let me list my lesser preferences, in order, and I caught four more days of the wildest aptitude tests I've ever heard of. I mean to say, what do they find out when a stenographer jumps on her chair and screams, "Snakes!" There was no snake, just a harmless piece of plastic hose.

The written and oral tests were mostly just as silly, but they seemed happy with them, so I took them. The thing I did most carefully was to list my preferences. Naturally I listed all of the Space Navy jobs (other than pilot) at the top; whether I went as power-room technician or as cook, I knew that I preferred any Navy job to any Army job—I wanted to travel.

Next I listed Intelligence—a spy gets around, too, and I figured that it couldn't possibly be dull. (I was wrong, but never mind.) After that came a long list; psychological warfare, chemical warfare, biological warfare, combat ecology (I didn't know what it was, but it sounded interesting), logistics corps (a simple mistake; I had studied logic for the debate team and "logistics" turns out to have two entirely separate meanings), and a dozen others. Clear at the bottom, with some hesitation, I put K-9 Corps, and Infantry.

I didn't bother to list the various non-combatant auxiliary corps because, if I wasn't picked for a combat corps, I didn't care whether they used me as an experimental animal or sent me as a laborer in the Terranizing of Venus—either one was a booby prize.

Mr. Weiss, the placement officer, sent for me a week after I was sworn in. He was actually a retired psychological-warfare major, on active duty for procurement, but he wore mufti and insisted on being called just "Mister" and you could relax and take it easy with him. He had my list of preferences and the reports on all my tests and I saw that he was holding my high school transcript—which pleased me, for I had done all right in school; I had stood high enough without standing so high as to be marked as a greasy grind, having never flunked any courses and dropped only one, and I had been rather a big man around school otherwise; swimming team, debate team, track squad, class treasurer, silver medal in the annual literary contest, chairman of the homecoming committee, stuff like that. A well-rounded record and it's all down in the transcript.

He looked up as I came in, said, "Sit down, Johnnie," and looked back at the transcript, then put it down. "You like dogs?"

"Huh? Yes, sir."

"How well do you like them? Did your dog sleep on your bed? By the way, where is your dog now?"

"Why, I don't happen to have a dog just at present. But when I did—well, no, he didn't sleep on my bed. You see, Mother didn't allow dogs in the house."

"But didn't you sneak him in?"

"Uh—" I thought of trying to explain Mother's not-angry-but-terribly-terribly-hurt routine when you tried to buck her on something she had her mind made up about. But I gave up. "No, sir."

"Mmm . . . have you ever seen a neodog?"

"Uh, once, sir. They exhibited one at the Macarthur Theater two years ago. But the S.P.C.A. made trouble for them."

"Let me tell you how it is with a K-9 team. A neodog is not just a dog that talks."

"I couldn't understand that neo at the Macarthur. Do they really talk?"

"They talk. You simply have to train your ear to their accent. Their mouths can't shape 'b,' 'm,' 'p,' or 'v' and you have to get used to their equivalents—something like the handicap of a split palate but with different letters. No matter, their speech is as clear as any human speech. But a neodog is not a talking dog; he is not a dog at all, he is an artificially mutated symbiote derived from dog stock. A neo, a trained Caleb, is about six times as bright as a dog, say about as intelligent as a human moron— except that the comparison is not fair to the neo; a moron is a defective, whereas a neo is a stable genius in his own line of work."

Mr. Weiss scowled. "Provided, that is, that he has his symbiote. That's the rub. Mmm . . . you're too young ever to have been married but you've seen marriage, your own parents at least. Can you imagine being married to a Caleb?"

"Huh? No. No, I can't."

"The emotional relationship between the dog-man and the man-dog in the K-9 team is a great deal closer and much more important than is the emotional relationship in most marriages. If the master is killed, we kill the neodog—at once! It is all that we can do for the poor thing. A mercy killing. If the neodog is killed . . . well, we can't kill the man even though it would be the simplest solution. Instead we restrain him and hospitalize him and slowly put him back together." He picked up a pen, made a mark. "I don't think we can risk assigning a boy to K-9 who didn't outwit his mother to have his dog sleep with him. So let's consider something else."

It was not until then that I realized that I must have already flunked every choice on my list above K-9 Corps—and now I had just flunked it, too. I was so startled that I almost missed his next remark. Major Weiss said meditatively, with no expression and as if he were talking about someone else, long dead and far away: "I was once half of a K-9 team. When my Caleb became a casualty, they kept me under sedation for six weeks, then rehabilitated me for other work. Johnnie, these courses you've taken—why didn't you study something useful?"

"Sir?"

"Too late now. Forget it. Mmm . . . your instructor in History and Moral Philosophy seems to think well of you."

"He does?" I was surprised. "What did he say?"

Weiss smiled. "He says that you are not stupid, merely ignorant and prejudiced by your environment. From him that is high praise—I know him."

It didn't sound like praise to me! That stuck-up stiff-necked old—

"And," Weiss went on, "a boy who gets a 'C-minus' in Appreciation of Television can't be all bad. I think we'll accept Mr. Dubois' recommendation. How would you like to be an infantryman?"

I came out of the Federal Building feeling subdued yet not really unhappy. At least I was a soldier; I had papers in my pocket to prove it. I hadn't been classed as too dumb and useless for anything but make-work.

It was a few minutes after the end of the working day and the building was empty save for a skeleton night staff and a few stragglers. I ran into a man in the rotunda who was just leaving; his face looked familiar but I couldn't place him.

But he caught my eye and recognized me. "Evening!" he said briskly. "You haven't shipped out yet?"

And then I recognized him—the Fleet Sergeant who had sworn us in. I guess my chin dropped; this man was in civilian clothes, was walking around on two legs and had two arms. "Uh, good evening, Sergeant," I mumbled.

He understood my expression perfectly, glanced down at himself and smiled easily. "Relax, lad. I don't have to put on my horror show after working hours— and I don't. You haven't been placed yet?"

"I just got my orders."

"For what?"

"Mobile Infantry."

His face broke in a big grin of delight and he shoved out his hand. "My outfit! Shake, son! We'll make a man of you—or kill you trying. Maybe both."

"It's a good choice?" I said doubtfully.

"'A good choice'? Son, it's the *only* choice. The Mobile Infantry *is* the Army. All the others are either button pushers or professors, along merely to hand us the saw; *we* do the work." He shook hands again and added, "Drop me a card—'Fleet Sergeant Ho, Federal

Building,' that'll reach me. Good luck!" And he was off, shoulders back, heels clicking, head up.

I looked at my hand. The hand he had offered me was the one that wasn't there—his right hand. Yet it had felt like flesh and had shaken mine firmly. I had read about these powered prosthetics, but it is startling when you first run across them.

I went back to the hotel where recruits were temporarily billeted during placement—we didn't even have uniforms yet, just plain coveralls we wore during the day and our own clothes after hours. I went to my room and started packing, as I was shipping out early in the morning—packing to send stuff home, I mean; Weiss had cautioned me not to take along anything but family photographs and possibly a musical instrument if I played one (which I didn't). Carl had shipped out three days earlier, having gotten the R&D assignment he wanted. I was just as glad, as he would have been just too confounded understanding about the billet I had drawn. Little Carmen had shipped out, too, with the rank of cadet midshipman (probationary)—she was going to be a pilot, all right, if she could cut it . . . and I suspected that she could.

My temporary roomie came in while I was packing. "Got your orders?" he asked.

"Yup."

"What?"

"Mobile Infantry."

"The *Infantry?* Oh, you poor stupid clown! I feel sorry for you, I really do."

I straightened up and said angrily, "Shut up! The Mobile Infantry is the best outfit in the Army—it *is* the Army! The rest of you jerks are just along to hand us the saw—*we* do the work."

He laughed. "You'll find out!"

"You want a mouthful of knuckles?"

CH:03

He shall rule them with a rod of iron.

—Revelations II:25

I did Basic at Camp Arthur Currie on the northern prairies, along with a couple of thousand other victims— and I do mean "Camp," as the only permanent buildings there were to shelter equipment. We slept and ate in tents; we lived outdoors—if you call that "living," which I didn't, at the time. I was used to a warm climate; it seemed to me that the North Pole was just five miles north of camp and getting closer. Ice Age returning, no doubt.

But exercise will keep you warm and they saw to it that we got plenty of that.

The first morning we were there they woke us up before daybreak. I had had trouble adjusting to the change in time zones and it seemed to me that I had just got to sleep; I couldn't believe that anyone seriously intended that I should get up in the middle of the night.

But they did mean it. A speaker somewhere was blaring out a military march, fit to wake the dead, and a hairy

nuisance who had come charging down the company
street yelling, "*Everybody out! Show a leg! On the bounce!*"
came marauding back again just as I had pulled the covers
over my head, tipped over my cot and dumped me on the
cold hard ground.

It was an impersonal attention; he didn't even wait to
see if I hit.

Ten minutes later, dressed in trousers, undershirt, and
shoes, I was lined up with the others in ragged ranks for
setting-up exercises just as the Sun looked over the east-
ern horizon. Facing us was a big broad-shouldered,
mean-looking man, dressed just as we were—except that
while I looked and felt like a poor job of embalming, his
chin was shaved blue, his trousers were sharply creased,
you could have used his shoes for mirrors, and his manner was alert, wide-awake, relaxed, and rested. You got
the impression that he never needed to sleep—just ten-
thousand-mile checkups and dust him off occasionally.

He bellowed, "C'p*nee! Atten . . . shut!* I am Career
Ship's Sergeant Zim, your company commander. When
you speak to me, you will salute and say, 'Sir'—you
will salute and 'sir' anyone who carries an instructor's
baton—" He was carrying a swagger cane and now made
a quick reverse moulinet with it to show what he meant
by an instructor's baton; I had noticed men carrying
them when we had arrived the night before and had intended to get one myself—they looked smart. Now I
changed my mind. "—because we don't have enough officers around here for you to practice on. You'll practice
on us. Who sneezed?"

No answer—

"WHO SNEEZED?"

"I did," a voice answered.

"'I did' *what*?"

"I sneezed."

"'I sneezed,' SIR!"

"I sneezed, sir. I'm cold, sir."

"Oho!" Zim strode up to the man who had sneezed, shoved the ferrule of the swagger cane an inch under his nose and demanded, "Name?"

"Jenkins . . . sir."

"Jenkins . . ." Zim repeated as if the word were somehow distasteful, even shameful. "I suppose some night on patrol you're going to sneeze just because you've got a runny nose. Eh?"

"I hope not, sir."

"So do I. But you're cold. Hmm . . . we'll fix that." He pointed with his stick. "See that armory over there?" I looked and could see nothing but prairie except for one building that seemed to be almost on the skyline.

"Fall out. Run around it. *Run,* I said. Fast! Bronski! Pace him."

"Right, Sarge." One of the five or six other baton carriers took out after Jenkins, caught up with him easily, cracked him across the tight of his pants with the baton. Zim turned back to the rest of us, still shivering at attention. He walked up and down, looked us over, and seemed awfully unhappy. At last he stepped out in front of us, shook his head, and said, apparently to himself but he had a voice that carried: "To think that this had to happen to *me*!"

He looked at us. "You apes—No, not 'apes'; you don't rate that much. You pitiful mob of sickly monkeys . . . you sunken-chested, slack-bellied, drooling refugees from apron strings. In my whole life I never saw such a disgraceful huddle of momma's spoiled little darlings in—you, there! Suck up the gut! Eyes front! I'm talking to *you*!"

I pulled in my belly, even though I was not sure he had addressed me. He went on and on and I began to forget my goose flesh in hearing him storm. He never once repeated himself and he never used either profanity or obscenity. (I learned later that he saved those for *very* special occasions, which this wasn't.) But he described our shortcomings, physical, mental, moral, and genetic, in great and insulting detail.

But somehow I was not insulted; I became greatly interested in studying his command of language. I wished that we had had him on our debate team.

At last he stopped and seemed about to cry. "I can't *stand* it," he said bitterly. "I've just got to work some of it off—I had a better set of wooden soldiers when I was six. ALL RIGHT! Is there any one of you jungle lice who thinks he can whip me? Is there a *man* in the crowd? Speak up!"

There was a short silence to which I contributed. I didn't have any doubt at all that he could whip me; I was convinced.

I heard a voice far down the line, the tall end. "Ah reckon ah can . . . suh."

Zim looked happy. "Good! Step out here where I can see you." The recruit did so and he was impressive, at least three inches taller than Sergeant Zim and broader across the shoulders. "What's your name, soldier?"

"Breckinridge, suh—and ah weigh two hundred and ten pounds an' theah ain't *any* of it 'slack-bellied.'"

"Any particular way you'd like to fight?"

"Suh, you jus' pick youah own method of dyin'. Ah'm not fussy."

"Okay, no rules. Start whenever you like." Zim tossed his baton aside.

It started—and it was over. The big recruit was sitting on the ground, holding his left wrist in his right hand. He didn't say anything.

Zim bent over him. "Broken?"

"Reckon it might be . . . suh."

"I'm sorry. You hurried me a little. Do you know where the dispensary is? Never mind—Jones! Take Breckinridge over to the dispensary." As they left Zim slapped him on the right shoulder and said quietly, "Let's try it again in a month or so. I'll show you what happened." I think it was meant to be a private remark but they were standing about six feet in front of where I was slowly freezing solid.

Zim stepped back and called out, "Okay, we've got one man in this company, at least. I feel better. Do we have another one? Do we have two more? Any two of you scrofulous toads think you can stand up to me?"

He looked back and forth along our ranks. "Chicken-livered, spineless—oh, oh! Yes? Step out."

Two men who had been side by side in ranks stepped out together; I suppose they had arranged it in whispers right there, but they also were far down the tall end, so I didn't hear. Zim smiled at them. "Names, for your next of kin, please."

"Heinrich."

"Heinrich *what*?"

"Heinrich, sir. Bitte." He spoke rapidly to the other recruit and added politely, "He doesn't speak much Standard English yet, sir."

"Meyer, mein Herr," the second man supplied.

"That's okay, lots of 'em don't speak much of it when they get here—I didn't myself. Tell Meyer not to worry, he'll pick it up. But he understands what we are going to do?"

"Jawohl," agreed Meyer.

"Certainly, sir. He understands Standard, he just can't speak it fluently."

"All right. Where did you two pick up those face scars? Heidelberg?"

"Nein—no, sir. Königsberg."

"Same thing." Zim had picked up his baton after fighting Breckinridge; he twirled it and asked, "Perhaps you would each like to borrow one of these?"

"It would not be fair to you, sir," Heinrich answered carefully. "Bare hands, if you please."

"Suit yourself. Though I might fool you. Königsberg, eh? Rules?"

"How can there be rules, sir, with three?"

"An interesting point. Well, let's agree that if eyes are gouged out they must be handed back when it's over. And tell your Korpsbruder that I'm ready now. Start when you like." Zim tossed his baton away; someone caught it.

"You joke, sir. We will not gouge eyes."

"No eye gouging, agreed. 'Fire when ready, Gridley.'"

"Please?"

"Come on and fight! Or get back into ranks!"

Now I am not sure that I saw it happen this way; I may have learned part of it later, in training. But here is what I think happened: The two moved out on each side of our company commander until they had him completely flanked but well out of contact. From this position there is a choice of four basic moves for the man working alone, moves that take advantage of his own mobility and of the superior co-ordination of one man as compared with two—Sergeant Zim says (correctly) that any group is weaker than a man alone unless they are perfectly trained to work together. For example, Zim could have feinted at one of them, bounced fast to the other with a disabler, such as a broken kneecap—then finished off the first at his leisure.

Instead he let them attack. Meyer came at him fast, intending to body check and knock him to the ground, I think, while Heinrich would follow through from above, maybe with his boots. That's the way it appeared to start.

And here's what I think I saw. Meyer never reached him with that body check. Sergeant Zim whirled to face him, while kicking out and getting Heinrich in the belly—and then Meyer was sailing through the air, his lunge helped along with a hearty assist from Zim.

But all I am sure of is that the fight started and then there were two German boys sleeping peacefully, almost end to end, one face down and one face up, and Zim was standing over them, not even breathing hard. "Jones," he said. "No, Jones left, didn't he? Mahmud! Let's have the water bucket, then stick them back into their sockets. Who's got my toothpick?"

A few moments later the two were conscious, wet, and back in ranks. Zim looked at us and inquired gently, "Anybody else? Or shall we get on with setting-up exercises?"

I didn't expect anybody else and I doubt if he did. But from down on the left flank, where the shorties hung out, a boy stepped out of ranks, came front and center. Zim looked down at him. "Just you? Or do you want to pick a partner?"

"Just myself, sir."

"As you say. Name?"

"Shujumi, sir."

Zim's eyes widened. "Any relation to Colonel Shujumi?"

"I have the honor to be his son, sir."

"Ah so! Well! Black Belt?"

"No, sir. Not yet."

"I'm glad you qualified that. Well, Shujumi, are we going to use contest rules, or shall I send for the ambulance?"

"As you wish, sir. But I think, if I may be permitted an opinion, that contest rules would be more prudent."

"I don't know just how you mean that, but I agree." Zim tossed his badge of authority aside, then, so help me, they backed off, faced each other, and bowed.

After that they circled around each other in a half crouch, making tentative passes with their hands, and looking like a couple of roosters.

Suddenly they touched—and the little chap was down on the ground and Sergeant Zim was flying through the air over his head. But he didn't land with the dull, breath-paralyzing thud that Meyer had; he lit rolling and was on his feet as fast as Shujumi was and facing him. "Banzai!" Zim yelled and grinned.

"Arigato," Shujumi answered and grinned back.

They touched again almost without a pause and I thought the Sergeant was going to fly again. He didn't; he slithered straight in, there was a confusion of arms and legs and when the motion slowed down you could see that Zim was tucking Shujumi's left foot in his right ear—a poor fit.

Shujumi slapped the ground with a free hand; Zim let him up at once. They again bowed to each other.

"Another fall, sir?"

"Sorry. We've got work to do. Some other time, eh? For fun . . . and honor. Perhaps I should have told you; your honorable father trained me."

"So I had already surmised, sir. Another time it is."

Zim slapped him hard on the shoulder. "Back in ranks, soldier. *C'pnee!*"

Then, for twenty minutes, we went through calisthenics that left me as dripping hot as I had been shivering cold. Zim led it himself, doing it all with us and shouting the count. He hadn't been mussed that I could see; he wasn't breathing hard as we finished. He never led the exercises after that morning (we never saw him again before breakfast; rank hath its privileges), but he did that morning, and when it was over and we were all bushed, he led us at a trot to the mess tent, shouting at us the whole way to "Step it up! On the bounce! You're dragging your tails!"

We *always* trotted *everywhere* at Camp Arthur Currie. I never did find out who Currie was, but he must have been a trackman.

Breckinridge was already in the mess tent, with a cast on his wrist but thumb and fingers showing. I heard him say, "Naw, just a greenstick fractchuh—ah've played a whole quahtuh with wuss. But you wait—ah'll fix him."

I had my doubts. Shujumi, maybe—but not that big ape. He simply didn't know when he was outclassed. I disliked Zim from the first moment I laid eyes on him. But he had style.

Breakfast was all right—all the meals were all right; there was none of that nonsense some boarding schools have of making your life miserable at the table. If you wanted to slump down and shovel it in with both hands, nobody bothered you—which was good, as meals were practically the only time somebody wasn't riding you. The menu for breakfast wasn't anything like what I had been used to at home and the civilians that waited on us slapped the food around in a fashion that would have made Mother grow pale and leave for her room—but it was hot and it was plentiful and the cooking was okay if plain. I ate about four times what I normally do and washed it down with mug after mug of coffee with cream and lots of sugar—I would have eaten a shark without stopping to skin him.

Jenkins showed up with Corporal Bronski behind him as I was starting on seconds. They stopped for a moment at a table where Zim was eating alone, then Jenkins slumped onto a vacant stool by mine. He looked mighty seedy—pale, exhausted, and his breath rasping. I said, "Here, let me pour you some coffee."

He shook his head.

"You better eat," I insisted. "Some scrambled eggs— they'll go down easily."

"Can't eat. Oh, that dirty, dirty so-and-so." He began cussing out Zim in a low, almost expressionless mono- tone. "All I asked him was to let me go lie down and skip breakfast. Bronski wouldn't let me—said I had to see the company commander. So I did and I *told* him I was sick, I *told* him. He just felt my cheek and counted my pulse and told me sick call was nine o'clock. Wouldn't let me go back to my tent. Oh, that rat! I'll catch him on a dark night, I will."

I spooned out some eggs for him anyway and poured coffee. Presently he began to eat. Sergeant Zim got up to leave while most of us were still eating, and stopped by our table. "Jenkins."

"Uh? Yes, sir."

"At oh-nine-hundred muster for sick call and see the doctor."

Jenkins' jaw muscles twitched. He answered slowly, "I don't need any pills—sir. I'll get by."

"Oh-nine-hundred. That's an order." He left.

Jenkins started his monotonous chant again. Finally he slowed down, took a bite of eggs and said somewhat more loudly, "I can't help wondering what kind of a mother produced *that*. I'd just like to have a look at her, that's all. Did he ever *have* a mother?"

It was a rhetorical question but it got answered. At the head of our table, several stools away, was one of the instructor-corporals. He had finished eating and was smoking and picking his teeth, simultaneously; he had evidently been listening. "Jenkins—"

"Uh—sir?"

"Don't you know about sergeants?"

"Well . . . I'm learning."

"They don't have mothers. Just ask any trained private." He blew smoke toward us. "They reproduce by fission . . . like all bacteria."

CH:04

And the LORD said unto Gideon, The people that are with thee are
too many . . . Now therefore go to, proclaim in the ears of the people,
saying, Whosoever is fearful and afraid, let him return . . . And
there returned of the people twenty and two thousand; and there re-
mained ten thousand. And the LORD said unto Gideon, The people
are yet too many; bring them down unto the water, and I will try them
for thee there . . . so he brought down the people unto the water: and
the LORD said unto Gideon, Every one that lappeth of the water with
his tongue, as a dog lappeth, him shalt thou set by himself; likewise
everyone that boweth down upon his knees to drink. And the num-
ber of them that drank, putting their hand to their mouth, were three
hundred men . . .

And the LORD said unto Gideon, By the three hundred . . . will
I save you . . . let all the other people go . . .

—Judges VII:2-7

Two weeks after we got there they took our cots away
from us. That is to say that we had the dubious pleasure
of folding them, carrying them four miles, and stowing
them in a warehouse. By then it didn't matter; the
ground seemed much warmer and quite soft—especially
when the alert sounded in the middle of the night and we
had to scramble out and play soldier. Which it did about
three times a week. But I could get back to sleep after one

of those mock exercises at once; I had learned to sleep any place, any time—sitting up, standing up, even marching in ranks. Why, I could even sleep through evening parade standing at attention, enjoy the music without being waked by it—and wake instantly at the command to pass in review.

I made a very important discovery at Camp Currie. Happiness consists in getting enough sleep. Just that, nothing more. All the wealthy, unhappy people you've ever met take sleeping pills; Mobile Infantrymen don't need them. Give a cap trooper a bunk and time to sack out in it and he's as happy as a worm in an apple—asleep.

Theoretically you were given eight full hours of sack time every night and about an hour and a half after evening chow for your own use. But in fact your night sack time was subject to alerts, to night duty, to field marches, and to acts of God and the whims of those over you, and your evenings, if not ruined by awkward squad or extra duty for minor offenses, were likely to be taken up by shining shoes, doing laundry, swapping haircuts (some of us got to be pretty fair barbers but a clean sweep like a billiard ball was acceptable and anybody can do that)—not to mention a thousand other chores having to do with equipment, person, and the demands of sergeants. For example we learned to answer morning roll call with: "Bathed!" meaning you had taken at least one bath since last reveille. A man might lie about it and get away with it (I did, a couple of times) but at least one in our company who pulled that dodge in the face of convincing evidence that he was not recently bathed got scrubbed with stiff brushes and floor soap by his squad mates while a corporal-instructor chaperoned and made helpful suggestions.

But if you didn't have more urgent things to do after supper, you could write a letter, loaf, gossip, discuss the myriad mental and moral shortcomings of sergeants and, dearest of all, talk about the female of the species (we became convinced that there were no such creatures, just mythology created by inflamed imaginations—one boy in our company claimed to have seen a girl, over at regimental headquarters; he was unanimously judged a liar and a braggart). Or you could play cards. I learned, the hard way, not to draw to an inside straight and I've never done it since. In fact I haven't played cards since.

Or, if you actually did have twenty minutes of your very own, you could sleep. This was a choice very highly thought of; we were always several weeks minus on sleep.

I may have given the impression that boot camp was made harder than necessary. This is not correct.

It was made *as hard as possible* and on purpose.

It was the firm opinion of every recruit that this was sheer meanness, calculated sadism, fiendish delight of witless morons in making other people suffer.

It was not. It was too scheduled, too intellectual, too efficiently and impersonally organized to be cruelty for the sick pleasure of cruelty; it was planned like surgery for purposes as unimpassioned as those of a surgeon. Oh, I admit that some of the instructors may have enjoyed it but I don't *know* that they did—and I *do* know (now) that the psych officers tried to weed out any bullies in selecting instructors. They looked for skilled and dedicated craftsmen to follow the art of making things as tough as possible for a recruit; a bully is too stupid, himself too emotionally involved and too likely to grow tired of his fun and slack off, to be efficient.

Still, there may have been bullies among them. But I've heard that some surgeons (and not necessarily bad ones) enjoy the cutting and the blood which accompanies the humane art of surgery.

That's what it was: surgery. Its immediate purpose was to get rid of, run right out of the outfit, those recruits who were too soft or too babyish ever to make Mobile Infantrymen. It accomplished that, in droves. (They darn near ran *me* out.) Our company shrank to platoon size in the first six weeks. Some of them were dropped without prejudice and allowed, if they wished, to sweat out their terms in the non-combatant services; others got Bad Conduct Discharges, or Unsatisfactory Performance Discharges, or Medical Discharges.

Usually you didn't know why a man left unless you saw him leave and he volunteered the information. But some of them got fed up, said so loudly, and resigned, forfeiting forever their chances of franchise. Some, especially the older men, simply couldn't stand the pace physically no matter how hard they tried. I remember one, a nice old geezer named Carruthers, must have been thirty-five; they carried him away in a stretcher while he was still shouting feebly that it wasn't fair!— and that he would be back.

It was sort of sad, because we liked Carruthers and he

did try—so we looked the other way and figured we would never see him again, that he was a cinch for a medical discharge and civilian clothes. Only I *did* see him again, long after. He had refused discharge (you don't have to accept a medical) and wound up as third cook in a troop transport. He remembered me and wanted to talk old times, as proud of being an alumnus of Camp Currie as Father is of his Harvard accent—he felt that he was a little bit better than the ordinary Navy man. Well, maybe he was.

But, much more important than the purpose of carving away the fat quickly and saving the government the training costs of those who would never cut it, was the prime purpose of making as sure as was humanly possible that no cap trooper ever climbed into a capsule for a combat drop unless he was prepared for it—fit, resolute, disciplined and skilled. If he is not, it's not fair to the Federation, it's certainly not fair to his teammates, and worst of all it's not fair to *him*.

But was boot camp more cruelly hard than was necessary?

All I can say to that is this: The next time I have to make a combat drop, I want the men on my flanks to be graduates of Camp Currie or its Siberian equivalent. Otherwise I'll refuse to enter the capsule.

But I certainly thought it was a bunch of crumby, vicious nonsense at the time. Little things— When we were there a week, we were issued undress maroons for parade to supplement the fatigues we had been wearing. (Dress and full-dress uniforms came much later.) I took my tunic back to the issue shed and complained to the supply sergeant. Since he was only a supply sergeant and rather fatherly in manner I thought of him as a semi-civilian—I didn't know how, as of then, to read the ribbons on his chest or I wouldn't have dared speak to him. "Sergeant, this tunic is too large. My company commander says it fits like a tent."

He looked at the garment, didn't touch it. "Really?"

"Yeah. I want one that fits."

He still didn't stir. "Let me wise you up, sonny boy. There are just two sizes in this army—too large and too small."

"But my company commander—"

"No doubt."

"But what am I going to *do*?"

"Oh, it's *advice* you want! Well, I've got that in stock—new issue, just today. Mmm . . . tell you what I'll do. Here's a needle and I'll even give you a spool of thread. You won't need a pair of scissors; a razor blade is better. Now you tight 'em plenty across the hips but leave cloth to loose 'em again across the shoulders; you'll need it later."

Sergeant Zim's only comment on my tailoring was: "You can do better than that. Two hours extra duty."

So I did better than that by next parade.

Those first six weeks were all hardening up and hazing, with lots of parade drill and lots of route march. Eventually, as files dropped out and went home or elsewhere, we reached the point where we could do fifty miles in ten hours on the level—which is good mileage for a good horse in case you've never used your legs. We rested, not by stopping, but by changing pace, slow march, quick march, and trot. Sometimes we went out the full distance, bivouacked and ate field rations, slept in sleeping bags and marched back the next day.

One day we started out on an ordinary day's march, no bed bags on our shoulders, no rations. When we didn't stop for lunch, I wasn't surprised, as I had already learned to sneak sugar and hard bread and such out of the mess tent and conceal it about my person, but when we kept on marching away from camp in the afternoon I began to wonder. But I had learned not to ask silly questions.

We halted shortly before dark, three companies, now somewhat abbreviated. We formed a battalion parade and marched through it, without music, guards were mounted, and we were dismissed. I immediately looked up Corporal-Instructor Bronski because he was a little easier to deal with than the others . . . and because I felt a certain amount of responsibility; I happened to be, at the time, a recruit-corporal myself. These boot chevrons didn't mean much—mostly the privilege of being chewed out for whatever your squad did as well as for what you did yourself—and they could vanish as quickly as they appeared. Zim had tried out all of the older men as temporary non-coms first and I had inherited a brassard with chevrons on it a couple of days before when our squad leader had folded up and gone to hospital.

I said, "Corporal Bronski, what's the straight word? When is chow call?"

He grinned at me. "I've got a couple of crackers on me. Want me to split 'em with you?"

"Huh? Oh, no, sir. Thank you." (I had considerably more than a couple of crackers; I was learning.) "No chow call?"

"They didn't tell me either, sonny. But I don't see any copters approaching. Now if I was you, I'd round up my squad and figure things out. Maybe one of you can hit a jack rabbit with a rock."

"Yes, sir. But— Well, are we staying here all night? We don't have our bedrolls."

His eyebrows shot up. "No bedrolls? Well, I do declare!" He seemed to think it over. "Mmm . . . ever see sheep huddle together in a snowstorm?"

"Oh, no, sir."

"Try it. They don't freeze, maybe you won't. Or if you don't care for company, you might walk around all night. Nobody'll bother you, as long as you stay inside the posted guards. You won't freeze if you keep moving. Of course you may be a little tired tomorrow." He grinned again.

I saluted and went back to my squad. We divvied up, share and share alike—and I came out with less food than I had started with; some of those idiots either hadn't sneaked out anything to eat, or had eaten all they had while we marched. But a few crackers and a couple of prunes will do a lot to quiet your stomach's sounding alert.

The sheep trick works, too; our whole section, three squads, did it together. I don't recommend it as a way to sleep; you are either in the outer layer, frozen on one side and trying to worm your way inside, or you are inside, fairly warm but with everybody else trying to shove his elbows, feet, and halitosis on you. You migrate from one condition to the other all night long in a sort of a Brownian movement, never quite waking up and never really sound asleep. All this makes a night about a hundred years long.

We turned out at dawn to the familiar shout of: "Up you come! On the bounce!" encouraged by instructors' batons applied smartly on fundaments sticking out of the piles . . . and then we did setting-up exercises. I felt like a corpse and didn't see how I could touch my toes. But I did, though it hurt, and twenty minutes later when we hit the trail I merely felt elderly. Sergeant Zim wasn't even mussed and somehow the scoundrel had managed to shave.

The Sun warmed our backs as we marched and Zim started us singing, oldies at first, like "Le Regiment de Sambre et Meuse" and "Caissons" and "Halls of Montezuma" and then our own "Cap Trooper's Polka" which moves you into quickstep and pulls you on into a trot. Sergeant Zim couldn't carry a tune in a sack; all he had was a loud voice. But Breckinridge had a sure, strong lead and could hold the rest of us in the teeth of Zim's terrible false notes. We all felt cocky and covered with spines.

But we didn't feel cocky fifty miles later. It had been a long night; it was an endless day—and Zim chewed us out for the way we looked on parade and several boots got gigged for failing to shave in the nine whole minutes between the time we fell out after the march and fell back in again for parade. Several recruits resigned that evening and I thought about it but didn't because I had those silly boot chevrons and hadn't been busted yet.

That night there was a two-hour alert.

But eventually I learned to appreciate the homey luxury of two or three dozen warm bodies to snuggle up to, because twelve weeks later they dumped me down raw naked in a primitive area of the Canadian Rockies and I had to make my way forty miles through mountains. I made it—and hated the Army every inch of the way.

I wasn't in too bad shape when I checked in, though. A couple of rabbits had failed to stay as alert as I was, so I didn't go entirely hungry . . . nor entirely naked; I had a nice warm thick coat of rabbit fat and dirt on my body and moccasins on my feet—the rabbits having no further use for their skins. It's amazing what you can do with a flake of rock if you have to—I guess our cave-man ancestors weren't such dummies as we usually think.

The others made it, too, those who were still around to try and didn't resign rather than take the test—all except two boys who died trying. Then we all went back into the mountains and spent thirteen days finding them, working with copters overhead to direct us and all the best communication gear to help us and our instructors in powered command suits to supervise and to check rumors—because the Mobile Infantry doesn't abandon its own while there is any thin shred of hope.

Then we buried them with full honors to the strains of "This Land Is Ours" and with the posthumous rank of PFC, the first of our boot regiment to go that high—because a cap trooper isn't necessarily expected to stay alive (dying is part of his trade) . . . but they care a lot about *how* you die. It has to be heads up, on the bounce, and still trying.

Breckinridge was one of them; the other was an Aussie boy I didn't know. They weren't the first to die in training; they weren't the last.

CH:05

He's bound *to be* guilty *'r he*
wouldn't *be* here!
Starboard gun . . . FIRE!

Shoo*ting's too* good *for 'im,*
kick *the louse* out!
Port gun . . . FIRE!

—Ancient chanty used to time
saluting guns

But that was after we had left Camp Currie and a lot
had happened in between. Combat training, mostly—
combat drill and combat exercises and combat maneuvers, using everything from bare hands to simulated
nuclear weapons. I hadn't known there were so many
different ways to fight. Hands and feet to start with—
and if you think those aren't weapons you haven't seen
Sergeant Zim and Captain Frankel, our battalion commander, demonstrate *la savate*, or had little Shujumi
work you over with just his hands and a toothy grin—Zim
made Shujumi an instructor for that purpose at once and
required us to take his orders, although we didn't have to
salute him and say "sir."

As our ranks thinned down Zim quit bothering with formations himself, except parade, and spent more and more time in personal instruction, supplementing the corporal-instructors. He was sudden death with anything but he loved knives, and made and balanced his own, instead of using the perfectly good general-issue ones. He mellowed quite a bit as a personal teacher, too, becoming merely unbearable instead of downright disgusting—he could be quite patient with silly questions.

Once, during one of the two-minute rest periods that were scattered sparsely through each day's work, one of the boys—a kid named Ted Hendrick—asked, "Sergeant? I guess this knife throwing is fun . . . but why do we have to learn it? What possible use is it?"

"Well," answered Zim, "suppose all you have is a knife? Or maybe not even a knife? What do you do? Just say your prayers and die? Or wade in and make him buy it anyhow? Son, this is *real*—it's not a checker game you can concede if you find yourself too far behind."

"But that's just what I mean, sir. Suppose you aren't armed at all? Or just one of these toadstickers, say? And the man you're up against has all sorts of dangerous weapons? There's nothing you can do about it; he's got you licked on showdown."

Zim said almost gently, "You've got it all wrong, son. There's no such thing as a 'dangerous weapon.'"

"Huh? Sir?"

"There are no dangerous weapons; there are only dangerous men. We're trying to teach you to be dangerous—

to the enemy. Dangerous even without a knife. Deadly as long as you still have one hand or one foot and are still alive. If you don't know what I mean, go read 'Horatius at the Bridge' or 'The Death of the Bon Homme Richard'; they're both in the Camp library. But take the case you first mentioned; I'm you and all you have is a knife. That target behind me—the one you've been missing, number three—is a sentry, armed with everything but an H-bomb. You've got to get him . . . quietly, at once, and without letting him call for help." Zim turned slightly—*thunk!*—a knife he hadn't even had in his hand was quivering in the center of target number three. "You see? Best to carry two knives—but get him you must, even barehanded."

"Uh—"

"Something still troubling you? Speak up. That's what I'm here for, to answer your questions."

"Uh, yes, sir. You said the sentry didn't have any H-bomb. But he *does* have an H-bomb; that's just the point. Well, at least we have, if we're the sentry . . . and any sentry we're up against is likely to have them, too. I don't mean the sentry, I mean the side he's on."

"I understood you."

"Well . . . you see, sir? If we can use an H-bomb—and, as you said, it's no checker game; it's real, it's war and nobody is fooling around—isn't it sort of ridiculous to go crawling around in the weeds, throwing knives and maybe getting yourself killed . . . and even losing the war . . . when you've got a real weapon you can use to win? What's the point in a whole lot of men risking their lives with obsolete weapons when one professor type can do so much more just by pushing a button?"

Zim didn't answer at once, which wasn't like him at all. Then he said softly, "Are you happy in the Infantry, Hendrick? You can resign, you know."

Hendrick muttered something; Zim said, "Speak up!"

"I'm not itching to resign, sir. I'm going to sweat out my term."

"I see. Well, the question you asked is one that a sergeant isn't really qualified to answer . . . and one that you shouldn't ask me. You're supposed to *know* the answer before you join up. Or you should. Did your school have a course in History and Moral Philosophy?"

"What? Sure—yes, sir."

"Then you've heard the answer. But I'll give you my own—unofficial—views on it. If you wanted to teach a baby a lesson, would you cut its head off?"

"Why . . . no, sir!"

"Of course not. You'd paddle it. There can be circumstances when it's just as foolish to hit an enemy city with an H-bomb as it would be to spank a baby with an ax. War is not violence and killing, pure and simple; war is *controlled* violence, for a purpose. The purpose of war is to support your government's decisions by force. The purpose is never to kill the enemy just to be killing him . . . but to make him do what you want him to do. Not killing . . . but controlled and purposeful violence. But it's not your business or mine to decide the purpose of the control. It's never a soldier's business to decide when or where or how—or *why*—he fights; that belongs to the statesmen and the generals. The statesmen decide why and how much; the generals take it from there and tell us where and when and how. *We* supply the violence; other people—'older and wiser heads,' as they say—supply the control. Which is as it should be. That's the best answer I can give you. If it doesn't satisfy you, I'll get you a chit to go talk to the regimental commander. If *he* can't convince you—then go home and be a civilian! Because in that case you will certainly never make a soldier."

Zim bounced to his feet. "I think you've kept me talking just to goldbrick. Up you come, soldiers! On the bounce! Man stations, on target—Hendrick, you first. This time I want you to throw that knife south of you. *South*, get it? Not north. The target is due south of you and I want that knife to go in a general southerly direction, at least. I know you won't hit the target but see if you can't scare it a little. Don't slice your ear off, don't let go of it and cut somebody behind you—just keep what tiny mind you have fixed on the idea of 'south'! Ready—on target! *Let fly!*"

Hendrick missed it again.

We trained with sticks and we trained with wire (lots of nasty things you can improvise with a piece of wire) and we learned what can be done with really modern weapons and how to do it and how to service and maintain the equipment—simulated nuclear weapons and infantry rockets and various sorts of gas and poison and incendiary and demolition. As well as other things maybe best not discussed. But we learned a lot of "obsolete" weapons, too. Bayonets on dummy guns for example, and guns that weren't dummies, too, but were almost identical with the infantry rifle of the XXth century—much like the sporting rifles used in hunting game, except that we fired nothing but solid slugs, alloy-jacketed lead bullets, both at targets on measured ranges and at surprise targets on booby-trapped skirmish runs. This was supposed to prepare us to learn to use any armed weapon and to train us to be on the bounce, alert, ready for anything. Well, I suppose it did. I'm pretty sure it did.

We used these rifles in field exercises to simulate a lot of deadlier and nastier aimed weapons, too. We used a lot of simulation; we had to. An "explosive" bomb or grenade, against matériel or personnel, would explode just enough to put out a lot of black smoke; another sort of gave off a gas that would make you sneeze and weep—that told you that you were dead or paralyzed . . . and was nasty enough to make you careful about anti-gas precautions, to say nothing of the chewing-out you got if you were caught by it.

We got still less sleep; more than half the exercises were held at night, with snoopers and radar and audio gear and such.

The rifles used to simulate aimed weapons were loaded with blanks except one in five hundred rounds at random, which was a real bullet. Dangerous? Yes and no. It's dangerous just to be alive . . . and a nonexplosive bullet probably won't kill you unless it hits you in the head or the heart and maybe not then. What that one-in-five-hundred "for real" did was to give us a deep interest in taking cover, especially as we knew that some of the rifles were being fired by instructors who were crack shots and actually trying their best to hit you—if the round happened not to be a blank. They assured us that they would not intentionally shoot a man in the head . . . but accidents do happen.

This friendly assurance wasn't very reassuring. That 500th bullet turned tedious exercises into large-scale Russian roulette; you stop being bored the very first time you hear a slug go *wheet!* past your ear before you hear the crack of the rifle.

But we did slack down anyhow and word came down from the top that if we didn't get on the bounce, the incidence of real ones would be changed to one in a hundred . . . and if that didn't work, to one in fifty. I don't know whether a change was made or not—no way to tell—but I do know we tightened up again, because a boy in the next company got creased across his buttocks with a live one, producing an amazing scar and a lot of half-witty comments and a renewed interest by all hands in taking cover. We laughed at this kid for getting shot where he did . . . but we all knew it could have been his head—or our *own* heads.

The instructors who were not firing rifles did not take cover. They put on white shirts and walked around upright with their silly canes, apparently calmly certain that even a recruit would not intentionally shoot an instructor—which may have been overconfidence on the part of some of them. Still, the chances were five hundred to one that even a shot aimed with murderous intent would not be live and the safety factor increased still higher because the recruit probably couldn't shoot that well anyhow. A rifle is not an easy weapon; it's got no target-seeking qualities at all—I understand that even back in the days when wars were fought and decided with just such rifles it used to take several thousand fired shots to average killing one man. This seems impossible but the military histories agree that it is true—apparently most shots weren't really aimed but simply acted to force the enemy to keep his head down and interfere with *his* shooting.

In any case we had no instructors wounded or killed by rifle fire. No trainees were killed, either, by rifle bullets; the deaths were all from other weapons or things—some of which could turn around and bite you if you didn't do things by the book. Well, one boy did manage to break his neck taking cover too enthusiastically when they first started shooting at him—but no bullet touched him.

However, by a chain reaction, this matter of rifle bullets and taking cover brought me to my lowest ebb at Camp Currie. In the first place I had been busted out of my boot chevrons, not over what I did but over something one of my squad did when I wasn't even around . . . which I pointed out. Bronski told me to button my lip. So I went to see Zim about it. He told me coldly that I was responsible for what my men did, regardless . . . and tacked on six hours of extra duty besides busting me for having spoken to him about it without Bronski's permission. Then I got a letter that upset me a lot; my mother finally wrote to me. Then I sprained a shoulder in my first drill with powered armor (they've got those practice suits rigged so that the instructor can cause casualties in the suit at will, by radio control; I got dumped and hurt my shoulder) and this put me on light duty with too much time to think at a time when I had many reasons, it seemed to me, to feel sorry for myself.

Because of "light duty" I was orderly that day in the battalion commander's office. I was eager at first. for I had never been there before and wanted to make a good impression. I discovered that Captain Frankel didn't want zeal; he wanted me to sit still, say nothing, and not bother him. This left me time to sympathize with myself, for I didn't dare go to sleep.

Then suddenly, shortly after lunch, I wasn't a bit sleepy; Sergeant Zim came in, followed by three men. Zim was smart and neat as usual but the expression on his face made him look like Death on a pale horse and he had a mark on his right eye that looked as if it might be shaping up into a shiner—which was impossible, of course. Of the other three, the one in the middle was Ted Hendrick. He was dirty—well, the company had been on a field exercise; they don't scrub those prairies and you spend a lot of your time snuggling up to the dirt. But his lip was split and there was blood on his chin and on his shirt and his cap was missing. He looked wild-eyed.

The men on each side of him were boots. They each had rifles; Hendrick did not. One of them was from my squad, a kid named Leivy. He seemed excited and pleased, and slipped me a wink when nobody was looking.

Captain Frankel looked surprised. "What is this, Sergeant?"

Zim stood frozen straight and spoke as if he were reciting something by rote. "Sir, H Company Commander reports to the Battalion Commander. Discipline. Article nine-one-oh-seven. Disregard of tactical command and doctrine, the team being in simulated combat. Article nine-one-two-oh. Disobedience of orders, same conditions."

Captain Frankel looked puzzled. "You are bringing this to *me*, Sergeant? Officially?"

I don't see how a man can manage to look as embarrassed as Zim looked and still have no expression of any sort in his face or voice. "Sir. If the Captain pleases. The man refused administrative discipline. He insisted on seeing the Battalion Commander."

"I see. A bedroll lawyer. Well, I still don't understand it, Sergeant, but technically that's his privilege. What was the tactical command and doctrine?"

"A 'freeze,' sir." I glanced at Hendrick, thinking: Oh, oh, he's going to catch it. In a "freeze" you hit dirt, taking any cover you can, fast, and then *freeze*— don't move at all, not even twitch an eyebrow, until released. Or you can freeze when you're already in cover. They tell stories about men who had been hit while in freeze . . . and had died slowly but without ever making a sound or a move.

Frankel's brows shot up. "Second part?"

"Same thing, sir. After breaking freeze, failing to return to it on being so ordered."

Captain Frankel looked grim. "Name?"

Zim answered. "Hendrick, T.C., sir. Recruit Private R-P-seven-nine-six-oh-nine-two-four."

"Very well. Hendrick, you are deprived of all privileges for thirty days and restricted to your tent when not on duty or at meals, subject only to sanitary necessities. You will serve three hours extra duty each day under the Corporal of the Guard, one hour to be served just before taps, one hour just before reveille, one hour at the time of the noonday meal and in place of it. Your evening meal will be bread and water—as much bread as you can eat. You will serve ten hours extra duty each Sunday, the time to be adjusted to permit you to attend divine services if you so elect."

(I thought: Oh my! He threw the book.)

Captain Frankel went on: "Hendrick, the only reason you are getting off so lightly is that I am not permitted to give you any more than that without convening a court-martial . . . and I don't want to spoil your company's record. Dismissed." He dropped his eyes back to the papers on his desk, the incident already forgotten—

—and Hendrick yelled, "You didn't hear *my* side of it!"

The Captain looked up. "Oh. Sorry. You have a side?"

"You're darn right I do! Sergeant Zim's got it in for

me! He's been riding me, riding me, riding me, all day long from the time I got here! He—"

"That's his job," the Captain said coldly. "Do you deny the two charges against you?"

"No, but—*He didn't tell you I was lying on an anthill.*"

Frankel looked disgusted. "Oh. So you would get yourself killed and perhaps your teammates as well because of a few little ants?"

"Not 'just a few'—there were hundreds of 'em. Stingers."

"So? Young man, let me put you straight. Had it been a nest of rattlesnakes you would still have been expected—and required—to freeze." Frankel paused. "Have you anything at all to say in your own defense?"

Hendrick's mouth was open. "I certainly do! He hit me! *He laid hands on me!* The whole bunch of 'em are always strutting around with those silly batons, whackin' you across the fanny, punchin' you between the shoulders and tellin' you to brace up—and I put up with it. But he hit me with his *hands*—he knocked me down to the ground and yelled, '*Freeze!* you stupid jackass!' How about *that*?"

Captain Frankel looked down at his hands, looked up again at Hendrick. "Young man, you are under a misapprehension very common among civilians. You think that your superior officers are not permitted to 'lay hands on you,' as you put it. Under purely social conditions, that is true—say if we happened to run across each other in a theater or a shop, I would have no more right, as long as you treated me with the respect due my rank, to slap your face than you have to slap mine. But in line of duty the rule is entirely different—"

The Captain swung around in his chair and pointed at some loose-leaf books. "There are the laws under which you live. You can search every article in those books, every court-martial case which has arisen under them, and you will not find *one word* which says, or implies, that your superior officer may not 'lay hands on you' or strike you in any other manner in line of duty. Hendrick, I could break your jaw . . . and I simply would be responsible to my own superior officers as to the appropriate necessity of the act. But I would not be responsible to *you*. I could do more than that. There are circumstances under which a superior officer, commissioned or not, is not only permitted but *required* to kill an officer or a man under him, without delay and perhaps without warning—and, far from being punished, be commended. To put a stop to pusillanimous conduct in the face of the enemy, for example."

The Captain tapped on his desk. "Now about those batons—They have two uses. First, they mark the men in authority. Second, we expect them to be used on you, to touch you up and keep you on the bounce. You can't possibly be hurt with one, not the way they are used; at most they sting a little. But they save thousands of words. Say you don't turn out on the bounce at reveille. No doubt the duty corporal could wheedle you, say 'pretty please with sugar on it,' inquire if you'd like breakfast in bed this morning—if we could spare one career corporal just to nursemaid you. We can't, so he gives your bedroll a whack and trots on down the line, applying the spur where needed. Of course he could simply kick you, which would be just as legal and nearly as effective. But the general in charge of training and discipline thinks that it is more dignified, both for the duty corporal and for you, to snap a late sleeper out of his fog with the impersonal rod of authority. And so do I. Not that it matters what you or I think about it; this is the way we do it."

Captain Frankel sighed. "Hendrick, I have explained these matters to you because it is useless to punish a man unless he knows why he is being punished. You've been a bad boy—I say 'boy' because you quite evidently aren't a man *yet*, although we'll keep trying—a surprisingly bad boy in view of the stage of your training. Nothing you have said is any defense, nor even any mitigation; you don't seem to know the score nor have any idea of your duty as a soldier. So tell me in your own words why you feel mistreated; I want to get you straightened out. There might even be something in your favor, though I confess that I cannot imagine what it could be."

I had sneaked a look or two at Hendrick's face while the Captain was chewing him out—somehow his quiet, mild words were a worse chewing-out than any Zim had ever given us. Hendrick's expression had gone from indignation to blank astonishment to sullenness.

"Speak up!" Frankel added sharply.

"Uh . . . well, we were ordered to freeze and I hit the dirt and I found I was on this anthill. So I got to my knees, to move over a couple of feet, and I was hit from behind and knocked flat and he yelled at me—and I bounced up and popped him one and he—"

"STOP!" Captain Frankel was out of his chair and standing ten feet tall, though he's hardly taller than I am. He stared at Hendrick.

"You . . . *struck* . . . your . . . company commander?"

"Huh? I said so. But he hit me first. From behind, I didn't even see him. I don't take that off of anybody. I popped him and then he hit me again and then—"

"Silence!"

Hendrick stopped. Then he added, "I just want out of this lousy outfit."

"I think we can accommodate you," Frankel said icily. "And quickly, too."

"Just gimme a piece of paper, I'm resigning."

"One moment. Sergeant Zim."

"Yes, sir." Zim hadn't said a word for a long time. He just stood, eyes front and rigid as a statue, nothing moving but his twitching jaw muscles. I looked at him now and saw that it certainly was a shiner—a beaut. Hendrick must have caught him just right. But he hadn't said anything about it and Captain Frankel hadn't asked—maybe he had just assumed Zim had run into a door and would explain it if he felt like it, later.

"Have the pertinent articles been published to your company, as required?"

"Yes, sir. Published and logged, every Sunday morning."

"I know they have. I asked simply for the record."

Just before church call every Sunday they lined us up and read aloud the disciplinary articles out of the Laws and Regulations of the Military Forces. They were posted on the bulletin board, too, outside the orderly tent. Nobody paid them much mind—it was just another drill; you could stand still and sleep through it. About the only thing we noticed, if we noticed anything, was what we called "the thirty-one ways to crash land." After all, the instructors see to it that you soak up all the regulations you need to know, through your skin. The "crash landings" were a worn-out joke, like "reveille oil" and "tent jacks" . . . they were the thirty-one capital offenses. Now and then somebody boasted, or accused somebody else, of having found a thirty-second way—always something preposterous and usually obscene.

"*Striking a superior officer—!*"

It suddenly wasn't amusing any longer. Popping Zim? *Hang* a man for that? Why, almost everybody in the company had taken a swing at Sergeant Zim and some of us had even landed . . . when he was instructing us in hand-to-hand combat. He would take us on after the other instructors had worked us over and we were beginning to feel cocky and pretty good at it—then he would put the polish on. Why, shucks, I once saw Shujumi knock him unconscious. Bronski threw water on him and Zim got up and grinned and shook hands—and threw Shujumi right over the horizon.

Captain Frankel looked around, motioned at me. "You. Flash regimental headquarters."

I did it, all thumbs, stepped back when an officer's face

came on and let the Captain take the call. "Adjutant," the face said.

Frankel said crisply, "Second Battalion Commander's respects to the Regimental Commander. I request and require an officer to sit as a court."

The face said, "When do you need him, Ian?"

"As quickly as you can get him here."

"Right away. I'm pretty sure Jake is in his HQ. Article and name?"

Captain Frankel identified Hendrick and quoted an article number. The face in the screen whistled and looked grim. "On the bounce, Ian. If I can't get Jake, I'll be over myself—just as soon as I tell the Old Man."

Captain Frankel turned to Zim. "This escort—are they witnesses?"

"Yes, sir."

"Did his section leader see it?"

Zim barely hesitated. "I think so, sir."

"Get him. Anybody out that way in a powered suit?"

"Yes, sir."

Zim used the phone while Frankel said to Hendrick, "What witnesses do you wish to call in your defense?"

"Huh? I don't need any witnesses, he knows what he did! Just hand me a piece of paper—I'm getting out of here."

"All in good time."

In very fast time, it seemed to me. Less than five minutes later Corporal Jones came bouncing up in a command suit, carrying Corporal Mahmud in his arms. He dropped Mahmud and bounced away just as Lieutenant

Spieksma came in. He said, "Afternoon, Cap'n. Accused and witnesses here?"

"All set. Take it, Jake."

"Recorder on?"

"It is now."

"Very well. Hendrick, step forward." Hendrick did so, looking puzzled and as if his nerve was beginning to crack. Lieutenant Spieksma said briskly: "Field Court-Martial, convened by order of Major F.X. Malloy, commanding Third Training Regiment, Camp Arthur Currie, under General Order Number Four, issued by the Commanding General, Training and Discipline Command, pursuant to the Laws and Regulations of the Military Forces, Terran Federation. Remanding officer: Captain Ian Frankel, M.I., assigned to and commanding Second Battalion, Third Regiment. The Court: Lieutenant Jacques Spieksma, M.I., assigned to and commanding First Battalion, Third Regiment. Accused: Hendrick, Theodore C., Recruit Private RP7960924. Article 9080. Charge: Striking his superior officer, the Terran Federation then being in a state of emergency."

The thing that got me was how *fast* it went. I found myself suddenly appointed an "officer of the court" and directed to "remove" the witnesses and have them ready. I didn't know how I would "remove" Sergeant Zim if he didn't feel like it, but he gathered Mahmud and the two boots up by eye and they all went outside, out of ear-shot. Zim separated himself from the others and simply waited; Mahmud sat down on the ground and rolled a cigarette—which he had to put out; he was the first one called. In less than twenty minutes all three of them had testified, all telling much the same story Hendrick had. Zim wasn't called at all.

Lieutenant Spieksma said to Hendrick, "Do you wish to cross-examine the witnesses? The Court will assist you, if you so wish."

"No."

"Stand at attention and say 'sir' when you address the Court."

"No, sir." He added, "I want a lawyer."

"The Law does not permit counsel in field courts-martial. Do you wish to testify in your own defense? You are not required to do so and, in view of the evidence thus far, the Court will take no judicial notice if you choose not to do so. But you are warned that any testimony that you give may be used against you and that you will be subject to cross-examination."

Hendrick shrugged. "I haven't anything to say. What good would it do me?"

"The Court repeats: Will you testify in your own defense?"

"Uh, no, sir."

"The Court must demand of you one technical question. Was the article under which you are charged published to you *before* the time of the alleged offense of which you stand accused? You may answer yes, or no, or stand mute—but you are responsible for your answer under Article 9167 which relates to perjury."

The accused stood mute.

"Very well, the Court will reread the article of the

charge aloud to you and again ask you that question. 'Article 9080: Any person in the Military Forces who strikes or assaults, or attempts to strike or assault—'"

"Oh, I suppose they did. They read a lot of stuff, every Sunday morning—a whole long list of things you couldn't do."

"Was or was not that particular article read to you?"

"Uh . . . yes, sir. It was."

"Very well. Having declined to testify, do you have any statement to make in mitigation or extenuation?"

"Sir?"

"Do you want to tell the Court anything about it? Any circumstance which you think might possibly affect the evidence already given? Or anything which might lessen the alleged offense? Such things as being ill, or under drugs or medication. You are not under oath at this point; you may say anything at all which you think may help you. What the Court is trying to find out is this: Does anything about this matter strike you as being unfair? If so, why?"

"Huh? Of course it is! Everything about it is unfair! He hit me first! You heard 'em!—*he hit me first!*"

"Anything more?"

"Huh? No, sir. Isn't that enough?"

"The trial is completed. Recruit Private Theodore C. Hendrick, stand forth!" Lieutenant Spieksma had been standing at attention the whole time; now Captain Frankel stood up. The place suddenly felt chilly.

"Private Hendrick, you are found guilty as charged."

My stomach did a flip-flop. They were going to do it to him . . . they were going to do the "Danny Deever" to Ted Hendrick. And I had eaten breakfast beside him just this morning.

"The Court sentences you," he went on, while I felt sick, "to ten lashes and Bad Conduct Discharge."

Hendrick gulped. "I want to resign!"

"The Court does not permit you to resign. The Court wishes to add that your punishment is light simply because this Court possesses no jurisdiction to assign greater punishment. The authority which remanded you specified a field court-martial—why it so chose, this Court will not speculate. But had you been remanded for general court-martial, it seems certain that the evidence before this Court would have caused a general court to sentence you to hang by the neck until dead. You are very lucky—and the remanding authority has been most merciful." Lieutenant Spieksma paused, then went on, "The sentence will be carried out at the earliest hour after the convening authority has reviewed and approved the record, if it does so approve. Court is adjourned. Remove and confine him."

The last was addressed to me, but I didn't actually have to do anything about it, other than phone the guard tent and then get a receipt for him when they took him away.

At afternoon sick call Captain Frankel took me off orderly and sent me to see the doctor, who sent me back to duty. I got back to my company just in time to dress and fall in for parade—and to get gigged by Zim for "spots on uniform." Well, he had a bigger spot over one eye but I didn't mention it.

Somebody had set up a big post in the parade ground just back of where the adjutant stood. When it came time to publish the orders, instead of "routine order of the day" or other trivia, they published Hendrick's court-martial.

Then they marched him out, between two armed guards, with his hands cuffed together in front of him.

I had never seen a flogging. Back home, while they do it in public of course, they do it back of the Federal Building—and Father had given me strict orders to stay away from there. I tried disobeying him on it once . . . but it was postponed and I never tried to see one again.

Once is too many.

The guards lifted his arms and hooked the manacles over a big hook high up on the post. Then they took his shirt off and it turned out that it was fixed so that it could come off and he didn't have an undershirt. The adjutant said crisply, "Carry out the sentence of the Court."

A corporal-instructor from some other battalion stepped forward with the whip. The Sergeant of the Guard made the count.

It's a slow count, five seconds between each one and it seems *much* longer. Ted didn't let out a peep until the third, then he sobbed.

The next thing I knew I was staring up at Corporal Bronski. He was slapping me and looking intently at me.

He stopped and asked, "Okay now? All right, back in ranks. On the bounce; we're about to pass in review." We did so and marched back to our company areas. I didn't eat much dinner but neither did a lot of them.

Nobody said a word to me about fainting. I found out later that I wasn't the only one—a couple of dozen of us had passed out.

CH:06

What we obtain too cheap, we esteem too lightly . . . it would be strange indeed if so celestial an article as FREEDOM *should not be highly rated.*

—Thomas Paine

It was the night after Hendrick was kicked out that I reached my lowest slump at Camp Currie. I couldn't sleep—and you have to have been through boot camp to understand just how far down a recruit has to sink before that can happen. But I hadn't had any real exercise all day so I wasn't physically tired, and my shoulder still hurt even though I had been marked "duty," and I had that letter from my mother preying on my mind, and every time I closed my eyes I would hear that *crack!* and see Ted slump against the whipping post.

I wasn't fretted about losing my boot chevrons. That no longer mattered at all because I was ready to resign, determined to. If it hadn't been the middle of the night and no pen and paper handy, I would have done so right then.

Ted had made a bad mistake, one that lasted all of half a second. And it really had been just a mistake, too, because, while he hated the outfit (who liked it?), he had been trying to sweat it out and win his franchise; he meant to go into politics—he talked a lot about how, when he got his citizenship, "There will be some changes made—you wait and see."

Well, he would never be in public office now; he had taken his finger off his number for a single instant and he was through.

If it could happen to him, it could happen to me. Suppose I slipped? Next day or next week? Not even allowed to resign . . . but drummed out with my back striped.

Time to admit that I was wrong and Father was right, time to put in that little piece of paper and slink home and tell Father that I was ready to go to Harvard and then go to work in the business—if he would still let me. Time to see Sergeant Zim, first thing in the morning, and tell him that I had had it. But not until morning, because you don't wake Sergeant Zim except for something you're certain that *he* will class as an emergency—believe me, you don't! Not Sergeant Zim.

Sergeant Zim—

He worried me as much as Ted's case did. After the court-martial was over and Ted had been taken away, he stayed behind and said to Captain Frankel, "May I speak with the Battalion Commander, sir?"

"Certainly. I was intending to ask you to stay behind for a word. Sit down."

Zim flicked his eyes my way and the Captain looked at me and I didn't have to be told to get out; I faded.

There was nobody in the outer office, just a couple of civilian clerks. I didn't dare go outside because the Captain might want me; I found a chair back of a row of files and sat down.

I could hear them talking, through the partition I had my head against. BHQ was a building rather than a tent, since it housed permanent communication and recording equipment, but it was a "minimum field building," a shack; the inner partitions weren't much. I doubt if the civilians could hear as they each were wearing transcriber phones and were bent over typers—besides, they didn't matter. I didn't mean to eavesdrop. Uh, well, maybe I did.

Zim said: "Sir, I request transfer to a combat team."

Frankel answered: "I can't hear you, Charlie. My tin ear is bothering me again."

Zim: "I'm quite serious, sir. This isn't my sort of duty."

Frankel said testily, "Quit bellyaching your troubles to me, Sergeant. At least wait until we've disposed of duty matters. What in the world happened?"

Zim said stiffly, "Captain, that boy doesn't rate ten lashes."

Frankel answered, "Of course he doesn't. You know who goofed—and so do I."

"Yes, sir. I know."

"Well? You know even better than I do that these kids are wild animals at this stage. You know when it's safe to turn your back on them and when it isn't. You know the doctrine and the standing orders about article nine-oh-eight-oh—you must *never* give them a *chance* to violate it. Of course some of them are going to try it— if they weren't aggressive they wouldn't be material for the M.I. They're docile in ranks; it's safe enough to turn your back when they're eating, or sleeping, or sitting on their tails and being lectured. But get them out in the field in a combat exercise, or anything that gets them keyed up and full of adrenaline, and they're as explosive as a hatful of mercury fulminate. You know that, all you instructors know that; you're trained—trained to watch for it, trained to snuff it out before it happens. Explain to me how it was possible for an untrained recruit to hang a mouse on your eye? He should never have laid a hand on you; you should have knocked him cold when you saw what he was up to. So why weren't you on the bounce? Are you slowing down?"

"I don't know," Zim answered slowly. "I guess I must be."

"Hmm! If true, a combat team is the last place for you. But it's not true. Or wasn't true the last time you and I worked out together, three days ago. So what slipped?"

Zim was slow in answering. "I think I had him tagged in my mind as one of the safe ones."

"There are no such."

"Yes, sir. But he was so earnest, so doggedly determined to sweat it out—he didn't have any aptitude but he kept on trying—that I must have done that, subconsciously." Zim was silent, then added, "I guess it was because I liked him."

Frankel snorted. "An instructor can't afford to like a man."

"I know it, sir. But I do. They're a nice bunch of kids. We've dumped all the real twerps by now—Hendrick's only shortcoming, aside from being clumsy, was that he thought he knew all the answers. I didn't mind that; I knew it all at that age myself. The twerps have gone home and those that are left are eager, anxious to please, and on the bounce—as cute as a litter of collie pups. A lot of them will make soldiers."

"So *that* was the soft spot. You liked him . . . so you failed to clip him in time. So he winds up with a court and the whip and a B.C.D. Sweet."

Zim said earnestly, "I wish to heaven there were some way for me to take that flogging myself, sir."

"You'd have to take your turn, I outrank you. What do you think I've been wishing the past hour? What do you think I was afraid of from the moment I saw you come in here sporting a shiner? I did my best to brush it off with administrative punishment and the young fool wouldn't let well enough alone. But I never thought he would be crazy enough to blurt it out that he'd hung one on you—he's *stupid*; you should have eased him out of the outfit weeks ago . . . instead of nursing him along until he got into trouble. But blurt it out he did, to me, in front of witnesses, forcing me to take official notice of it—and that licked us. No way to get it off the record, no way to avoid a court . . . just go through the whole dreary mess and take our medicine, and wind up with one more civilian who'll be against us the rest of his days. Because he *has* to be flogged; neither you nor I can take it for him, even though the fault was ours. Because the regiment has to see what happens when nine-oh-eight-oh is violated. Our fault . . . but his lumps."

"*My* fault, Captain. That's why I want to be transferred. Uh, sir, I think it's best for the outfit."

"You do, eh? But I decide what's best for my battalion, not you, Sergeant. Charlie, who do you think pulled your name out of the hat? And why? Think back twelve years. You were a corporal, remember? Where were you?"

"Here, as you know quite well, Captain. Right here on this same godforsaken prairie—and I wish I had never come back to it!"

"Don't we all. But it happens to be the most important and the most delicate work in the Army—turning unspanked young cubs into soldiers. Who was the worst unspanked young cub in your section?"

"Mmm . . ." Zim answered slowly. "I wouldn't go so far as to say you were the worst, Captain."

"You wouldn't, eh? But you'd have to think hard to name another candidate. I hated your guts, 'Corporal' Zim."

Zim sounded surprised, and a little hurt. "You did, Captain? I didn't hate you—I rather liked you."

"So? Well, 'hate' is the other luxury an instructor can never afford. We must not hate them, we must not like them; we must teach them. But if you liked me then—mmm, it seemed to me that you had very strange ways of showing it. Do you still like me? Don't answer that; I don't care whether you do or not—or, rather, I don't want to know, whichever it is. Never mind; I despised you then and I used to dream about ways to get you. But you were always on the bounce and never gave me a chance to buy a nine-oh-eight-oh court of my own. So here I am, thanks to you. Now to handle your request: You used to have one order that you gave to me over and over again when I was a boot. I got so I loathed it almost more than anything else you did or said. Do you remember it? *I* do and now I'll give it back to you: 'Soldier, shut up and soldier!'"

"Yes, sir."

"Don't go yet. This weary mess isn't all loss; any regiment of boots needs a stern lesson in the meaning of nine-oh-eight-oh, as we both know. They haven't yet learned to think, they won't read, and they rarely listen—but they can *see* . . . and young Hendrick's misfortune may save one of his mates, someday, from swinging by the neck until he's dead, dead, dead. But I'm sorry the object lesson had to come from my battalion and I certainly don't intend to let this battalion supply another one. You get your instructors together and warn them. For about twenty-four hours those kids will be in a state of shock. Then they'll turn sullen and the tension will build. Along about Thursday or Friday some boy who is about to flunk out anyhow will start thinking over the fact that Hendrick didn't get so very much, not even the number of lashes for drunken driving . . . and he's going to start brooding that it might be worth it, to take a swing at the instructor he hates worst. Sergeant—*that blow must never land!* Understand me?"

"Yes, sir."

"I want them to be eight times as cautious as they have been. I want them to keep their distance, I want them to have eyes in the backs of their heads. I want them to be as alert as a mouse at a cat show. Bronski—you have a special word with Bronski; he has a tendency to fraternize."

"I'll straighten Bronski out, sir."

"See that you do. Because when the next kid starts swinging, it's *got* to be stop-punched—not muffed, like today. The boy has got to be knocked cold and the instructor must do so without ever being touched himself—or I'll damned well break him for incompetence. Let them know that. They've got to teach those kids that it's not merely expensive but *impossible* to violate nine-oh-eight-oh . . . that even trying it wins a short nap, a bucket of water in the face, and a very sore jaw— and nothing else."

"Yes, sir. It'll be done."

"It had better be done. I will not only break the instructor who slips, I will personally take him 'way out on the prairie and give him lumps . . . because *I will not have another one of my boys strung up to that whipping post* through sloppiness on the part of his teachers. Dismissed."

"Yes, sir. Good afternoon, Captain."

"What's good about it? Charlie—"

"Yes, sir."

"If you're not too busy this evening, why don't you bring your soft shoes and your pads over to officers'

row and we'll go waltzing Matilda? Say about eight
o'clock."

"Yes, sir."

"That's not an order, that's an invitation. If you really
are slowing down, maybe I'll be able to kick your shoulder blades off."

"Uh, would the Captain care to put a small bet on it?"

"Huh? With me sitting here at this desk getting
swivel-chair spread? I will not! Not unless you agree to
fight with one foot in a bucket of cement. Seriously,
Charlie, we've had a miserable day and it's going to be
worse before it gets better. If you and I work up a good
sweat and swap a few lumps, maybe we'll be able to sleep
tonight despite all of mother's little darlings."

"I'll be there, Captain. Don't eat too much dinner—I
need to work off a couple of matters myself."

"I'm not going to dinner; I'm going to sit right
here and sweat out this quarterly report . . . which
the Regimental Commander is graciously pleased to see
right after *his* dinner . . . and which somebody whose
name I won't mention has put me two hours behind on.
So I may be a few minutes late for our waltz. Go 'way
now, Charlie, and don't bother me. See you later."

Sergeant Zim left so abruptly that I barely had time
to lean over and tie my shoe and thereby be out of sight
behind the file case as he passed through the outer
office. Captain Frankel was already shouting, "Orderly!
Orderly! ORDERLY!—do I have to call you three times?
What's your name? Put yourself down for an hour's extra
duty, full kit. Find the company commanders of E, F,

and G, my compliments and I'll be pleased to see them before parade. Then bounce over to my tent and fetch me a clean dress uniform, cap, side arms, shoes, ribbons—no medals. Lay it out for me here. Then make afternoon sick call—if you can scratch with that arm, as I've seen you doing, your shoulder can't be too sore. You've got thirteen minutes until sick call—on the bounce, soldier!"

I made it . . . by catching two of them in the senior instructors' shower (an orderly can go anywhere) and the third at his desk; the orders you get aren't impossible, they merely seem so because they nearly are. I was laying out Captain Frankel's uniform for parade as sick call sounded. Without looking up he growled, "Belay that extra duty. Dismissed." So I got home just in time to catch extra duty for "Uniform, Untidy in, Two Particulars" and see the sickening end of Ted Hendrick's time in the M.I.

So I had plenty to think about as I lay awake that night. I had known that Sergeant Zim worked hard, but it had never occurred to me that he could possibly be other than completely and smugly self-satisfied with what he did. He *looked* so smug, so self-assured, so at peace with the world and with himself.

The idea that this invincible robot could feel that he had failed, could feel so deeply and personally disgraced that he wanted to run away, hide his face among strangers, and offer the excuse that his leaving would be "best for the outfit," shook me up as much, and in a way even more, than seeing Ted flogged.

To have Captain Frankel agree with him—as to the

seriousness of the failure, I mean—and then rub his nose in it, chew him out. Well! I mean really. Sergeants don't get chewed out; sergeants do the chewing. A law of nature.

But I had to admit that what Sergeant Zim had taken, and swallowed, was so completely humiliating and withering as to make the worst I had ever heard or overheard from a sergeant sound like a love song. And yet the Captain hadn't even raised his voice.

The whole incident was so preposterously unlikely that I was never even tempted to mention it to anyone else.

And Captain Frankel himself— Officers we didn't see very often. They showed up for evening parade, sauntering over at the last moment and doing nothing that would work up a sweat; they inspected once a week, making private comments to sergeants, comments that invariably meant grief for somebody else, not them; and they decided each week what company had won the honor of guarding the regimental colors. Aside from that, they popped up occasionally on surprise inspections, creased, immaculate, remote, and smelling faintly of cologne—and went away again.

Oh, one or more of them did always accompany us on route marches and twice Captain Frankel had demonstrated his virtuosity at *la savate*. But officers didn't work, not real work, and they had no worries because sergeants were *under* them, not *over* them.

But it appeared that Captain Frankel worked so hard that he skipped meals, was kept so busy with something or other that he complained of lack of exercise and would waste his own free time just to work up a sweat.

As for worries, he had honestly seemed to be even more upset at what had happened to Hendrick than Zim had been. And yet he hadn't even known Hendrick by sight; he had been forced to ask his name.

I had an unsettling feeling that I had been completely mistaken as to the very nature of the world I was in, as if every part of it was something wildly different from what it appeared to be—like discovering that your own mother isn't anyone you've ever seen before, but a stranger in a rubber mask.

But I was sure of one thing: I didn't even want to find out what the M.I. really was. If it was so tough that even the gods-that-be—sergeants and officers—were made unhappy by it, it was certainly too tough for Johnnie! How could you keep from making mistakes in an outfit you didn't understand? *I* didn't want to swing by my neck till I was dead, dead, dead! I didn't even want to risk being flogged . . . even though the doctor stands by to make certain that it doesn't do you any permanent injury. Nobody in our family had *ever* been flogged (except paddlings in school, of course, which isn't at all the same thing). There were no criminals in our family on either side, none who had even been accused of crime. We were a proud family; the only thing we lacked was citizenship and Father regarded that as no real honor, a vain and useless thing. But if I were flogged—Well, he'd probably have a stroke.

And yet Hendrick hadn't done anything that I hadn't

thought about doing a thousand times. Why hadn't I? Timid, I guess. I *knew* that those instructors, any one of them, could beat the tar out of me, so I had buttoned my lip and hadn't tried it. No guts, Johnnie. At least Ted Hendrick had had guts. I didn't have . . . and a man with no guts has no business in the Army in the first place.

Besides that, Captain Frankel hadn't even considered it to be Ted's fault. Even if I didn't buy a 9080, through lack of guts, what day would I do something other than a 9080—something not my fault—and wind up slumped against the whipping post anyhow?

Time to get out, Johnnie, while you're still ahead.

My mother's letter simply confirmed my decision. I had been able to harden my heart to my parents as long as they were refusing me—but when they softened, I couldn't stand it. Or when Mother softened, at least. She had written:

—but I am afraid I must tell you that your father will still not permit your name to be mentioned. But, dearest, that is his way of grieving, since he cannot cry. You must understand, my darling baby, that he loves you more than life itself—more than he does me—and that you have hurt him very deeply. He tells the world that you are a grown man, capable of making your own decisions, and that he is proud of you. But that is his own pride speaking, the bitter hurt of a proud man who has been wounded deep in his heart by the one he loves best. You must understand, Juanito, that he does not speak of you and has not written to you because he cannot—not yet, not till his grief becomes bearable. When it has, I will know it, and then I will intercede for you—and we will all be together again.

Myself? How could anything her baby boy does anger his mother? You can hurt me, but you cannot make me love you the less. Wherever you are, whatever you choose to do, you are always my little boy who bangs his knee and comes running to my lap for comfort. My lap has shrunk, or perhaps you have grown (though I have never believed it), but nonetheless it will always be waiting, when you need it. Little boys never get over needing their mother's laps—do they, darling? I hope not. I hope that you will write and tell me so.

But I must add that, in view of the terribly long time that you have not written, it is probably best (until I let you know otherwise) for you to write to me care of your Aunt Eleanora. She will pass it on to me at once— and without causing any more upset. You understand?

> A thousand kisses to my baby,
> Your Mother

I understood, all right—and if Father could not cry, I could. I did.

And at last I got to sleep . . . and was awakened at once by an alert. We bounced out to the bombing range, the whole regiment, and ran through a simulated exercise, without ammo. We were wearing full unarmored kit otherwise, including ear-plug receivers, and we had no more than extended when the word came to freeze.

We held that freeze for at least an hour—and I mean we held it, barely breathing. A mouse tiptoeing past would have sounded noisy. Something did go past and ran right over me, a coyote I think. I never twitched. We got awfully cold holding that freeze, but I didn't care; I knew it was my last.

I didn't even hear reveille the next morning; for the first time in weeks I had to be whacked out of my sack and barely made formation for morning jerks. There was no point in trying to resign before breakfast anyhow, since I had to see Zim as the first step. But he wasn't at breakfast. I did ask Bronski's permission to see the C.C. and he said, "Sure. Help yourself," and didn't ask me why.

But you can't see a man who isn't there. We started a route march after breakfast and I still hadn't laid eyes on him. It was an out-and-back, with lunch fetched out to us by copter—an unexpected luxury, since failure to issue field rations before marching usually meant practice starvation except for whatever you had cached . . . and I hadn't; too much on my mind.

Sergeant Zim came out with the rations and he held mail call in the field—which was not an unexpected lux-ury. I'll say this for the M.I.; they might chop off your food, water, sleep, or anything else, without warning, but they never held up a person's mail a minute longer than circumstances required. That was yours, and they got it to you by the first transportation available and you could read it at your earliest break, even on maneuvers.

This hadn't been too important for me, as (aside from a couple of letters from Carl) I hadn't had anything but junk mail until Mother wrote to me.

I didn't even gather around when Zim handed it out; I figured now on not speaking to him until he got in—no point in giving him reason to notice me until we were actually in reach of headquarters. So I was surprised when he called my name and held up a letter. I bounced over and took it.

And was surprised again—it was from Mr. Dubois, my high school instructor in History and Moral Philosophy. I would sooner have expected a letter from Santa Claus.

Then, when I read it, it still seemed like a mistake. I had to check the address and the return address to convince myself that he had written it and had meant it for me.

My dear boy,

I would have written to you much sooner to express my delight and my pride in learning that you had not only volunteered to serve but also had chosen my own service. But not to express surprise; it is what I expected of you—except, possibly, the additional and very personal bonus that you chose the M.I. This is the sort of consummation, which does not happen too often, that nevertheless makes all of a teacher's efforts worth while. We necessarily sift a great many pebbles, much sand, for each nugget—but the nuggets are the reward.

By now the reason I did not write at once is obvious to you. Many young men, not necessarily through any reprehensible fault, are dropped during recruit training. I have waited (I have kept in touch through my own connections) until you had "sweated it out" past the hump (how well we all know that hump!) and were certain, barring accidents or illness, of completing your training and your term.

You are now going through the hardest part of your service—not the hardest physically (though physical hardship will never trouble you again; you now have its measure), but the hardest spiritually . . . the deep, soul-turning readjustments and re-evaluations necessary to metamorphize a potential citizen into one in being. Or, rather I should say: you have already gone through the hardest part, despite all the tribulations you still have ahead of you and all the hurdles, each higher than the last, which you still must clear. But it is that "hump" that counts—and, knowing you, lad, I know that I have waited long enough to be sure that you are past your "hump"—or you would be home now.

When you reached that spiritual mountaintop you felt something, a new something. Perhaps you haven't words for it (I know I didn't, when I was a boot). So perhaps you will permit an older comrade to lend you the words, since it often helps to have discrete words. Simply this: The noblest fate that a man can endure is to place his own mortal body between his loved home and the war's desolation. The words are not mine, of course, as you will recognize. Basic truths cannot change and once a man of insight expresses one of them it is never necessary, no matter how much the world changes, to reformulate them. This is an immutable, true everywhere, throughout all time, for all men and all nations.

Let me hear from you, please, if you can spare an old man some of your precious sack time to write an occasional letter. And if you should happen to run across any of my former mates, give them my warmest greetings.

Good luck, trooper! You've made me proud.

Jean V. Dubois
Lt.-Col., M.I., rtd.

The signature was as amazing as the letter itself. Old Sour Mouth was a short colonel? Why, our regional commander was only a major. Mr. Dubois had never used any sort of rank around school. We had supposed (if we thought about it at all) that he must have been a corporal or some such who had been let out when he lost his hand and had been fixed up with a soft job teaching a course that didn't have to be passed, or even taught— just audited. Of course we had known that he was a veteran since History and Moral Philosophy must be taught by a citizen. But an M.I.? He didn't look it. Prissy, faintly scornful, a dancing-master type—not one of us apes.

But that was the way he had signed himself.

I spent the whole long hike back to camp thinking about that amazing letter. It didn't sound in the least like anything he had ever said in class. Oh, I don't mean it contradicted anything he had told us in class; it was just entirely different in tone. Since when does a short colonel call a recruit private "comrade"?

When he was plain "Mr. Dubois" and I was one of the kids who had to take his course he hardly seemed to see me—except once when he got me sore by implying that I had too much money and not enough sense. (So my old man could have bought the school and given it to me for Christmas—is that a crime? It was none of his business.)

He had been droning along about "value," comparing the Marxist theory with the orthodox "use" theory. Mr. Dubois had said, "Of course, the Marxian definition of value is ridiculous. All the work one cares to add will not turn a mud pie into an apple tart; it remains a mud pie, value zero. By corollary, unskillful work can easily subtract value; an untalented cook can turn wholesome dough and fresh green apples, valuable already, into an inedible mess, value zero. Conversely, a great chef can fashion of those same materials a confection of greater value than a commonplace apple tart, with no more effort than an ordinary cook uses to prepare an ordinary sweet.

"These kitchen illustrations demolish the Marxian theory of value—the fallacy from which the entire magnificent fraud of communism derives—and illustrate the truth of the common-sense definition as measured in terms of use."

Dubois had waved his stump at us. "Nevertheless—wake up, back there!—nevertheless the disheveled old mystic of *Das Kapital*, turgid, tortured, confused, and neurotic, unscientific, illogical, this pompous fraud Karl Marx, *nevertheless* had a glimmering of a very important truth. If he had possessed an analytical mind, he might have formulated the first adequate definition of value . . . and this planet might have been saved endless grief.

"Or might not," he added. "You!"

I had sat up with a jerk.

"If you can't listen, perhaps you can tell the class whether 'value' is a relative, or an absolute?"

I had been listening; I just didn't see any reason not to listen with eyes closed and spine relaxed. But his question caught me out; I hadn't read that day's assignment. "An absolute," I answered, guessing.

"Wrong," he said coldly. "'Value' has no meaning other than in relation to living beings. The value of a thing is always relative to a particular person, is completely personal and different in quantity for each living human—'market value' is a fiction, merely a rough guess at the average of personal values, all of which must be quantitatively different or trade would be impossible." (I had wondered what Father would have said if he had heard "market value" called a "fiction"—snort in disgust, probably.)

"This very personal relationship, 'value,' has two factors for a human being: first, what he can do with a thing, its *use* to him . . . and second, what he must do to get it, its *cost* to him. There is an old song which asserts 'the best things in life are free.' Not true! Utterly false! This was the tragic fallacy which brought on the decadence and collapse of the democracies of the twentieth century;

those noble experiments failed because the people had been led to believe that they could simply vote for whatever they wanted . . . and get it, without toil, without sweat, without tears.

"Nothing of value is free. Even the breath of life is purchased at birth only through gasping effort and pain." He had been still looking at me and added, "If you boys and girls had to sweat for your toys the way a newly born baby has to struggle to live you would be happier . . . and much richer. As it is, with some of you, I pity the poverty of your wealth. You! I've just awarded you the prize for the hundred-meter dash. Does it make you happy?"

"Uh, I suppose it would."

"No dodging, please. You have the prize—here, I'll write it out: 'Grand prize for the championship, one hundred-meter sprint.'" He had actually come back to my seat and pinned it on my chest. "There! Are you happy? You value it—or don't you?"

I was sore. First that dirty crack about rich kids—a typical sneer of those who haven't got it—and now this farce. I ripped it off and chucked it at him.

Mr. Dubois had looked surprised. "It doesn't make you happy?"

"You know darn well I placed fourth!"

"*Exactly!* The prize for first place is worthless to you . . . because you haven't earned it. But you enjoy a modest satisfaction in placing fourth; you earned it. I trust that some of the somnambulists here understood this little morality play. I fancy that the poet who wrote that song meant to imply that the best things in life must be purchased other than with money—which is *true*—just as the literal meaning of his words is false. The best things in life are beyond money; their price is agony and sweat and devotion . . . and the price demanded for the most precious of all things in life is life itself—ultimate cost for perfect value."

I mulled over things I had heard Mr. Dubois—*Colonel* Dubois—say, as well as his extraordinary letter, while we went swinging back toward camp. Then I stopped thinking because the band dropped back near our position in column and we sang for a while, a French group— "Marseillaise," of course, and "Madelon" and "Sons of Toil and Danger," and then "Legion Étrangère" and "Mademoiselle from Armentières."

It's nice to have the band play; it picks you right up when your tail is dragging the prairie. We hadn't had anything but canned music at first and that only for parade and calls. But the powers-that-be had found out early who could play and who couldn't; instruments were provided and a regimental band was organized, all our own—even the director and the drum major were boots.

It didn't mean they got out of anything. Oh no! It just meant they were allowed and encouraged to do it on their own time, practicing evenings and Sundays and such—and that they got to strut and countermarch and show off at parade instead of being in ranks with their platoons. A lot of things that we did were run that way. Our chaplain, for example, was a boot. He was older than most of us and had been ordained in some obscure little sect I had never heard of. But he put a lot of passion into his preaching whether his theology was orthodox or not (don't ask me) and he was certainly in a position to understand the problems of a recruit. And the singing was fun. Besides, there was nowhere else to go on Sunday morning between morning police and lunch.

The band suffered a lot of attrition but somehow they always kept it going. The camp owned four sets of pipes and some Scottish uniforms, donated by Lochiel of Cameron whose son had been killed there in training— and one of us boots turned out to be a piper; he had learned it in the Scottish Boy Scouts. Pretty soon we had four pipers, maybe not good but loud. Pipes seem very odd when you first hear them, and a tyro practicing can set your teeth on edge—it sounds and looks as if he had a cat under his arm, its tail in his mouth, and biting it.

But they grow on you. The first time our pipers kicked their heels out in front of the band, skirling away at "Alamein Dead," my hair stood up so straight it lifted my cap. It gets you—makes tears.

We couldn't take a parade band out on route march, of course, because no special allowances were made for the band. Tubas and bass drums had to stay behind because a boy in the band had to carry a full kit, same as everybody, and could only manage an instrument small enough to add to his load. But the M.I. has band instruments which I don't believe anybody else has, such as a little box hardly bigger than a harmonica, an electric gadget which does an amazing job of faking a big horn and is played the same way. Comes band call when you are headed for the horizon, each bandsman sheds his kit without stopping, his squad mates split it up, and he trots to the column position of the color company and starts blasting.

It helps.

The band drifted aft, almost out of earshot, and we stopped singing because your own singing drowns out the beat when it's too far away.

I suddenly realized I felt good.

I tried to think why I did. Because we would be in after a couple of hours and I could resign?

No. When I had decided to resign, it had indeed given me a measure of peace, quieted down my awful jitters and let me go to sleep. But this was something else—and no reason for it, that I could see.

Then I knew. *I had passed my hump!*

I was over the "hump" that Colonel Dubois had written about. I actually walked over it and started down, swinging easily. The prairie through there was flat as a griddle-cake, but just the same I had been plodding wearily uphill all the way out and about halfway back. Then, at some point—I think it was while we were singing—I had passed the hump and it was all downhill. My kit felt lighter and I was no longer worried.

When we got in, I didn't speak to Sergeant Zim; I no longer needed to. Instead he spoke to me, motioned me to him as we fell out.

"Yes, sir?"

"This is a personal question . . . so don't answer it unless you feel like it." He stopped, and I wondered if he suspected that I had overheard his chewing-out, and shivered.

"At mail call today," he said, "you got a letter. I noticed—purely by accident, none of my business—the name on the return address. It's a fairly common name, some places, but—this is the personal question you need not answer—by any chance does the person who wrote that letter have his left hand off at the wrist?"

I guess my chin dropped. "How did you know? Sir?"

"I was nearby when it happened. It *is* Colonel Dubois? Right?"

"Yes, sir." I added, "He was my high school instructor in History and Moral Philosophy."

I think that was the only time I ever impressed Sergeant Zim, even faintly. His eyebrows went up an eighth of an inch and his eyes widened slightly. "So? You were extraordinarily fortunate." He added, "When you answer his letter—if you don't mind—you might say that Ship's Sergeant Zim sends his respects."

"Yes, sir. Oh . . . I think maybe he sent you a message, sir."

"*What?*"

"Uh, I'm not certain." I took out the letter, read just: "'—if you should happen to run across any of my former mates, give them my warmest greetings.' Is that for you, sir?"

Zim pondered it, his eyes looking through me, some-where else. "Eh? Yes, it is. For me among others. Thanks very much." Then suddenly it was over and he said briskly, "Nine minutes to parade. And you still have to shower and change. On the bounce, soldier."

CH:07

The young recruit is silly—'e thinks o' suicide.
'E's lost 'is gutter-devil; 'e 'asin't got 'is pride;
But day by day they kicks 'im, which 'elps 'im on a bit,
Till 'e finds 'isself one mornin' with a full an' proper kit.
Gettin' clear o' dirtiness, gettin' done with mess,
Gettin' shut o' doin' things rather-more-or-less.

—Rudyard Kipling

I'm not going to talk much more about my boot training. Mostly it was simply work, but I was squared away—enough said.

But I do want to mention a little about powered suits, partly because I was fascinated by them and also because that was what led me into trouble. No complaints—I rated what I got.

An M.I. lives by his suit the way a K-9 man lives by and with and on his doggie partner. Powered armor is one-half the reason we call ourselves "mobile infantry" instead of just "infantry." (The other half are the space-ships that drop us and the capsules we drop in.) Our suits give us better eyes, better ears, stronger backs (to carry heavier weapons and more ammo), better legs, more intelligence ("intelligence" in the military meaning; a man in a suit can be just as stupid as anybody else—only he had better not be), more firepower, greater endurance, less vulnerability.

A suit isn't a space suit—although it can serve as one. It is not primarily armor—although the Knights of the Round Table were not armored as well as we are. It isn't a tank—but a single M.I. private could take on a squadron of those things and knock them off unassisted if anybody was silly enough to put tanks against M.I. A suit is not a ship but it can fly, a little—on the other hand neither spaceships nor atmosphere craft can fight against a man in a suit except by saturation bombing of the area he is in (like burning down a house to get one flea!). Contrariwise we can do many things that no ship—air, submersible, or space—can do.

There are a dozen different ways of delivering destruction in impersonal wholesale, via ships and missiles of one sort or another, catastrophes so widespread, so unselective, that the war is over because that nation or planet has ceased to exist. What we do is entirely different. We make war as personal as a punch in the nose. We can be selective, applying precisely the required amount of pressure at the specified point at a designated time— we've never been told to go down and kill or capture all left-handed redheads in a particular area, but if they tell us to, we can. We will.

We are the boys who go to a particular place, at H-hour, occupy a designated terrain, stand on it, dig the enemy out of their holes, force them then and there to

surrender or die. We're the bloody infantry, the doughboy, the duckfoot, the foot soldier who goes where the
enemy is and takes him on in person. We've been doing
it, with changes in weapons but very little change in our
trade, at least since the time five thousand years ago
when the foot sloggers of Sargon the Great forced the
Sumerians to cry "Uncle!"

Maybe they'll be able to do without us someday.
Maybe some mad genius with myopia, a bulging forehead, and a cybernetic mind will devise a weapon that
can go down a hole, pick out the opposition, and force
it to surrender or die—without killing that gang of
your own people they've got imprisoned down there. I
wouldn't know; I'm not a genius, I'm an M.I. In the
meantime, until they build a machine to replace us, my
mates can handle that job—and I might be some help on
it, too.

Maybe someday they'll get everything nice and tidy
and we'll have that thing we sing about, when "we ain't
a-gonna study war no more." Maybe. Maybe the same
day the leopard will take off his spots and get a job as a
Jersey cow, too. But again, I wouldn't know; I am not a
professor of cosmopolitics; I'm an M.I. When the government sends me, I go. In between, I catch a lot of sack
time.

But, while they have not yet built a machine to replace us, they've surely thought up some honeys to help
us. The suit, in particular.

No need to describe what it looks like, since it has
been pictured so often. Suited up, you look like a big

steel gorilla, armed with gorilla-sized weapons. (This may be why a sergeant generally opens his remarks with "You apes—" However, it seems more likely that Caesar's sergeants used the same honorific.)

But the suits are considerably stronger than a gorilla. If an M.I. in a suit swapped hugs with a gorilla, the gorilla would be dead, crushed; the M.I. and the suit wouldn't be mussed.

The "muscles," the pseudo-musculature, get all the publicity but it's the control of all that power which merits it. The real genius in the design is that you *don't* have to control the suit; you just wear it, like your clothes, like skin. Any sort of ship you have to learn to pilot; it takes a long time, a new full set of reflexes, a different and artificial way of thinking. Even riding a bicycle demands an acquired skill, very different from walking, whereas a spaceship—oh, brother! I won't live that long. Spaceships are for acrobats who are also mathematicians.

But a suit you just wear.

Two thousand pounds of it, maybe, in full kit—yet the very first time you are fitted into one you can immediately walk, run, jump, lie down, pick up an egg without breaking it (takes a trifle of practice, but anything improves with practice), dance a jig (if you can dance a jig, that is, *without* a suit)—and jump right over the house next door and come down to a feather landing.

The secret lies in negative feedback and amplification.

Don't ask me to sketch the circuitry of a suit; I can't. But I understand that some very good concert violinists can't build a violin, either. I can do field maintenance and field repairs and check off the three hundred and forty-seven items from "cold" to ready to wear, and that's all a dumb M.I. is expected to do. But if my suit gets really sick, I call the doctor—a doctor of science (electromechanical engineering) who is a staff Naval officer, usually a lieutenant (read "captain" for our ranks), and is part of the ship's company of the troop transport—or who is reluctantly assigned to a regimental headquarters at Camp Currie, a fate-worse-than-death to a Navy man.

But if you really are interested in the prints and stereos and schematics of a suit's physiology, you can find most of it, the unclassified part, in any fairly large public library. For the small amount that is classified, you must look up a reliable enemy agent—"reliable" I say, because spies are a tricky lot; he's likely to sell you the parts you could get free from the public library.

But here is how it works, minus the diagrams. The inside of the suit is a mass of pressure receptors, hundreds of them. You push with the heel of your hand; the suit feels it, amplifies it, pushes with you to take the pressure off the receptors that gave the order to push. That's confusing, but negative feedback is always a confusing idea the first time, even though your body has been doing it ever since you quit kicking helplessly as a baby. Young children are still learning it; that's why they are clumsy. Adolescents and adults do it without knowing they ever learned it—and a man with Parkinson's disease has damaged his circuits for it.

The suit has feedback which causes it to match *any* motion you make, exactly—but with great force.

Controlled force . . . force controlled without your having to think about it. You jump, that heavy suit jumps, but higher than you can jump in your skin. Jump really hard and the suit's jets cut in, amplifying what the suit's leg "muscles" did, giving you a three-jet shove, the axis of pressure of which passes through your center of mass. So you jump over that house next door. Which makes you come down as fast as you went up . . . which the suit notes through your proximity & closing gear (a sort of simple-minded radar resembling a proximity fuse) and therefore cuts in the jets again just the right amount to cushion your landing without your having to think about it.

And *that* is the beauty of a powered suit: you don't have to think about it. You don't have to drive it, fly it, conn it, operate it; you just wear it and it takes orders directly from your muscles and does for you what your muscles are trying to do. This leaves you with your whole mind free to handle your weapons and notice what is going on around you . . . which is *supremely* important to an infantryman who wants to die in bed. If you load a mud foot down with a lot of gadgets that he has to watch, somebody a lot more simply equipped—say with a stone ax—will sneak up and bash his head in while he is trying to read a vernier.

Your "eyes" and your "ears" are rigged to help you without cluttering up your attention, too. Say you have three audio circuits, common in a marauder suit. The frequency control to maintain tactical security is very complex, at least two frequencies for each circuit, both of which are necessary for any signal at all and each of which wobbles under the control of a cesium clock timed to a micromicrosecond with the other end—but all this is no problem of yours. You want circuit A to your squad leader, you bite down once—for circuit B, bite down twice—and so on. The mike is taped to your throat, the plugs are in your ears and can't be jarred out; just talk. Besides that, outside mikes on each side of your helmet give you binaural hearing for your immediate surroundings just as if your head were bare—or you can suppress any noisy neighbors and not miss what your platoon leader is saying simply by turning your head.

Since your head is the one part of your body not involved in the pressure receptors controlling the suit's muscles, you use your head—your jaw muscles, your chin, your neck—to switch things for you and thereby leave your hands free to fight. A chin plate handles all visual displays the way the jaw switch handles the audios. All displays are thrown on a mirror in front of your forehead from where the work is actually going on above and back of your head. All this helmet gear makes you look like a hydrocephalic gorilla but, with luck, the enemy won't live long enough to be offended by your appearance, and it is a very convenient arrangement; you can flip through your several types of radar displays quicker than you can change channels to avoid a commercial—catch a range & bearing, locate your boss, check your flank men, whatever.

If you toss your head like a horse bothered by a fly,

your infrared snoopers go up on your forehead—toss it again, they come down. If you let go of your rocket launcher, the suit snaps it back until you need it again. No point in discussing water nipples, air supply, gyros, etc.— the point to *all* the arrangements is the same: to leave you free to follow your trade, slaughter.

Of course these things do require practice and you do practice until picking the right circuit is as automatic as brushing your teeth, and so on. But simply wearing the suit, moving in it, requires almost no practice. You practice jumping because, while you do it with a completely natural motion, you jump higher, faster, farther, and stay up longer. The last alone calls for a new orientation; those seconds in the air can be used—seconds are jewels beyond price in combat. While off the ground in a jump, you can get a range & bearing, pick a target, talk & receive, fire a weapon, reload, decide to jump again without landing and override your automatics to cut in the jets again. You can do *all* of these things in one bounce, with practice.

But, in general, powered armor doesn't require practice; it simply does it for you, just the way you were doing it, only better. All but one thing—you *can't* scratch where it itches. If I ever find a suit that will let me scratch between my shoulder blades, I'll marry it.

There are three main types of M.I. armor: marauder, command, and scout. Scout suits are very fast and very long-range, but lightly armed. Command suits are heavy on go juice and jump juice, are fast and can jump high; they have three times as much comm & radar gear as other suits, and a dead-reckoning tracker, inertial. Marauders are for those guys in ranks with the sleepy look—the executioners.

As I may have said, I fell in love with powered armor, even though my first crack at it gave me a strained shoulder. Any day thereafter that my section was allowed to practice in suits was a big day for me. The day I goofed I had simulated sergeant's chevrons as a simulated section leader and was armed with simulated A-bomb rockets to use in simulated darkness against a simulated enemy. That was the trouble; everything was simulated—but you are required to behave as if it is all real.

We were retreating—"advancing toward the rear," I mean—and one of the instructors cut the power on one of my men, by radio control, making him a helpless casualty. Per M.I. doctrine, I ordered the pickup, felt rather cocky that I had managed to get the order out before my number two cut out to do it anyhow, turned to do the next thing I had to do, which was to lay down a simulated atomic ruckus to discourage the simulated enemy overtaking us.

Our flank was swinging; I was supposed to fire it sort of diagonally but with the required spacing to protect my own men from blast while still putting it in close enough to trouble the bandits. On the bounce, of course. The movement over the terrain and the problem itself had been discussed ahead of time; we were still green—the only variations supposed to be left in were casualties.

Doctrine required me to locate *exactly*, by radar beacon, my own men who could be affected by the blast. But this all had to be done fast and I wasn't too sharp at reading those little radar displays anyhow. I cheated just a touch—flipped my snoopers up and looked, bare eyes in broad daylight. I left plenty of room. Shucks, I could *see* the only man affected, half a mile away, and all I had was just a little bitty H.E. rocket, intended to make a lot of smoke and not much else. So I picked a spot by eye, took the rocket launcher and let fly.

Then I bounced away, feeling smug—no seconds lost.

And had my power cut in the air. This doesn't hurt you; it's a delayed action, executed by your landing. I grounded and there I stuck, squatting, held upright by gyros but unable to move. You do not repeat *not* move when surrounded by a ton of metal with your power dead.

Instead I cussed to myself—I hadn't thought that they would make me a casualty when I was supposed to be leading the problem. Shucks and other comments.

I should have known that Sergeant Zim would be monitoring the section leader.

He bounced over to me, spoke to me privately on the face-to-face. He suggested that I might be able to get a job sweeping floors since I was too stupid, clumsy, and careless to handle dirty dishes. He discussed my past and probable future and several other things that I did not want to hear about. He ended by saying tonelessly, "How would you like to have Colonel Dubois see what you've done?"

Then he left me. I waited there, crouched over, for two hours until the drill was over. The suit, which had been feather-light, real seven-league boots, felt like an Iron Maiden. At last he returned for me, restored power, and we bounded together at top speed to BHQ.

Captain Frankel said less but it cut more.

Then he paused and added in that flat voice officers use when quoting regulations: "You may demand trial by court-martial if such be your choice. How say you?"

I gulped and said, "No, sir!" Until that moment I hadn't fully realized just how *much* trouble I was in.

Captain Frankel seemed to relax slightly. "Then we'll see what the Regimental Commander has to say. Sergeant, escort the prisoner." We walked rapidly over to RHQ and for the first time I met the Regimental Commander face to face—and by then I was sure that I was going to catch a court no matter what. But I remembered sharply how Ted Hendrick had talked himself into one; I said nothing.

Major Malloy said a total of five words to me. After hearing Sergeant Zim, he said three of them: "Is that correct?"

I said, "Yes, sir," which ended my part of it.

Major Malloy said, to Captain Frankel: "Is there any possibility of salvaging this man?"

Captain Frankel answered, "I believe so, sir."

Major Malloy said, "Then we'll try administrative punishment," turned to me and said:

"Five lashes."

Well, they certainly didn't keep me dangling. Fifteen minutes later the doctor had completed checking my heart and the Sergeant of the Guard was outfitting me with that special shirt which comes off without having to be pulled over the hands—zippered from the neck down the arms. Assembly for parade had just sounded. I was feeling detached, unreal . . . which I have learned is one way of being scared right out of your senses. The nightmare hallucination—

Zim came into the guard tent just as the call ended. He glanced at the Sergeant of the Guard—Corporal Jones—and Jones went out. Zim stepped up to me, slipped something into my hand. "Bite on that," he said quietly. "It helps. I know."

It was a rubber mouthpiece such as we used to avoid broken teeth in hand-to-hand combat drill. Zim left. I put it in my mouth. Then they handcuffed me and marched me out.

The order read: "—in simulated combat, gross negligence which would in action have caused the death of a teammate." Then they peeled off my shirt and strung me up.

Now here is a very odd thing: A flogging isn't as hard to *take* as it is to *watch*. I don't mean it's a picnic. It hurts worse than anything else I've ever had happen to me, and the waits between strokes are worse than the strokes themselves. But the mouthpiece did help and the only yelp I let out never got past it.

Here's the second odd thing: Nobody even mentioned it to me, not even other boots. So far as I could see, Zim and the instructors treated me exactly the same afterwards as they had before. From the instant the doctor painted the marks and told me to go back to duty it was all done with, completely. I even managed to eat a little at dinner that night and pretend to take part in the jawing at the table.

Another thing about administrative punishment: There is no permanent black mark. Those records are destroyed at the end of boot training and you start clean. The only record is one where it counts most.

You *don't* forget it.

CH:08

Train up a child in the way he should go; and when he is old he will not depart from it.

—Proverbs XXII:6

There were other floggings but darn few. Hendrick was the only man in our regiment to be flogged by sentence of court-martial; the others were administrative punishment, like mine, and for lashes it was necessary to go all the way up to the Regimental Commander—which a subordinate commander finds distasteful, to put it faintly. Even then, Major Malloy was much more likely to kick the man out, "Undesirable Discharge," than to have the whipping post erected. In a way, an administrative flogging is the mildest sort of a compliment; it means that your superiors think that there is a faint possibility that you just might have the character eventually to make a soldier and a citizen, unlikely as it seems at the moment.

I was the only one to get the maximum administrative punishment; none of the others got more than three lashes. Nobody else came as close as I did to putting on civilian clothes but still squeaked by. This is a social distinction of sorts. I don't recommend it.

But we had another case, much worse than mine or Ted Hendrick's—a really sick-making one. Once they erected gallows.

Now, look, get this straight. This case didn't really have anything to do with the Army. The crime didn't take place at Camp Currie and the placement officer who accepted this boy for M.I. should turn in his suit.

He deserted, only two days after we arrived at Currie. Ridiculous, of course, but nothing about the case made sense—why didn't he resign? Desertion, naturally, is one of the "thirty-one crash landings" but the Army doesn't invoke the death penalty for it unless there are special circumstances, such as "in the face of the enemy" or something else that turns it from a highly informal way of resigning into something that can't be ignored.

The Army makes no effort to find deserters and bring them back. This makes the hardest kind of sense. We're all volunteers; we're M.I. because we want to be, we're proud to be M.I. and the M.I. is proud of us. If a man doesn't feel that way about it, from his callused feet to his hairy ears, I don't want him on my flank when trouble starts. If I buy a piece of it, I want men around me who will pick me up because they're M.I. and I'm M.I. and my skin means as much to them as their own. I don't want any ersatz soldiers, dragging their tails and ducking out when the party gets rough. It's a whole lot safer to have a blank file on your flank than to have an alleged soldier who is nursing the "conscript" syndrome. So if they run, let 'em run; it's a waste of time and money to fetch them back.

Of course most of them do come back, though it may take them years—in which case the Army tiredly lets them have their fifty lashes instead of hanging them, and turns them loose. I suppose it must wear on a man's nerves to be a fugitive when everybody else is either a citizen or a legal resident, even when the police aren't trying to find him. "The wicked flee when no man pursueth." The temptation to turn yourself in, take your lumps, and breathe easily again must get to be overpowering.

But this boy didn't turn himself in. He was gone four months and I doubt if his own company remembered him, since he had been with them only a couple of days; he was probably just a name without a face, the "Dillinger, N.L." who had to be reported, day after day, as absent without leave on the morning muster.

Then he killed a baby girl.

He was tried and convicted by a local tribunal but identity check showed that he was an undischarged soldier; the Department had to be notified and our commanding general at once intervened. He was returned to us, since military law and jurisdiction take precedence over civil code.

Why did the general bother? Why didn't he let the local sheriff do the job?

In order to "teach us a lesson"?

Not at all. I'm quite sure that our general did not think that any of his boys needed to be nauseated in order not to kill any baby girls. By now I believe that he would have spared us the sight—had it been possible.

We did learn a lesson, though nobody mentioned it at the time and it is one that takes a long time to sink in until it becomes second nature:

The M.I. take care of their own—no matter what.

Dillinger belonged to us, he was still on our rolls. Even though we didn't want him, even though we should never have had him, even though we would have been happy to disclaim him, he was a member of our regiment. We couldn't brush him off and let a sheriff a thousand miles away handle it. If it has to be done, a man—a real man—shoots his own dog himself; he doesn't hire a proxy who may bungle it.

The regimental records said that Dillinger was ours, so taking care of him was our duty.

That evening we marched to the parade grounds at slow march, sixty beats to the minute (hard to keep step, when you're used to a hundred and forty), while the band played "Dirge for the Unmourned." Then Dillinger was marched out, dressed in M.I. full dress just as we were, and the band played "Danny Deever" while they stripped off every trace of insignia, even buttons and cap, leaving him in a maroon and light blue suit that was no longer a uniform. The drums held a sustained roll and it was all over.

We passed in review and on home at a fast trot. I don't think anybody fainted and I don't think anybody quite got sick, even though most of us didn't eat much dinner that night and I've never heard the mess tent so quiet. But, grisly as it was (it was the first time I had seen death, first time for most of us), it was not the shock that Ted Hendrick's flogging was—I mean, you couldn't put yourself in Dillinger's place; you didn't have any feeling of: "It could have been *me*." Not counting the technical matter of desertion, Dillinger had committed at least four capital crimes; if his victim had lived, he still would have danced Danny Deever for any one of the other three—kidnaping, demand of ransom, criminal neglect, etc.

I had no sympathy for him and still haven't. That old saw about "To understand all is to forgive all" is a lot of tripe. Some things, the more you understand the more you loathe them. My sympathy is reserved for Barbara Anne Enthwaite whom I had never seen, and for her parents, who would never again see their little girl.

As the band put away their instruments that night we started thirty days of mourning for Barbara and of disgrace for us, with our colors draped in black, no music at parade, no singing on route march. Only once did I hear anybody complain and another boot promptly asked him how he would like a full set of lumps? Certainly, it hadn't been our fault—but our business was to guard little girls, not kill them. Our regiment had been dishonored; we had to clean it. We were disgraced and we *felt* disgraced.

That night I tried to figure out how such things could be kept from happening. Of course, they hardly ever do nowadays—but even once is 'way too many. I never did reach an answer that satisfied me. This Dillinger—he looked like anybody else, and his behavior and record couldn't have been too odd or he would never have reached Camp Currie in the first place. I suppose he was one of those pathological personalities you read about—no way to spot them.

Well, if there was no way to keep it from happening once, there was only one sure way to keep it from happening twice. Which we had used.

If Dillinger had understood what he was doing (which seemed incredible) then he got what was coming to him . . . except that it seemed a shame that he hadn't suffered as much as had little Barbara Anne—he practically hadn't suffered at all.

But suppose, as seemed more likely, that he was so crazy that he had never been aware that he was doing anything wrong? What then?

Well, we shoot mad dogs, don't we?

Yes, but being crazy that way is a sickness—

I couldn't see but two possibilities. Either he couldn't be made well—in which case he was better dead for his own sake and for the safety of others—or he could be treated and made sane. In which case (it seemed to me) if he ever became sane enough for civilized society . . . and thought over what he had done while he was "sick"—what could be left for him but suicide? How could he *live* with himself?

And suppose he escaped *before* he was cured and did the same thing again? And maybe *again*? How do you explain *that* to bereaved parents? In view of his record?

I couldn't see but one answer.

I found myself mulling over a discussion in our class in

History and Moral Philosophy. Mr. Dubois was talking about the disorders that preceded the breakup of the North American republic, back in the XXth century. According to him, there was a time just before they went down the drain when such crimes as Dillinger's were as common as dog-fights. The Terror had not been just in North America—Russia and the British Isles had it, too, as well as other places. But it reached its peak in North America shortly before things went to pieces.

"Law-abiding people," Dubois had told us, "hardly dared go into a public park at night. To do so was to risk attack by wolf packs of children, armed with chains, knives, homemade guns, bludgeons . . . to be hurt at least, robbed most certainly, injured for life probably—or even killed. This went on for years, right up to the war between the Russo-Anglo-American Alliance and the Chinese Hegemony. Murder, drug addiction, larceny, assault, and vandalism were commonplace. Nor were parks the only places—these things happened also on the streets in daylight, on school grounds, even inside school buildings. But parks were so notoriously unsafe that honest people stayed clear of them after dark."

I had tried to imagine such things happening in our schools. I simply couldn't. Nor in our parks. A park was a place for fun, not for getting hurt. As for getting killed in one—"Mr. Dubois, didn't they have police? Or courts?"

"They had many more police than we have. And more courts. All overworked."

"I guess I don't get it." If a boy in our city had done

anything half that bad . . . well, he and his father would have been flogged side by side. But such things just didn't happen.

Mr. Dubois then demanded of me, "Define a 'juvenile delinquent.'"

"Uh, one of those kids—the ones who used to beat up people."

"Wrong."

"Huh? But the book said—"

"My apologies. Your textbook does so state. But calling a tail a leg does not make the name fit. 'Juvenile delinquent' is a contradiction in terms, one which gives a clue to their problem and their failure to solve it. Have you ever raised a puppy?"

"Yes, sir."

"Did you housebreak him?"

"Err . . . yes, sir. Eventually." It was my slowness in this that caused my mother to rule that dogs must stay out of the house.

"Ah, yes. When your puppy made mistakes, were you angry?"

"What? Why, he didn't know any better; he was just a puppy."

"What did you do?"

"Why, I scolded him and rubbed his nose in it and paddled him."

"Surely he could not understand your words?"

"No, but he could tell I was sore at him!"

"But you just said that you were not angry."

Mr. Dubois had an infuriating way of getting a person

mixed up. "No, but I had to make him *think* I was. He had to learn, didn't he?"

"Conceded. But, having made it clear to him that you disapproved, how could you be so cruel as to spank him as well? You said the poor beastie didn't know that he was doing wrong. Yet you inflicted pain. Justify yourself! Or are you a sadist?"

I didn't then know what a sadist was—but I knew pups. "Mr. Dubois, you *have* to! You scold him so that he knows he's in trouble, you rub his nose in it so that he will know what trouble you mean, you paddle him so that he darn well won't do it again—and you have to do it right away! It doesn't do a bit of good to punish him later; you'll just confuse him. Even so, he won't learn from one lesson, so you watch and catch him again and paddle him still harder. Pretty soon he learns. But it's a waste of breath just to scold him." Then I added, "I guess you've never raised pups."

"Many. I'm raising a dachshund now—by your methods. Let's get back to those juvenile criminals. The most vicious averaged somewhat younger than you here in this class . . . and they often started their lawless careers much younger. Let us never forget that puppy. These children were often caught; police arrested batches each day. Were they scolded? Yes, often scathingly. Were their noses rubbed in it? Rarely. News organs and officials usually kept their names secret—in many places the law so required for criminals under eighteen. Were they spanked? Indeed not! Many had never been spanked even as small children; there was a widespread belief that spanking, or any punishment involving pain, did a child permanent psychic damage."

(I had reflected that my father must never have heard of that theory.)

"Corporal punishment in schools was forbidden by law," he had gone on. "Flogging was lawful as sentence of court only in one small province, Delaware, and there only for a few crimes and was rarely invoked; it was regarded as 'cruel and unusual punishment.'" Dubois had mused aloud, "I do not understand objections to 'cruel and unusual' punishment. While a judge should be benevolent in purpose, his awards should cause the criminal to suffer, else there is no punishment—and pain is the basic mechanism built into us by millions of years of evolution which safeguards us by warning when something threatens our survival. Why should society refuse to use such a highly perfected survival mechanism? However, that period was loaded with pre-scientific pseudo-psychological nonsense.

"As for 'unusual,' punishment *must* be unusual or it serves no purpose." He then pointed his stump at another boy. "What would happen if a puppy were spanked every hour?"

"Uh . . . probably drive him crazy!"

"Probably. It certainly will not teach him anything. How long has it been since the principal of this school last had to switch a pupil?"

"Uh, I'm not sure. About two years. The kid that swiped—"

"Never mind. Long enough. It means that such punishment is so unusual as to be significant, to deter, to instruct. Back to these young criminals—They probably were not spanked as babies; they certainly were not flogged for their crimes. The usual sequence was: for a first offense, a warning—a scolding, often without trial. After several offenses a sentence of confinement but with sentence suspended and the youngster placed on probation. A boy might be arrested many times and convicted several times before he was punished—and then it would be merely confinement, with others like him from whom he learned still more criminal habits. If he kept out of major trouble while confined, he could usually evade most of even that mild punishment, be given probation—'paroled' in the jargon of the times.

"This incredible sequence could go on for years while his crimes increased in frequency and viciousness, with no punishment whatever save rare dull-but-comfortable confinements. Then suddenly, usually by law on his eighteenth birthday, this so-called 'juvenile delinquent' becomes an adult criminal—and sometimes wound up in only weeks or months in a death cell awaiting execution for murder. *You*—"

He had singled me out again. "Suppose you merely scolded your puppy, never punished him, let him go on making messes in the house . . . and occasionally locked him up in an outbuilding but soon let him back into the house with a warning not to do it again. Then one day you notice that he is now a grown dog and *still* not housebroken—whereupon you whip out a gun and shoot him dead. Comment, please?"

"Why . . . that's the craziest way to raise a dog I ever heard of!"

"I agree. Or a child. Whose fault would it be?"

"Uh . . . why, mine, I guess."

"Again I agree. But I'm not guessing."

"Mr. Dubois," a girl blurted out, "but *why*? Why didn't they spank little kids when they needed it and use a good dose of the strap on any older ones who deserved it—the sort of lesson they wouldn't forget! I mean ones who did things really *bad*. Why not?"

"I don't know," he had answered grimly, "except that the time-tested method of instilling social virtue and respect for law in the minds of the young did not appeal to a pre-scientific pseudo-professional class who called themselves 'social workers' or sometimes 'child psychologists.' It was too simple for them, apparently, since anybody could do it, using only the patience and firmness needed in training a puppy. I have sometimes wondered if they cherished a vested interest in disorder—but that is unlikely; adults almost always act from conscious 'highest motives' no matter what their behavior."

"But—good heavens!" the girl answered. "I didn't like being spanked any more than any kid does, but when I needed it, my mama delivered. The only time I ever got a switching in school I got another one when I got home—and that was years and years ago. I don't ever expect to be hauled up in front of a judge and sentenced to a flogging; you behave yourself and such things don't happen. I don't see anything wrong with our system; it's a lot better than not being able to walk outdoors for fear of your life—why, that's *horrible*!"

"I agree. Young lady, the tragic wrongness of what those well-meaning people did, contrasted with what they *thought* they were doing, goes very deep. They had no scientific theory of morals. They did have a theory of morals and they tried to live by it (I should not have sneered at their motives), but their theory was *wrong*— half of it fuzzy-headed wishful thinking, half of it rationalized charlatanry. The more earnest they were, the farther it led them astray. You see, they assumed that Man has a moral instinct."

"Sir? I thought— But he does! *I* have."

"No, my dear, you have a cultivated conscience, a most carefully trained one. Man has *no moral instinct*. He is not born with moral sense. You were not born with it, I was not—and a puppy has none. We *acquire* moral sense, when we do, through training, experience, and hard sweat of the mind. These unfortunate juvenile criminals were born with none, even as you and I, and they had no chance to acquire any; their experiences did not permit it. What *is* 'moral sense'? It is an elaboration of the instinct to survive. The instinct to survive is human nature itself, and every aspect of our personalities derives from it. Anything that conflicts with the survival instinct acts sooner or later to eliminate the individual and thereby fails to show up in future generations. This truth is mathematically demonstrable, everywhere verifiable; it is the single eternal imperative controlling everything we do.

"But the instinct to survive," he had gone on, "can be cultivated into motivations more subtle and much more complex than the blind, brute urge of the individual to stay alive. Young lady, what you miscalled your 'moral instinct' was the instilling in you by your elders of the truth that survival can have stronger imperatives than that of your own personal survival. Survival of your family, for example. Of your children, when you have them. Of your nation, if you struggle that high up the scale. And so on up. A scientifically verifiable theory of morals must be rooted in the individual's instinct to survive—*and nowhere else!*—and must correctly describe the hierarchy of survival, note the motivations at each level, and resolve all conflicts.

"We have such a theory now; we can solve any moral problem, on any level. Self-interest, love of family, duty to country, responsibility toward the human race—we are even developing an exact ethic for extra-human relations. But all moral problems can be illustrated by one misquotation: 'Greater love hath no man than a mother cat dying to defend her kittens.' Once you understand the problem facing that cat and how she solved it, you will then be ready to examine yourself and learn how high up the moral ladder you are capable of climbing.

"These juvenile criminals hit a low level. Born with only the instinct for survival, the highest morality they achieved was a shaky loyalty to a peer group, a street gang. But the do-gooders attempted to 'appeal to their better natures,' to 'reach them,' to 'spark their moral sense.' *Tosh!* They *had* no 'better natures'; experience taught them that what they were doing was the way to survive. The puppy never got his spanking; therefore what he did with pleasure and success must be 'moral.'

"The basis of all morality is duty, a concept with the same relation to group that self-interest has to individual. Nobody preached duty to these kids in a way they could understand—that is, with a spanking. But the society they were in told them endlessly about their 'rights.'

"The results should have been predictable, since a human being has *no natural rights of any nature*."

Mr. Dubois had paused. Somebody took the bait. "Sir? How about 'life, liberty, and the pursuit of happiness'?"

"Ah, yes, the 'unalienable rights.' Each year someone quotes that magnificent poetry. Life? What 'right' to life has a man who is drowning in the Pacific? The ocean will not hearken to his cries. What 'right' to life has a man who must die if he is to save his children? If he chooses to save his own life, does he do so as a matter of 'right'? If two men are starving and cannibalism is the only alternative to death, which man's right is 'unalienable'? And is it 'right'? As to liberty, the heroes who signed the great document pledged themselves to *buy* liberty with their lives. Liberty is *never* unalienable; it must be redeemed regularly with the blood of patriots or it *always* vanishes. Of all the so-called natural human rights that have ever been invented, liberty is least likely to be cheap and is *never* free of cost.

"The third 'right'?—the 'pursuit of happiness'? It is

indeed unalienable but it is not a right; it is simply a universal condition which tyrants cannot take away nor patriots restore. Cast me into a dungeon, burn me at the stake, crown me king of kings, I can 'pursue happiness' as long as my brain lives—but neither gods nor saints, wise men nor subtle drugs, can insure that I will catch it."

Mr. Dubois then turned to me. "I told you that 'juvenile delinquent' is a contradiction in terms. 'Delinquent' means 'failing in duty.' But *duty* is an *adult* virtue—indeed a juvenile becomes an adult when, and only when, he acquires a knowledge of duty and embraces it as dearer than the self-love he was born with. There never was, there cannot *be*, a 'juvenile delinquent.' But for every juvenile criminal there are always one or more adult delinquents—people of mature years who either do not know their duty, or who, knowing it, fail.

"And *that* was the soft spot which destroyed what was in many ways an admirable culture. The junior hoodlums who roamed their streets were symptoms of a greater sickness; their citizens (all of them counted as such) glorified their mythology of 'rights' . . . and lost track of their duties. No nation, so constituted, can endure."

I wondered how Colonel Dubois would have classed Dillinger. Was he a juvenile criminal who merited pity even though you had to get rid of him? Or was he an adult delinquent who deserved nothing but contempt?

I didn't know, I would never know. The one thing I was sure of was that he would never again kill any little girls.

That suited me. I went to sleep.

CH:09

We've got no place in this outfit for good losers. We want tough hombres who will go in there and win!

—Admiral Jonas Ingram, 1926

When we had done all that a mud foot can do in flat country, we moved into some rough mountains to do still rougher things—the Canadian Rockies between Good Hope Mountain and Mount Waddington. Camp Sergeant Spooky Smith was much like Camp Currie (aside from its rugged setting) but it was much smaller. Well, the Third Regiment was much smaller now, too—less than four hundred whereas we had started out with more than two thousand. H Company was now organized as a single platoon and the battalion paraded as if it were a company. But we were still called "H Company" and Zim was "Company Commander," not platoon leader.

What the sweat-down meant, really, was much more personal instruction; we had more corporal-instructors than we had squads and Sergeant Zim, with only fifty men on his mind instead of the two hundred and sixty he had started with, kept his Argus eyes on each one of us all the time—even when he wasn't there. At least, if you goofed, it turned out he was standing right behind you.

However, the chewing-out you got had almost a friendly quality, in a horrid sort of way, because we had changed, too, as well as the regiment—the one-in-five who was left was almost a soldier and Zim seemed to be trying to make him into one, instead of running him over the hill.

We saw a lot more of Captain Frankel, too; he now spent most of his time teaching us, instead of behind a desk, and he knew all of us by name and face and seemed to have a card file in his mind of exactly what progress each man had made on every weapon, every piece of equipment—not to mention your extra-duty status, medical record, and whether you had had a letter from home lately.

He wasn't as severe with us as Zim was; his words were milder and it took a really stupid stunt to take that friendly grin off his face—but don't let that fool you; there was beryl armor under the grin. I never did figure out which one was the better soldier, Zim or Captain Frankel—I mean, if you took away the insignia and thought of them as privates. Unquestionably they were both better soldiers than any of the other instructors—but which was best? Zim did everything with precision and style, as if he were on parade; Captain Frankel did the same thing with dash and gusto, as if it were a game. The results were about the same—and it never turned out to be as easy as Captain Frankel made it look.

We needed the abundance of instructors. Jumping

a suit (as I have said) was easy on flat ground. Well, the suit jumps just as high and just as easily in the mountains—but it makes a lot of difference when you have to jump up a vertical granite wall, between two close-set fir trees, and override your jet control at the last instant. We had three major casualties in suit practice in broken country, two dead and one medical retirement.

But that rock wall is even tougher without a suit, tackled with lines and pitons. I didn't really see what use alpine drill was to a cap trooper but I had learned to keep my mouth shut and try to learn what they shoved at us. I learned it and it wasn't too hard. If anybody had told me, a year earlier, that I could go up a solid chunk of rock, as flat and as perpendicular as a blank wall of a building, using only a hammer, some silly little steel pins, and a chunk of clothesline, I would have laughed in his face; I'm a sea-level type. Correction: I was a sea-level type. There had been some changes made.

Just how much I had changed I began to find out. At Camp Sergeant Spooky Smith we had liberty—to go to town, I mean. Oh, we had "liberty" after the first month at Camp Currie, too. This meant that, on a Sunday afternoon, if you weren't in the duty platoon, you could check out at the orderly tent and walk just as far away from camp as you wished, bearing in mind that you had to be back for evening muster. But there was nothing within walking distance, if you don't count jack rabbits—no girls, no theaters, no dance halls, et cetera.

Nevertheless, liberty, even at Camp Currie, was no mean privilege; sometimes it can be very important in-

deed to be able to go so far away that you can't see a
tent, a sergeant, nor even the ugly faces of your best
friends among the boots . . . not have to be on the
bounce about anything, have time to take out your soul
and look at it. You could lose that privilege in several degrees; you could be restricted to camp . . . or you could
be restricted to your own company street, which meant
that you couldn't go to the library nor to what was mis-
leadingly called the "recreation" tent (mostly some
parcheesi sets and similar wild excitements) . . . or you
could be under close restriction, required to stay in your
tent when your presence was not required elsewhere.

This last sort didn't mean much in itself since it was
usually added to extra duty so demanding that you didn't
have any time in your tent other than for sleep anyhow;
it was a decoration added like a cherry on top of a dish of
ice cream to notify you and the world that you had
pulled not some everyday goof-off but something unbe-
coming of a member of the M.I. and were thereby unfit
to associate with other troopers until you had washed
away the stain.

But at Camp Spooky we could go into town—duty
status, conduct status, etc., permitting. Shuttles ran to
Vancouver every Sunday morning, right after divine services (which were moved up to thirty minutes after
breakfast) and came back again just before supper and
again just before taps. The instructors could even spend
Saturday night in town, or cop a three-day pass, duty
permitting.

I had no more than stepped out of the shuttle, my

first pass, than I realized in part that I had changed. Johnnie didn't fit in any longer. Civilian life, I mean. It all seemed amazingly complex and unbelievably untidy.

I'm not running down Vancouver. It's a beautiful city in a lovely setting; the people are charming and they are used to having the M.I. in town and they make a trooper welcome. There is a social center for us downtown, where they have dances for us every week and see to it that junior hostesses are on hand to dance with, and senior hostesses to make sure that a shy boy (*me*, to my amazement—but you try a few months with nothing female around but lady jack rabbits) gets introduced and has a partner's feet to step on.

But I didn't go to the social center that first pass. Mostly I stood around and gawked—at beautiful buildings, at display windows filled with all manner of unnecessary things (and not a weapon among them), at all those people running around, or even strolling, doing exactly as they pleased and no two of them dressed alike—and at girls.

Especially at girls. I hadn't realized just how wonderful they were. Look, I've approved of girls from the time I first noticed that the difference was more than just that they dress differently. So far as I remember I never did go through that period boys are supposed to go through when they know that girls are different but dislike them; I've *always* liked girls.

But that day I realized that I had long been taking them for granted.

Girls are simply wonderful. Just to stand on a corner and watch them going past is delightful. They don't *walk*. At least not what we do when we talk. I don't know how to describe it, but it's much more complex and utterly delightful. They don't move just their feet; everything moves and in different directions . . . and all of it graceful.

I might have been standing there yet if a policeman hadn't come by. He sized us up and said, "Howdy, boys. Enjoying yourselves?"

I quickly read the ribbons on his chest and was impressed. "Yes, *sir*!"

"You don't have to say 'sir' to me. Not much to do here. Why don't you go to the hospitality center?" He gave us the address, pointed the direction and we started that way—Pat Leivy, "Kitten" Smith, and myself. He called after us, "Have a good time, boys . . . and stay out of trouble." Which was exactly what Sergeant Zim had said to us as we climbed into the shuttle.

But we didn't go there. Pat Leivy had lived in Seattle when he was a small boy and wanted to take a look at his old home town. He had money and offered to pay our shuttle fares if we would go with him. I didn't mind and it was all right; shuttles ran every twenty minutes and our passes were not restricted to Vancouver. Smith decided to go along, too.

Seattle wasn't so very different from Vancouver and the girls were just as plentiful; I enjoyed it. But Seattle wasn't quite as used to having M.I. around in droves and we picked a poor spot to eat dinner, one where we weren't quite so welcome—a bar-restaurant, down by the docks.

Now, look, we weren't drinking. Well, Kitten Smith had had one repeat *one* beer with his dinner but he was never anything but friendly and nice. That is how he got his name; the first time we had hand-to-hand combat drill Corporal Jones had said to him disgustedly: "A kitten would have hit me harder than *that*!" The nickname stuck.

We were the only uniforms in the place; most of the other customers were merchant marine sailors—Seattle handles an awful lot of surface tonnage. I hadn't known it at the time but merchant sailors don't like us. Part of it has to do with the fact that their guilds have tried and tried to get their trade classed as equivalent to Federal Service, without success—but I understand that some of it goes way back in history, centuries.

There were some young fellows there, too, about our age—the right age to serve a term, only they weren't—long-haired and sloppy and kind of dirty-looking. Well, say about the way I looked, I suppose, before I joined up.

Presently we started noticing that at the table behind us, two of these young twerps and two merchant sailors (to judge by clothes) were passing remarks that were intended for us to overhear. I won't try to repeat them.

We didn't say anything. Presently, when the remarks were even more personal and the laughs louder and everybody else in the place was keeping quiet and listening, Kitten whispered to me, "Let's get out of here."

I caught Pat Leivy's eye; he nodded. We had no score to settle; it was one of those pay-as-you-get-it places. We got up and left.

They followed us out.

Pat whispered to me, "Watch it." We kept on walking, didn't look back.

They charged us.

I gave my man a side-neck chop as I pivoted and let him fall past me, swung to help my mates. But it was over. Four in, four down. Kitten had handled two of them and Pat had sort of wrapped the other one around a lamppost from throwing him a little too hard.

Somebody, the proprietor I guess, must have called the police as soon as we stood up to leave, since they arrived almost at once while we were still standing around wondering what to do with the meat—two policemen; it was that sort of a neighborhood.

The senior of them wanted us to prefer charges, but none of us was willing—Zim had told us to "stay out of trouble." Kitten looked blank and about fifteen years old and said, "I guess they stumbled."

"So I see," agreed the police officer and toed a knife away from the outflung hand of my man, put it against the curb and broke the blade. "Well, you boys had better run along . . . farther uptown."

We left. I was glad that neither Pat nor Kitten wanted to make anything of it. It's a mighty serious thing, a civilian assaulting a member of the Armed Forces, but what the deuce?—the books balanced. They jumped us, they got their lumps. All even.

But it's a good thing we *never* go on pass armed . . .
and have been trained to disable without killing. Because
every bit of it happened by reflex. I didn't believe that
they would jump us until they already had, and I didn't
do any thinking at all until it was over.

But that's how I learned for the first time just how
much I had changed.

We walked back to the station and caught a shuttle to
Vancouver.

We started practice drops as soon as we moved to Camp
Spooky—a platoon at a time, in rotation (a full platoon,
that is—a company), would shuttle down to the field
north of Walla Walla, go aboard, space, make a drop, go
through an exercise, and home on a beacon. A day's work.
With eight companies that gave us not quite a drop each
week, and then it gave us a little more than a drop
each week as attrition continued, whereupon the drops
got tougher—over mountains, into the arctic ice, into the
Australian desert, and, before we graduated, onto the face
of the Moon, where your capsule is placed only a hundred
feet up and explodes as it ejects—and you have to look
sharp and land with only your suit (no air, no parachute)
and a bad landing can spill your air and kill you.

Some of the attrition was from casualties, deaths or
injuries, and some of it was just from refusing to enter
the capsule—which some did, and that was that; they
weren't even chewed out; they were just motioned aside
and that night they were paid off. Even a man who had

made several drops might get the panic and refuse . . . and the instructors were just gentle with him, treated him the way you do a friend who is ill and won't get well.

I never quite refused to enter the capsule—but I certainly learned about the shakes. I always got them, I was scared silly every time. I still am.

But you're not a cap trooper unless you drop.

They tell a story, probably not true, about a cap trooper who was sight-seeing in Paris. He visited Les Invalides, looked down at Napoleon's coffin and said to a French guard there: "Who's he?"

The Frenchman was properly scandalized. "Monsieur does not *know*? This is the tomb of *Napoleon*! Napoleon Bonaparte—the greatest soldier who ever lived!"

The cap trooper thought about it. Then he asked, "So? Where were his drops?"

It is almost certainly not true, because there is a big sign outside there that tells you exactly who Napoleon was. But that is how cap troopers feel about it.

Eventually we graduated.

I can see that I've left out almost everything. Not a word about most of our weapons, nothing about the time we dropped everything and fought a forest fire for three days, no mention of the practice alert that was a real one, only we didn't know it until it was over, nor about the day the cook tent blew away—in fact not any mention of weather and, believe me, weather is important to a doughboy, rain and mud especially. But though weather is important while it happens it seems to me to be pretty dull to look back on. You can take descriptions of most any sort of weather out of an almanac and stick them in just anywhere; they'll probably fit.

The regiment had started with 2009 men; we graduated 187—of the others, fourteen were dead (one executed and his name struck) and the rest resigned, dropped, transferred, medical discharge, etc. Major Malloy made a short speech, we each got a certificate, we passed in review for the last time, and the regiment was disbanded, its colors to be cased until they would be needed (three weeks later) to tell another couple of thousand civilians that they were an outfit, not a mob.

I was a "trained soldier," entitled to put "TP" in front of my serial number instead of "RP." Big day.

The biggest I ever had.

CH:10

The tree of Liberty must be refreshed from time to time with the blood of patriots . . .

—Thomas Jefferson, 1787

That is, I thought I was a "trained soldier" until I reported to my ship. Any law against having a wrong opinion?

I see that I didn't make any mention of how the Terran Federation moved from "peace" to a "state of emergency" and then on into "war." I didn't notice it too closely myself. When I enrolled, it was "peace," the normal condition, at least so people think (who ever expects anything else?). Then, while I was at Currie, it became a "state of emergency" but I still didn't notice it, as what Corporal Bronski thought about my haircut, uniform, combat drill, and kit was much more important—and what Sergeant Zim thought about such matters was overwhelmingly important. In any case, "emergency" is still "peace."

"Peace" is a condition in which no civilian pays any attention to military casualties which do not achieve page-one, lead-story prominence—unless that civilian is a close relative of one of the casualties. But, if there ever was a time in history when "peace" meant that there was no fighting going on, I have been unable to find out about it. When I reported to my first outfit, "Willie's Wildcats," sometimes known as Company K, Third Regiment, First M.I. Division, and shipped with them in the *Valley Forge* (with that misleading certificate in my kit), the fighting had already been going on for several years.

The historians can't seem to settle whether to call this one "The Third Space War" (or the "Fourth"), or whether "The First Interstellar War" fits it better. We just call it "The Bug War" if we call it anything, which we usually don't, and in any case the historians date the beginning of "war" after the time I joined my first outfit and ship. Everything up to then and still later were "incidents," "patrols," or "police actions." However, you are just as dead if you buy a farm in an "incident" as you are if you buy it in a declared war.

But, to tell the truth, a soldier doesn't notice a war much more than a civilian does, except his own tiny piece of it and that just on the days it is happening. The rest of the time he is much more concerned with sack time, the vagaries of sergeants, and the chances of wheedling the cook between meals. However, when Kitten Smith and Al Jenkins and I joined them at Luna Base, each of Willies' Wildcats had made more than one combat drop; they were soldiers and we were not. We weren't hazed for it—at least I was not—and the sergeants and corporals were amazingly easy to deal with after the calculated frightfulness of instructors.

It took a little while to discover that this comparatively gentle treatment simply meant that we were nobody, hardly worth chewing out, until we had proved in a drop—a real drop—that we might possibly replace real Wildcats who had fought and bought it and whose bunks we now occupied.

Let me tell you how green I was. While the *Valley Forge* was still at Luna Base, I happened to come across my section leader just as he was about to hit dirt, all slicked up in dress uniform. He was wearing in his left ear lobe a rather small earring, a tiny gold skull beautifully made and under it, instead of the conventional crossed bones of the ancient Jolly Roger design, was a whole bundle of little gold bones, almost too small to see.

Back home, I had always worn earrings and other jewelry when I went out on a date—I had some beautiful ear clips, rubies as big as the end of my little finger which had belonged to my mother's grandfather. I like jewelry and had rather resented being required to leave it all behind when I went to Basic . . . but here was a type of jewelry which was apparently okay to wear with uniform. My ears weren't pierced—my mother didn't approve of it, for boys—but I could have the jeweler mount it on a clip . . . and I still had some money left from pay call at graduation and was anxious to spend it before it mildewed. "Unh, Sergeant? Where do you get earrings like that one? Pretty neat."

He didn't look scornful, he didn't even smile. He just said, "You like it?"

"I certainly do!" The plain raw gold pointed up the

gold braid and piping of the uniform even better than gems would have done. I was thinking that a pair would be still handsomer, with just crossbones instead of all that confusion at the bottom. "Does the base PX carry them?"

"No, the PX here never sells them." He added, "At least I don't think you'll ever be able to buy one here—I hope. But I tell you what—when we reach a place where you can buy one of your own, I'll see to it you know about it. That's a promise."

"Uh, thanks!"

"Don't mention it."

I saw several of the tiny skulls thereafter, some with more "bones," some with fewer; my guess had been correct, this was jewelry permitted with uniform, when on pass at least. Then I got my own chance to "buy" one almost immediately thereafter and discovered that the prices were unreasonably high, for such plain ornaments.

It was Operation Bughouse, the First Battle of Klendathu in the history books, soon after Buenos Aires was smeared. It took the loss of B.A. to make the ground-hogs realize that anything was going on, because people who haven't been out don't really believe in other planets, not down deep where it counts. I know I hadn't and I had been space-happy since I was a pup.

But B.A. really stirred up the civilians and inspired loud screams to bring all our forces home, from everywhere—orbit them around the planet practically shoulder to shoulder and interdict the space Terra occupies. This is silly, of course; you don't win a war by defense but by attack—no "Department of Defense" ever won a war; see the histories. But it seems to be a standard civilian reaction to scream for defensive tactics as soon as they do notice a war. They then want to run the war—like a passenger trying to grab the controls away from the pilot in an emergency.

However, nobody asked my opinion at the time; I was told. Quite aside from the impossibility of dragging the troops home in view of our treaty obligations and what it would do to the colony planets in the Federation and to our allies, we were awfully busy doing something else, to wit: carrying the war to the Bugs. I suppose I noticed the destruction of B.A. much less than most civilians did. We were already a couple of parsecs away under Cherenkov drive and the news didn't reach us until we got it from another ship after we came out of drive.

I remember thinking, "Gosh, that's terrible!" and feeling sorry for the one Porteño in the ship. But B.A. wasn't my home and Terra was a long way off and I was very busy, as the attack on Klendathu, the Bugs' home planet, was mounted immediately after that and we spent the time to rendezvous strapped in our bunks, doped and unconscious, with the internal-gravity field of the *Valley Forge* off, to save power and give greater speed.

The loss of Buenos Aires did mean a great deal to me; it changed my life enormously, but this I did not know until many months later.

When it came time to drop onto Klendathu, I was assigned to PFC Dutch Bamburger as a supernumerary. He managed to conceal his pleasure at the news and as soon as the platoon sergeant was out of earshot, he said, "Listen, boot, you stick close behind me and stay out of my way. You go slowing me down, I break your silly neck."

I just nodded. I was beginning to realize that this was not a practice drop.

Then I had the shakes for a while and then we were down—

Operation Bughouse should have been called "Operation Madhouse." Everything went wrong. It had been planned as an all-out move to bring the enemy to their knees, occupy their capital and the key points of their home planet, and end the war. Instead it darn near lost the war.

I am not criticizing General Diennes. I don't know whether it's true that he demanded more troops and more support and allowed himself to be overruled by the Sky Marshal-in-Chief—or not. Nor was it any of my business. Furthermore I doubt if some of the smart second-guessers know all the facts.

What I do know is that the General dropped with us and commanded us on the ground and, when the situation became impossible, he personally led the diversionary attack that allowed quite a few of us (including me) to be retrieved—and, in so doing, bought his farm. He's radioactive debris on Klendathu and it's much too late to court-martial him, so why talk about it?

I do have one comment to make to any armchair strategist who has never made a drop. Yes, I agree that the Bugs' planet possibly could have been plastered with

H-bombs until it was surfaced with radioactive glass. But would that have won the war? The Bugs are not like us. The Pseudo-Arachnids aren't even like spiders. They are arthropods who happen to look like a madman's conception of a giant, intelligent spider, but their organization, psychological and economic, is more like that of ants or termites; they are communal entities, the ultimate dictatorship of the hive. Blasting the surface of their planet would have killed soldiers and workers; it would not have killed the brain caste and the queens—I doubt if anybody can be certain that even a direct hit with a burrowing H-rocket would kill a queen; we don't know how far down they are. Nor am I anxious to find out; none of the boys who went down those holes came up again.

So suppose we did ruin the productive surface of Klendathu? They still would have ships and colonies and other planets, same as we have, and their HQ is still intact—so unless they surrender, the war isn't over. We didn't have nova bombs at that time; we couldn't crack Klendathu open. If they absorbed the punishment and didn't surrender, the war was still on.

If they *can* surrender—

Their soldiers can't. Their workers can't fight (and you can waste a lot of time and ammo shooting up workers who wouldn't say *boo!*) and their soldier caste can't surrender. But don't make the mistake of thinking that the Bugs are just stupid insects because they look the way they do and don't know how to surrender. Their warriors are smart, skilled, and aggressive—smarter than you are, by the only universal rule, if the Bug shoots first.

You can burn off one leg, two legs, three legs, and he just keeps on coming; burn off four on one side and he topples over—but keeps on shooting. You have to spot the nerve case and get it . . . whereupon he will trot right on past you, shooting at nothing, until he crashes into a wall or something.

The drop was a shambles from the start. Fifty ships were in our piece of it and they were supposed to come out of Cherenkov drive and into reaction drive so perfectly co-ordinated that they could hit orbit and drop us, in formation and where we were supposed to hit, without even making one planet circuit to dress up their own formation. I suppose this is difficult. Shucks, I *know* it is. But when it slips, it leaves the M.I. holding the sack.

We were lucky at that, because the *Valley Forge* and every Navy file in her bought it before we ever hit the ground. In that tight, fast formation (4.7 miles/sec. orbital speed is not a stroll) she collided with the *Ypres* and both ships were destroyed. We were lucky to get out of her tubes—those of us who did get out, for she was still firing capsules as she was rammed. But I wasn't aware of it; I was inside my cocoon, headed for the ground. I suppose our company commander knew that the ship had been lost (and half his Wildcats with it) since he was out first and would know when he suddenly lost touch, over the command circuit, with the ship's captain.

But there is no way to ask him, because he wasn't retrieved. All I ever had was a gradually dawning realization that things were in a mess.

The next eighteen hours were a nightmare. I shan't

tell much about it because I don't remember much, just snatches, stop-motion scenes of horror. I have never liked spiders, poisonous or otherwise; a common house spider in my bed can give me the creeps. Tarantulas are simply unthinkable, and I can't eat lobster, crab, or anything of that sort. When I got my first sight of a Bug, my mind jumped right out of my skull and started to yammer. It was seconds later that I realized that I had killed it and could stop shooting. I suppose it was a worker; I doubt if I was in any shape to tackle a warrior and win.

But, at that, I was in better shape than was the K-9 Corps. They were to be dropped (if the drop had gone perfectly) on the periphery of our entire target and the neodogs were supposed to range outward and provide tactical intelligence to interdiction squads whose business it was to secure the periphery. Those Calebs aren't armed, of course, other than their teeth. A neodog is supposed to hear, see, and smell and tell his partner what he finds by radio; all he carries is a radio and a destruction bomb with which he (or his partner) can blow the dog up in case of bad wounds or capture.

Those poor dogs didn't wait to be captured; apparently most of them suicided as soon as they made contact. They felt the way I do about the Bugs, only worse. They have neodogs now that are indoctrinated from puppyhood to observe and evade without blowing their tops at the mere sight or smell of a Bug. But these weren't.

But that wasn't all that went wrong. Just name it, it was fouled up. I didn't know what was going on, of course; I just stuck close behind Dutch, trying to shoot or flame anything that moved, dropping a grenade down a hole whenever I saw one. Presently I got so that I could kill a Bug without wasting ammo or juice, although I did not learn to distinguish between those that were harmless and those that were not. Only about one in fifty is a warrior—but he makes up for the other forty-nine. Their personal weapons aren't as heavy as ours but they are lethal just the same—they've got a beam that will penetrate armor and slice flesh like cutting a hard-boiled egg, and they co-operate even better than we do . . . because the brain that is doing the heavy thinking for a "squad" isn't where you can reach it; it's down one of the holes.

Dutch and I stayed lucky for quite a long time, milling around over an area about a mile square, corking up holes with bombs, killing what we found above surface, saving our jets as much as possible for emergencies. The idea was to secure the entire target and allow the reinforcements and the heavy stuff to come down without important opposition; this was not a raid, this was a battle to establish a beachhead, stand on it, hold it, and enable fresh troops and heavies to capture or pacify the entire planet.

Only we didn't.

Our own section was doing all right. It was in the wrong pew and out of touch with the other section— the platoon leader and sergeant were dead and we never re-formed. But we had staked out a claim, our special-weapons squad had set up a strong point, and we were ready to turn our real estate over to fresh troops as soon as they showed up.

Only they didn't. They dropped in where we should have dropped, found unfriendly natives and had their own troubles. We never saw them. So we stayed where we were, soaking up casualties from time to time and passing them out ourselves as opportunity offered—while we ran low on ammo and jump juice and even power to keep the suits moving. This seemed to go on for a couple of thousand years.

Dutch and I were zipping along close to a wall, headed for our special-weapons squad in answer to a yell for help, when the ground suddenly opened in front of Dutch, a Bug popped out, and Dutch went down.

I flamed the Bug and tossed a grenade and the hole closed up, then turned to see what had happened to Dutch. He was down but he didn't look hurt. A platoon sergeant can monitor the physicals of every man in his platoon, sort out the dead from those who merely can't make it unassisted and must be picked up. But you can do the same thing manually from switches right on the belt of a man's suit.

Dutch didn't answer when I called to him. His body temperature read ninety-nine degrees, his respiration, heartbeat, and brain wave read zero—which looked bad but maybe his suit was dead rather than he himself. Or so I told myself, forgetting that the temperature indicator would give no reading if it were the suit rather than the man. Anyhow, I grabbed the can-opener wrench from my own belt and started to take him out of his suit while trying to watch all around me.

Then I heard an all-hands call in my helmet that I

never want to hear again. "*Sauve qui peut!* Home! Home! Pickup and *home*! Any beacon you can hear. Six minutes! All hands, save yourselves, pick up your mates. Home on any beacon! *Sauve qui—*"

I hurried.

His head came off as I tried to drag him out of his suit, so I dropped him and got out of there. On a later drop I would have had sense enough to salvage his ammo, but I was far too sluggy to think; I simply bounced away from there and tried to rendezvous with the strong point we had been heading for.

It was already evacuated and I felt lost . . . lost and deserted. Then I heard recall, not the recall it should have been: "Yankee Doodle" (if it had been a boat from the *Valley Forge*)—but "Sugar Bush," a tune I didn't know. No matter, it was a beacon; I headed for it, using the last of my jump juice lavishly—got aboard just as they were about to button up and shortly thereafter was in the *Voortrek*, in such a state of shock that I couldn't remember my serial number.

I've heard it called a "strategic victory"—but I was there and I claim we took a terrible licking.

Six weeks later (and feeling about sixty years older) at Fleet Base on Sanctuary I boarded another ground boat and reported for duty to Ship's Sergeant Jelal in the *Rodger Young*. I was wearing, in my pierced left ear lobe, a broken skull with one bone. Al Jenkins was with me and was wearing one exactly like it (Kitten never made it out of the tube). The few surviving Wildcats were distributed elsewhere around the Fleet; we had lost half our strength, about, in the collision between the *Valley Forge* and the *Ypres*; that disastrous mess on the ground had run our casualties up over 80 per cent and the powers-that-be decided that it was impossible to put the outfit back together with the survivors—close it out, put the records in the archives, and wait until the scars had healed before reactivating Company K (Wildcats) with new faces but old traditions.

Besides, there were a lot of empty files to fill in other outfits.

Sergeant Jelal welcomed us warmly, told us that we were joining a smart outfit, "best in the Fleet," in a taut ship, and didn't seem to notice our ear skulls. Later that day he took us forward to meet the Lieutenant, who smiled rather shyly and gave us a fatherly little talk. I noticed that Al Jenkins wasn't wearing his gold skull. Neither was I—because I had already noticed that nobody in Rasczak's Roughnecks wore the skulls.

They didn't wear them because, in Rasczak's Roughnecks, it didn't matter in the least how many combat drops you had made, nor which ones; you were either a Roughneck or you weren't—and if you were not, they didn't care who you were. Since we had come to them not as recruits but as combat veterans, they gave us all possible benefit of doubt and made us welcome with no more than that unavoidable trace of formality anybody necessarily shows to a house guest who is not a member of the family.

But, less than a week later when we had made one combat drop with them, we were full-fledged Roughnecks, members of the family, called by our first names, chewed out on occasion without any feeling on either side that we were less than blood brothers thereby, borrowed from and lent to, included in bull sessions and privileged to express our own silly opinions with complete freedom—and have them slapped down just as freely. We even called non-coms by their first names on any but strictly duty occasions. Sergeant Jelal was always on duty, of course, unless you ran across him dirtside, in which case he was "Jelly" and went out of his way to behave as if his lordly rank meant nothing between Roughnecks.

But the Lieutenant was always "The Lieutenant"—never "Mr. Rasczak," nor even "Lieutenant Rasczak." Simply "The Lieutenant," spoken to and of in the third person. There was no god but the Lieutenant and Sergeant Jelal was his prophet. Jelly could say "No" in his own person and it might be subject to further argument, at least from junior sergeants, but if he said, "The Lieutenant wouldn't like it," he was speaking *ex cathedra* and the matter was dropped permanently. Nobody ever tried to check up on whether or not the Lieutenant would or would not like it; the Word had been spoken.

The Lieutenant was father to us and loved us and spoiled us and was nevertheless rather remote from us aboard ship—and even dirtside . . . unless we reached dirt via a drop. But in a drop—well, you wouldn't think that an officer could worry about every man of a platoon spread over a hundred square miles of terrain. But he can. He can worry himself sick over each one of them. How he could keep track of us all I can't describe, but in the midst of a ruckus his voice would sing out over the command circuit: "Johnson! Check squad six! Smitty's in trouble," and it was better than even money that the Lieutenant had noticed it before Smith's squad leader.

Besides that, you knew with utter and absolute certainty that, as long as you were still alive, the Lieutenant would not get into the retrieval boat without you. There have been prisoners taken in the Bug War, but none from Rasczak's Roughnecks.

Jelly was mother to us and was close to us and took care of us and didn't spoil us at all. But he didn't report us to the Lieutenant—there was never a court-martial among the Roughnecks and no man was *ever* flogged. Jelly didn't even pass out extra duty very often; he had other ways of paddling us. He could look you up and down at daily inspection and simply say, "In the Navy you might look good. Why don't you transfer?"—and get results, it being an article of faith among us that the Navy crew members slept in their uniforms and never washed below their collar lines.

But Jelly didn't have to maintain discipline among privates because he maintained discipline among his non-coms and expected them to do likewise. My squad leader, when I first joined, was "Red" Greene. After a couple of drops, when I knew how *good* it was to be a Roughneck, I got to feeling gay and a bit too big for my clothes—and talked back to Red. He didn't report me to

Jelly; he just took me back to the washroom and gave me a medium set of lumps, and we got to be pretty good friends. In fact, he recommended me for lance, later on.

Actually we didn't know whether the crew members slept in their clothes or not; we kept to our part of the ship and the Navy men kept to theirs, because they were made to feel unwelcome if they showed up in our country other than on duty—after all, one has social standards one must maintain, mustn't one? The Lieutenant had his stateroom in male officers' country, a Navy part of the ship, but we never went there, either, except on duty and rarely. We did go forward for guard duty, because the *Rodger Young* was a mixed ship, female captain and pilot officers, some female Navy ratings; forward of bulkhead thirty was ladies' country—and two armed M.I. day and night stood guard at the one door cutting it. (At battle stations that door, like all other gastight doors, was secured; nobody missed a drop.)

Officers were privileged to go forward of bulkhead thirty on duty and all officers, including the Lieutenant, ate in a mixed mess just beyond it. But they didn't tarry there; they ate and got out. Maybe other corvette transports were run differently, but that was the way the *Rodger Young* was run—both the Lieutenant and Captain Deladrier wanted a taut ship and got it.

Nevertheless guard duty was a privilege. It was a rest to stand beside that door, arms folded, feet spread, doping off and thinking about nothing . . . but always warmly aware that any moment you might see a feminine creature even though you were not privileged to speak to her other than on duty. Once I was called all the way into the Skipper's office and she spoke to me—she looked right at me and said, "Take this to the Chief Engineer, please."

My daily shipside job, aside from cleaning, was servicing electronic equipment under the close supervision of "Padre" Migliaccio, the section leader of the first section, exactly as I used to work under Carl's eye. Drops didn't happen too often and everybody worked every day. If a man didn't have any other talent he could always scrub bulkheads; nothing was ever quite clean enough to suit Sergeant Jelal. We followed the M.I. rule; everybody fights, everybody works. Our first cook was Johnson, the second section's sergeant, a big friendly boy from Georgia (the one in the western hemisphere, not the other one) and a very talented chef. He wheedled pretty well, too; he liked to eat between meals himself and saw no reason why other people shouldn't.

With the Padre leading one section and the cook leading the other, we were well taken care of, body and soul—but suppose one of them bought it? Which one would you pick? A nice point that we never tried to settle but could always discuss.

The *Rodger Young* kept busy and we made a number of drops, all different. Every drop has to be different so that they never can figure out a pattern on you. But no more pitched battles; we operated alone, patrolling, harrying, and raiding. The truth was that the Terran Federation was not then able to mount a large battle; the foul-up with Operation Bughouse had cost too many ships, 'way too many trained men. It was necessary to take time to heal up, train more men.

In the meantime, small fast ships, among them the *Rodger Young* and other corvette transports, tried to be everywhere at once, keeping the enemy off balance, hurting him and running. We suffered casualties and filled our holes when we returned to Sanctuary for more capsules. I still got the shakes every drop, but actual drops didn't happen too often nor were we ever down long—and between times there were days and days of shipboard life among the Roughnecks.

It was the happiest period of my life although I was never quite consciously aware of it—I did my full share of beefing just as everybody else did, and enjoyed that, too.

We weren't *really* hurt until the Lieutenant bought it.

I guess that was the worst time in all my life. I was already in bad shape for a personal reason: My mother had been in Buenos Aires when the Bugs smeared it.

I found out about it one time when we put in at Sanctuary for more capsules and some mail caught up with us—a note from my Aunt Eleanora, one that had not been coded and sent fast because she had failed to mark for that; the letter itself came. It was about three bitter lines. Somehow she seemed to blame me for my mother's death. Whether it was my fault because I was in the Armed Services and should have therefore prevented the raid, or whether she felt that my mother had made a trip to Buenos Aires because I wasn't home where I should have been, was not quite clear; she managed to imply both in the same sentence.

I tore it up and tried to walk away from it. I thought that both my parents were dead—since Father would never send Mother on a trip that long by herself. Aunt Eleanora had not said so, but she wouldn't have mentioned Father in any case; her devotion was entirely to her sister. I was almost correct—eventually I learned that Father had planned to go with her but something had come up and he stayed over to settle it, intending to come along the next day. But Aunt Eleanora did not tell me this.

A couple of hours later the Lieutenant sent for me and asked me very gently if I would like to take leave at Sanctuary while the ship went out on her next patrol—he pointed out that I had plenty of accumulated R&R and might as well use some of it. I don't know how he knew that I had lost a member of my family, but he obviously did. I said no, thank you, sir; I preferred to wait until the outfit all took R&R together.

I'm glad I did it that way, because if I hadn't, I wouldn't have been along when the Lieutenant bought it . . . and that would have been just too much to be borne. It happened very fast and just before retrieval. A man in the third squad was wounded, not badly but he was down; the assistant section leader moved in to pick up—and bought a small piece of it himself. The Lieutenant, as usual, was watching everything at once—no doubt he had checked physicals on each of them by remote, but we'll never know. What he did was to make sure that the assistant section leader was still alive; then made pickup on both of them himself, one in each arm of his suit.

He threw them the last twenty feet and they were passed into the retrieval boat—and with everybody else in, the shield gone and no interdiction, was hit and died instantly.

I haven't mentioned the names of the private and of the assistant section leader on purpose. The Lieutenant was making pickup on *all* of us, with his last breath. Maybe I was the private. It doesn't matter who he was. What did matter was that our family had had its head chopped off. The head of the family from which we took our name, the father who made us what we were.

After the Lieutenant had to leave us Captain Deladrier invited Sergeant Jelal to eat forward, with the other heads of departments. But he begged to be excused. Have you ever seen a widow with stern character keep her family together by behaving as if the head of the family had simply stepped out and would return at any moment? That's what Jelly did. He was just a touch more strict with us than ever and if he ever had to say: "The Lieutenant wouldn't like that," it was almost more than a man could take. Jelly didn't say it very often.

He left our combat team organization almost unchanged; instead of shifting everybody around, he moved the assistant section leader of the second section over into the (nominal) platoon sergeant spot, leaving his section leaders where they were needed—with their sections—and he moved me from lance and assistant squad leader into acting corporal as a largely ornamental assistant section leader. Then he himself behaved as if the Lieutenant were merely out of sight and that he was just passing on the Lieutenant's orders, as usual.

It saved us.

CH:11

I have nothing to offer but blood, toil, tears, and sweat.
— W. Churchill, XXth century
soldier-statesman

As we came back into the ship after the raid on the
Skinnies—the raid in which Dizzy Flores bought it,
Sergeant Jelal's first drop as platoon leader—a ship's
gunner who was tending the boat lock spoke to me:

"How'd it go?"

"Routine," I answered briefly. I suppose his remark
was friendly but I was feeling very mixed up and in no
mood to talk—sad over Dizzy, glad that we had made
pickup anyhow, mad that the pickup had been useless,
and all of it tangled up with that washed-out but happy
feeling of being back in the ship again, able to muster
arms and legs and note that they are all present. Besides,
how can you talk about a drop to a man who has never
made one?

"So?" he answered. "You guys have got it soft. Loaf
thirty days, work thirty minutes. Me, I stand a watch in
three *and* turn to."

"Yeah, I guess so," I agreed and turned away. "Some of us are born lucky."

"Soldier, you ain't peddlin' vacuum," he said to my back.

And yet there was much truth in what the Navy gunner had said. We cap troopers are like aviators of the earlier mechanized wars; a long and busy military career could contain only a few hours of actual combat facing the enemy, the rest being: train, get ready, go out—then come back, clean up the mess, get ready for another one, and practice, practice, practice, in between. We didn't make another drop for almost three weeks and that on a different planet around another star—a Bug colony. Even with Cherenkov drive, stars are far apart.

In the meantime I got my corporal's stripes, nominated by Jelly and confirmed by Captain Deladrier in the absence of a commissioned officer of our own. Theoretically the rank would not be permanent until approved against vacancy by the Fleet M.I. repple-depple, but that meant nothing, as the casualty rate was such that there were always more vacancies in the T.O. than there were warm bodies to fill them. I was a corporal when Jelly said I was a corporal; the rest was red tape.

But the gunner was not quite correct about "loafing"; there were fifty-three suits of powered armor to check, service, and repair between each drop, not to mention weapons and special equipment. Sometimes Migliaccio would down-check a suit, Jelly would confirm it, and the ship's weapons engineer, Lieutenant Farley, would decide that he couldn't cure it short of base facilities—whereupon a new suit would have to be broken out of stores and brought from "cold" to "hot," an exacting process requiring twenty-six man-hours not counting the time of the man to whom it was being fitted.

We kept busy.

But we had fun, too. There were always several competitions going on, from acey-deucy to Honor Squad, and we had the best jazz band in several cubic light-years (well, the only one, maybe), with Sergeant Johnson on the trumpet leading them mellow and sweet for hymns or tearing the steel right off the bulkheads, as the occasion required. After that masterful (or should it be "mistressful"?) retrieval rendezvous without a programmed ballistic, the platoon's metalsmith, PFC Archie Campbell, made a model of the *Rodger Young* for the Skipper and we all signed and Archie engraved our signatures on a base plate: *To Hot Pilot Yvette Deladrier, with thanks from Rasczak's Roughnecks,* and we invited her aft to eat with us and the Roughneck Downbeat Combo played during dinner and then the junior private presented it to her. She got tears and kissed him—and kissed Jelly as well and he blushed purple.

After I got my chevrons I simply had to get things straight with Ace, because Jelly kept me on as assistant section leader. This is not good. A man ought to fill each spot on his way up; I should have had a turn as squad leader instead of being bumped from lance and assistant squad leader to corporal and assistant section leader. Jelly knew this, of course, but I know perfectly well that he was trying to keep the outfit as much as possible the way it had been when the Lieutenant was alive—which meant that he left his squad leaders and section leaders unchanged.

But it left me with a ticklish problem; all three of the corporals under me as squad leaders were actually senior to me—but if Sergeant Johnson bought it on the next drop, it would not only lose us a mighty fine cook, it would leave me leading the section. There mustn't be any shadow of doubt when you give an order, not in combat; I had to clear up any possible shadow before we dropped again.

Ace was the problem. He was not only senior of the three, he was a career corporal as well and older than I was. If Ace accepted me, I wouldn't have any trouble with the other two squads.

I hadn't really had any trouble with him aboard. After we made pickup on Flores together he had been civil enough. On the other hand we hadn't had anything to have trouble over; our shipside jobs didn't put us together, except at daily muster and guard mount, which is all cut and dried. But you can feel it. He was not treating me as somebody he took orders from.

So I looked him up during off hours. He was lying in his bunk, reading a book, *Space Rangers against the Galaxy*—a pretty good yarn, except that I doubt if a military outfit ever had so many adventures and so few goof-offs. The ship had a good library.

"Ace. Got to see you."

He glanced up. "So? I just left the ship, I'm off duty."

"I've got to see you now. Put your book down."

"What's so aching urgent? I've got to finish this chapter."

"Oh, come off it, Ace. If you can't wait, I'll tell you how it comes out."

"You do and I'll clobber you." But he put the book down, sat up, and listened.

I said, "Ace, about this matter of the section organization—you're senior to me, you ought to be assistant section leader."

"Oh, so it's *that* again!"

"Yep. I think you and I ought to go see Johnson and get him to fix it up with Jelly."

"You do, eh?"

"Yes, I do. That's how it's got to be."

"So? Look, Shortie, let me put you straight. I got nothing against you at all. Matter of fact, you were on the bounce that day we had to pick up Dizzy; I'll hand you that. But if you want a squad, you go dig up one of your own. Don't go eyeing mine. Why, my boys wouldn't even peel potatoes for you."

"That's your final word?"

"That's my first, last, and only word."

I sighed. "I thought it would be. But I had to make sure. Well, that settles that. But I've got one thing on my mind. I happened to notice that the washroom needs cleaning . . . and I think maybe you and I ought to attend to it. So put your book aside . . . as Jelly says, non-coms are always on duty."

He didn't stir at once. He said quietly, "You really think it's necessary, Shortie? As I said, I got nothing against you."

"Looks like."

"Think you can do it?"

"I can sure try."

"Okay. Let's take care of it."

We went aft to the washroom, chased out a private who was about to take a shower he didn't really need, and locked the door. Ace said, "You got any restrictions in mind, Shortie?"

"Well . . . I hadn't planned to kill you."

"Check. And no broken bones, nothing that would keep either one of us out of the next drop—except maybe by accident, of course. That suit you?"

"Suits," I agreed. "Uh, I think maybe I'll take my shirt off."

"Wouldn't want to get blood on your shirt." He relaxed. I started to peel it off and he let go a kick for my kneecap. No wind up. Flat-footed and not tense.

Only my kneecap wasn't there—I had learned.

A real fight ordinarily can last only a second or two, because that is all the time it takes to kill a man, or knock him out, or disable him to the point where he can't fight. But we had agreed to avoid inflicting permanent damage; this changes things. We were both young, in top condition, highly trained, and used to absorbing punishment. Ace was bigger, I was maybe a touch faster. Under such conditions the miserable business simply has to go on until one or the other is too beaten down to continue—unless a fluke settles it sooner. But neither one of us was allowing any flukes; we were professionals and wary.

So it did go on, for a long, tedious, painful time. Details would be trivial and pointless; besides, I had no time to take notes.

A long time later I was lying on my back and Ace was flipping water in my face. He looked at me, then hauled me to my feet, shoved me against a bulkhead, steadied me. "Hit me!"

"Huh?" I was dazed and seeing double.

"Johnnie . . . hit me."

His face was floating in the air in front of me; I zeroed in on it and slugged it with all the force in my body, hard enough to mash any mosquito in poor health. His eyes closed and he slumped to the deck and I had to grab at a stanchion to keep from following him.

He got slowly up. "Okay, Johnnie," he said, shaking his head, "I've had my lesson. You won't have any more lip out of me . . . nor out of anybody in the section. Okay?"

I nodded and my head hurt.

"Shake?" he asked.

We shook on it, and that hurt, too.

Almost anybody else knew more about how the war was going than we did, even though we were in it. This was the period, of course, after the Bugs had located our home planet, through the Skinnies, and had raided it, destroying Buenos Aires and turning "contact troubles" into all-out war, but *before* we had built up our forces and before the Skinnies had changed sides and become our co-belligerents and de facto allies. Partly effective interdiction for Terra had been set up from Luna (we didn't know it), but speaking broadly, the Terran Federation was losing the war.

We didn't know that, either. Nor did we know that strenuous efforts were being made to subvert the alliance against us and bring the Skinnies over to our side; the nearest we came to being told about that was when we got instructions, before the raid in which Flores was killed, to go easy on the Skinnies, destroy as much property as possible but to kill inhabitants only when unavoidable.

What a man doesn't know he can't spill if he is captured; neither drugs, nor torture, nor brainwash, nor endless lack of sleep can squeeze out a secret he doesn't possess. So we were told only what we had to know for tactical purposes. In the past, armies have been known to fold up and quit because the men didn't know what they were fighting for, or why, and therefore lacked the will to fight. But the M.I. does not have that weakness. Each one of us was a volunteer to begin with, each for some reason or other—some good, some bad. But now we fought because we were M.I. We were professionals, with *esprit de corps*. We were Rasczak's Roughnecks, the best unprintable outfit in the whole expurgated M.I.; we climbed into our capsules because Jelly told us it was time to do so and we fought when we got down there because that is what Rasczak's Roughnecks do.

We certainly didn't know that we were losing.

Those Bugs lay eggs. They not only lay them, they hold them in reserve, hatch them as needed. If we killed a warrior—or a thousand, or ten thousand—his or their replacements were hatched and on duty almost before we could get back to base. You can imagine, if you like, some Bug supervisor of population flashing a phone to somewhere down inside and saying, "Joe, warm up ten thousand warriors and have 'em ready by Wednesday . . . and tell engineering to activate reserve incubators N, O, P, Q, and R; the demand is picking up."

I don't say they did exactly that, but those were the results. But don't make the mistake of thinking that they acted purely from instinct, like termites or ants; their actions were as intelligent as ours (stupid races don't build spaceships!) and were much better co-ordinated. It takes a minimum of a year to train a private to fight and to mesh his fighting in with his mates; a Bug warrior is *hatched* able to do this.

Every time we killed a thousand Bugs at a cost of one M.I. it was a net victory for the Bugs. We were learning, expensively, just how efficient a total communism can be when used by a people actually adapted to it by evolution; the Bug commissars didn't care any more about expending soldiers than we cared about expending ammo. Perhaps we could have figured this out about the Bugs by noting the grief the Chinese Hegemony gave the Russo-Anglo-American Alliance; however the trouble with "lessons from history" is that we usually read them best after falling flat on our chins.

But we were learning. Technical instructions and tactical doctrine orders resulted from every brush with them, spread through the Fleet. We learned to tell the workers from the warriors—if you had time, you could tell from the shape of the carapace, but the quick rule of thumb was: If he comes at you, he's a warrior; if he runs, you can turn your back on him. We learned not to waste ammo even on warriors except in self-protection; instead we went after their lairs. Find a hole, drop down it first a gas bomb which explodes gently a few seconds later, releasing an oily liquid which evaporates as a nerve gas tailored to Bugs (it is harmless to us) and which is heavier than air and keeps on going down—then you use a second grenade of H.E. to seal the hole.

We still didn't know whether we were getting deep enough to kill the queens—but we did know that the Bugs didn't like these tactics; our intelligence through the Skinnies and on back into the Bugs themselves was definite on this point. Besides, we cleaned their colony off Sheol completely this way. Maybe they managed to evacuate the queens and the brains . . . but at least we were learning to hurt them.

But so far as the Roughnecks were concerned, these gas bombings were simply another drill, to be done according to orders, by the numbers, and on the bounce.

*   *   *

Eventually we had to go back to Sanctuary for more capsules. Capsules are expendable (well, so were we) and when they are gone, you must return to base, even if the Cherenkov generators could still take you twice around the Galaxy. Shortly before this a dispatch came through breveting Jelly to lieutenant, vice Rasczak. Jelly tried to keep it quiet but Captain Deladrier published it and then required him to eat forward with the other officers. He still spent all the rest of his time aft.

But we had taken several drops by then with him as platoon leader and the outfit had gotten used to getting along without the Lieutenant—it still hurt but it was routine now. After Jelal was commissioned the word was slowly passed around among us and chewed over that it was time for us to name ourselves for our boss, as with other outfits.

Johnson was senior and took the word to Jelly; he picked me to go along with him as moral support. "Yeah?" growled Jelly.

"Uh, Sarge—I mean Lieutenant, we've been thinking—"

"With what?"

"Well, the boys have sort of been talking it over and they think—well, they say the outfit ought to call itself: 'Jelly's Jaguars.'"

"They do, eh? How many of 'em favor that name?"

"It's unanimous," Johnson said simply.

"So? Fifty-two ayes . . . and one no. The noes have it." Nobody ever brought up the subject again.

Shortly after that we orbited at Sanctuary. I was glad to

be there, as the ship's internal pseudo-gravity field had been off for most of two days before that, while the Chief Engineer tinkered with it, leaving us in free fall—which I hate. I'll never be a real spaceman. Dirt underfoot felt good. The entire platoon went on ten days' rest & recreation and transferred to accommodation barracks at the Base.

I never have learned the co-ordinates of Sanctuary, nor the name or catalogue number of the star it orbits—because what you don't know, you can't spill; the location is ultra-top-secret, known only to ships' captains, piloting officers, and such . . . and, I understand, with each of them under orders and hypnotic compulsion to suicide if necessary to avoid capture. So I don't want to know. With the possibility that Luna Base might be taken and Terra herself occupied, the Federation kept as much of its beef as possible at Sanctuary. so that a disaster back home would not necessarily mean capitulation.

But I can tell you what sort of a planet it is. Like Earth, but retarded.

Literally retarded, like a kid who takes ten years to learn to wave bye-bye and never does manage to master patty-cake. It is a planet as near like Earth as two planets can be, same age according to the planetologists and its star is the same age as the Sun and the same type, so say the astrophysicists. It has plenty of flora and fauna, the same atmosphere as Earth, near enough, and much the same weather; it even has a good-sized moon and Earth's exceptional tides.

With all these advantages it barely got away from the

starting gate. You see, it's short on mutations; it does not enjoy Earth's high level of natural radiation.

Its typical and most highly developed plant life is a very primitive giant fern; its top animal life is a proto-insect which hasn't even developed colonies. I am not speaking of transplanted Terran flora and fauna—*our* stuff moves in and brushes the native stuff aside.

With its evolutionary progress held down almost to zero by lack of radiation and a consequent most unhealthily low mutation rate, native life forms on Sanctuary just haven't had a decent chance to evolve and aren't fit to compete. Their gene patterns remain fixed for a relatively long time; they aren't adaptable—like being forced to play the same bridge hand over and over again, for eons, with no hope of getting a better one.

As long as they just competed with each other, this didn't matter too much—morons among morons, so to speak. But when types that had evolved on a planet enjoying high radiation and fierce competition were introduced, the native stuff was outclassed.

Now all the above is perfectly obvious from high school biology . . . but the high forehead from the research station there who was telling me about this brought up a point I would never have thought of.

What about the human beings who have colonized Sanctuary?

Not transients like me, but the colonists who live there, many of whom were born there, and whose descendants will live there, even unto the umpteenth generation—what about those descendants? It doesn't do a person any harm not to be radiated; in fact it's a bit safer—leukemia and some types of cancer are almost unknown there. Besides that, the economic situation is at present all in their favor; when they plant a field of (Terran) wheat, they don't even have to clear out the weeds. Terran wheat displaces anything native.

But the descendants of those colonists won't evolve. Not much, anyhow. This chap told me that they could improve a little through mutation from other causes, from new blood added by immigration, and from natural selection among the gene patterns they already own—but that is all very minor compared with the evolutionary rate on Terra and on any usual planet. So what happens? Do they stay frozen at their present level while the rest of the human race moves on past them, until they are living fossils, as out of place as a pithecanthropus in a spaceship?

Or will they worry about the fate of their descendants and dose themselves regularly with X-rays or maybe set off lots of dirty-type nuclear explosions each year to build up a fallout reservoir in their atmosphere? (Accepting, of course, the immediate dangers of radiation to themselves in order to provide a proper genetic heritage of mutation for the benefit of their descendants.)

This bloke predicted that they would not do anything. He claims that the human race is too individualistic, too self-centered, to worry that much about future generations. He says that the genetic impoverishment of distant generations through lack of radiation is something most people are simply incapable of worrying about. And of course it is a far-distant threat; evolution works so slowly, even on Terra, that the development of a new species is a matter of many, many thousands of years.

I don't know. Shucks, I don't know what I myself will do more than half the time; how can I predict what a colony of strangers will do? But I'm sure of this: Sanctuary is going to be fully settled, either by us or by the Bugs. Or by somebody. It is a potential utopia, and, with desirable real estate so scarce in this end of the Galaxy, it will not be left in the possession of primitive life forms that failed to make the grade.

Already it is a delightful place, better in many ways for a few days R&R than is most of Terra. In the second place, while it has an awful lot of civilians, more than a million, as civilians go they aren't bad. They know there is a war on. Fully half of them are employed either at the Base or in war industry; the rest raise food and sell it to the Fleet. You might say they have a vested interest in war, but, whatever their reasons, they respect the uniform and don't resent the wearers thereof. Quite the contrary. If an M.I. walks into a shop there, the proprietor calls him "Sir," and really seems to mean it, even while he's trying to sell something worthless at too high a price.

But in the *first* place, half of those civilians are female.

You have to have been out on a long patrol to appreciate this properly. You need to have looked forward to your day of guard duty, for the privilege of standing two hours out of each six with your spine against bulkhead thirty and your ears cocked for just the *sound* of a female voice. I suppose it's actually easier in the all-stag ships . . . but I'll take the *Rodger Young*. It's good to know that the ultimate reason you are fighting actually exists and that they are not just a figment of the imagination.

Besides the civilian wonderful 50 per cent, about 40 per cent of the Federal Service people on Sanctuary are female. Add it all up and you've got the most beautiful scenery in the explored universe.

Besides these unsurpassed natural advantages, a great deal has been done artificially to keep R&R from being wasted. Most of the civilians seem to hold two jobs; they've got circles under their eyes from staying up all night to make a service man's leave pleasant. Churchill Road from the Base to the city is lined both sides with enterprises intended to separate painlessly a man from money he really hasn't any use for anyhow, to the pleasant accompaniment of refreshment, entertainment, and music.

If you are able to get past these traps, through having already been bled of all valuta, there are still other places in the city almost as satisfactory (I mean there are girls there, too) which are provided free by a grateful populace—much like the social center in Vancouver, these are, but even more welcome.

Sanctuary, and especially Espiritu Santo, the city, struck me as such an ideal place that I toyed with the notion of asking for my discharge there when my term was up—after all, I didn't really care whether my descendants (if any) twenty-five thousand years hence had long green tendrils like everybody else, or just the equipment I had been forced to get by with. That professor type from the Research Station couldn't frighten me with that no radiation scare talk; it seemed to me (from what I could see around me) that the human race had reached its ultimate peak anyhow.

No doubt a gentleman wart hog feels the same way about a lady wart hog—but, if so, both of us are very sincere.

There are other opportunities for recreation there, too. I remember with particular pleasure one evening when a table of Roughnecks got into a friendly discussion with a group of Navy men (not from the *Rodger Young*) seated at the next table. The debate was spirited, a bit noisy, and some Base police came in and broke it up with stun guns just as we were warming to our rebuttal. Nothing came of it, except that we had to pay for the furniture—the Base Commandant takes the position that a man on R&R should be allowed a little freedom as long as he doesn't pick one of the "thirty-one crash landings."

The accommodation barracks are all right, too—not fancy, but comfortable and the chow line works twenty-five hours a day with civilians doing all the work. No reveille, no taps, you're actually on leave and you don't have to go to the barracks at all. I did, however, as it seemed downright preposterous to spend money on hotels when there was a clean, soft sack free and so many better ways to spend accumulated pay. That extra hour in each day was nice, too, as it meant nine hours solid and the day still untouched—I caught up sack time clear back to Operation Bughouse.

It might as well have been a hotel; Ace and I had a room all to ourselves in visiting non-com quarters. One morning, when R&R was regrettably drawing to a close, I was just turning over about local noon when Ace shook my bed. "On the bounce, soldier! The Bugs are attacking."

I told him what to do with the Bugs.

"Let's hit dirt," he persisted.

"No dinero." I had had a date the night before with a chemist (female, of course, and charmingly so) from the Research Station. She had known Carl on Pluto and Carl had written to me to look her up if I ever got to Sanctuary. She was a slender redhead, with expensive tastes. Apparently Carl had intimated to her that I had more money than was good for me, for she decided that the night before was just the time for her to get acquainted with the local champagne. I didn't let Carl down by admitting that all I had was a trooper's honorarium; I bought it for her while I drank what they said was (but wasn't) fresh pineapple squash. The result was that I had to walk home, afterwards—the cabs aren't free. Still, it had been worth it. After all, what is money?—I'm speaking of Bug money, of course.

"No ache," Ace answered. "I can juice you—I got

lucky last night. Ran into a Navy file who didn't know percentages."

So I got up and shaved and showered and we hit the chow line for half a dozen shell eggs and sundries such as potatoes and ham and hot cakes and so forth and then we hit dirt to get something to eat. The walk up Churchill Road was hot and Ace decided to stop in a cantina. I went along to see if their pineapple squash was real. It wasn't, but it was cold. You can't have everything.

We talked about this and that and Ace ordered another round. I tried their strawberry squash—same deal. Ace stared into his glass, then said, "Ever thought about greasing for officer?"

I said, "*Huh*? Are you crazy?"

"Nope. Look, Johnnie, this war may run on quite a piece. No matter what propaganda they put out for the folks at home, you and I know that the Bugs aren't ready to quit. So why don't you plan ahead? As the man says, if you've got to play in the band, it's better to wave the stick than to carry the big drum."

I was startled by the turn the talk had taken, especially from Ace. "How about you? Are you planning to buck for a commission?"

"Me?" he answered. "Check your circuits, son— you're getting wrong answers. I've got no education and I'm ten years older than you are. But you've got enough education to hit the selection exams for O.C.S. *and* you've got the I.Q. they like. I guarantee that if you go career, you'll make sergeant before I do . . . and get picked for O.C.S. the day after."

"Now I know you're crazy!"

"You listen to your pop. I hate to tell you this, but you are just stupid and eager and sincere enough to make the kind of officer that men love to follow into some silly predicament. But me—well, I'm a natural non-com, with the proper pessimistic attitude to offset the enthusiasm of the likes of you. Someday I'll make sergeant . . . and presently I'll have my twenty years in and retire and get one of the reserved jobs—cop, maybe—and marry a nice fat wife with the same low tastes I have, and I'll follow the sports and fish and go pleasantly to pieces."

Ace stopped to wet his whistle. "But *you*," he went on. "You'll stay in and probably make high rank and die gloriously and I'll read about it and say proudly, 'I knew him when. Why, I used to lend him money—we were corporals together.' Well?"

"I've never thought about it," I said slowly. "I just meant to serve my term."

He grinned sourly. "Do you see any term enrollees being paid off today? You expect to make it on two years?"

He had a point. As long as the war continued, a "term" didn't end—at least not for cap troopers. It was mostly a difference in attitude, at least for the present. Those of us on "term" could at least feel like short-timers; we could talk about: "When this flea-bitten war is over." A career man didn't say that; he wasn't going anywhere, short of retirement—or buying it.

On the other hand, neither were we. But if you

went "career" and then didn't finish twenty . . . well, they could be pretty sticky about your franchise even though they wouldn't keep a man who didn't want to stay.

"Maybe not a two-year term," I admitted. "But the war won't last forever."

"It won't?"

"How can it?"

"Blessed if I know. They don't tell me these things. But I know that's not what is troubling you, Johnnie. You got a girl waiting?"

"No. Well, I had," I answered slowly, "but she 'Dear-Johned' me." As a lie, this was no more than a mild decoration, which I tucked in because Ace seemed to expect it. Carmen wasn't my girl and she never waited for anybody—but she *did* address letters with "Dear Johnnie" on the infrequent occasions when she wrote to me.

Ace nodded wisely. "They'll do it every time. They'd rather marry civilians and have somebody around to chew out when they feel like it. Never you mind, son— you'll find plenty of them more than willing to marry when you're retired . . . and you'll be better able to handle one at that age. Marriage is a young man's disaster and an old man's comfort." He looked at my glass. "It nauseates me to see you drinking that slop."

"I feel the same way about the stuff you drink," I told him.

He shrugged. "As I say, it takes all kinds. You think it over."

"I will."

Ace got into a card game shortly after, and lent me some money and I went for a walk; I needed to think.

Go career? Quite aside from that noise about a commission, did I want to go career? Why, I had gone through all this to get my franchise, hadn't I?—and if I went career, I was just as far away from the privilege of voting as if I had never enrolled . . . because as long as you were still in uniform you weren't entitled to vote. Which was the way it should be, of course—why, if they let the Roughnecks vote the idiots might vote not to make a drop. Can't have that.

Nevertheless I had signed up in order to win a vote.

Or had I?

Had I ever cared about voting? No, it was the prestige, the pride, the status . . . of being a citizen.

Or was it?

I couldn't to save my life remember *why* I had signed up.

Anyhow, it wasn't the process of voting that made a citizen—the Lieutenant had been a citizen in the truest sense of the word, even though he had not lived long enough ever to cast a ballot. He had "voted" every time he made a drop.

And so had I!

I could hear Colonel Dubois in my mind: "Citizenship is an attitude, a state of mind, an emotional conviction that the whole is greater than the part . . . and that the part should be humbly proud to sacrifice itself that the whole may live."

I still didn't know whether I yearned to place my one-and-only body "between my loved home and the war's desolation"—I still got the shakes every drop and that "desolation" could be pretty desolate. But nevertheless I knew at last what Colonel Dubois had been talking about. The M.I. was mine and I was theirs. If that was what the M.I. did to break the monotony, then that was what I did. Patriotism was a bit esoteric for me, too large-scale to see. But the M.I. was my gang, I belonged. They were all the family I had left; they were the brothers I had never had, closer than Carl had ever been. If I left them, I'd be lost.

So why shouldn't I go career?

All right, all right—but how about this nonsense of greasing for a commission? That was something else again. I could see myself putting in twenty years and then taking it easy, the way Ace had described, with ribbons on my chest and carpet slippers on my feet . . . or evenings down at the Veterans Hall, rehashing old times with others who belonged. But O.C.S.? I could hear Al Jenkins, in one of the bull sessions we had about such things: "I'm a private! I'm going to stay a private! When you're a private they don't expect anything of you. Who wants to be an officer? Or even a sergeant? You're breathing the same air, aren't you? Eating the same food. Going the same places, making the same drops. But no worries."

Al had a point. What had chevrons ever gotten me?— aside from lumps.

Nevertheless I knew I would take sergeant if it was ever offered to me. You don't refuse, a cap trooper doesn't refuse anything; he steps up and takes a swing at it. Commission, too, I supposed.

Not that it would happen. Who was I to think that I could ever be what Lieutenant Rasczak had been?

My walk had taken me close to the candidates' school, though I don't believe I intended to come that way. A company of cadets were out on their parade ground, drilling at trot, looking for all the world like boots in Basic. The sun was hot and it looked not nearly as comfortable as a bull session in the drop room of the *Rodger Young*—why, I hadn't marched farther than bulkhead thirty since I had finished Basic; that breaking-in nonsense was past.

I watched them a bit, sweating through their uniforms; I heard them being chewed out—by sergeants, too. Old Home Week. I shook my head and walked away from there—

—went back to the accommodation barracks, over to the B.O.Q. wing, found Jelly's room.

He was in it, his feet up on a table and reading a magazine. I knocked on the frame of the door. He looked up and growled, "Yeah?"

"Sarge—I mean, Lieutenant—"

"Spit it out!"

"Sir, I want to go career."

He dropped his feet to the desk. "Put up your right hand."

He swore me, reached into the drawer of the table and pulled out papers.

He had my papers already made out, waiting for me ready to sign. And I hadn't even told Ace. How about that?

CH:12

It is by no means enough that an officer should be capable. . . . He should be as well a gentleman of liberal education, refined manners, punctilious courtesy, and the nicest sense of personal honor. . . . No meritorious act of a subordinate should escape his attention, even if the reward be only one word of approval. Conversely, he should not be blind to a single fault in any subordinate.

True as may be the political principles for which we are now contending . . . the ships themselves must be ruled under a system of absolute despotism.

I trust that I have now made clear to you the tremendous responsibilities. . . . We must do the best we can with what we have.

—John Paul Jones, September 14, 1775;
excerpts from a letter to the naval
committee of the N.A. insurrectionists

The *Rodger Young* was again returning to Base for replacements, both capsules and men. Al Jenkins had bought his farm, covering a pickup—and that one had cost us the Padre, too. And besides that, I had to be replaced. I was wearing brand-new sergeant's chevrons (vice Migliaccio) but I had a hunch that Ace would be wearing them as soon as I was out of the ship—they were mostly honorary, I knew; the promotion was Jelly's way of giving me a good send-off as I was detached for O.C.S.

But it didn't keep me from being proud of them. At the Fleet landing field I went through the exit gate with my nose in the air and strode up to the quarantine desk to have my orders stamped. As this was being done I heard a polite, respectful voice behind me: "Excuse me, Sergeant, but that boat that just came down—is it from the *Rodger*—"

I turned to see the speaker, flicked my eyes over his sleeves, saw that it was a small, slightly stoop-shouldered corporal, no doubt one of our—

"*Father!*"

Then the corporal had his arms around me. "Juan! Juan! Oh, my little Johnnie!"

I kissed him and hugged him and started to cry. Maybe that civilian clerk at the quarantine desk had never seen two non-coms kiss each other before. Well, if I had noticed him so much as lifting an eyebrow, I would have pasted him. But I didn't notice him; I was busy. He had to remind me to take my orders with me.

By then we had blown our noses and quit making an open spectacle of ourselves. I said, "Father, let's find a corner somewhere and sit down and talk. I want to know . . . well, *everything!*" I took a deep breath. "I thought you were dead."

"No. Came close to buying it once or twice, maybe. But, Son . . . Sergeant—I really do have to find out about that landing boat. You see—"

"Oh, that. It's from the *Rodger Young*. I just—"

He looked terribly disappointed. "Then I've got to bounce, right now. I've got to report in." Then he added eagerly, "But you'll be back aboard soon, won't you, Juanito? Or are you going on R&R?"

"Uh, no." I thought fast. Of all the ways to have things roll! "Look, Father, I know the boat schedule. You can't go aboard for at least an hour and a bit. That boat is not on a fast retrieve; she'll make a minimum-fuel rendezvous when the *Rog* completes this pass—if the pilot doesn't have to wait over for the next pass after that; they've got to load first."

He said dubiously, "My orders read to report at once to the pilot of the first available ship's boat."

"Father, Father! Do you have to be so confounded regulation? The girl who's pushing that heap won't care whether you board the boat now, or just as they button up. Anyhow they'll play the ship's recall over the speakers in here ten minutes before boost and announce it. You *can't* miss it."

He let me lead him over to an empty corner. As we sat down he added, "Will you be going up in the same boat, Juan? Or later?"

"Uh—" I showed him my orders; it seemed the simplest way to break the news. Ships that pass in the night, like the Evangeline story—cripes, what a way for things to break!

He read them and got tears in his eyes and I said hastily, "Look, Father, I'm going to try to come back—I wouldn't want any other outfit than the Roughnecks. And with you in them . . . oh, I know it's disappointing but—"

"It's not disappointment, Juan."

"Huh?"

"It's pride. My boy is going to be an officer. My little Johnnie—Oh, it's disappointment, too; I had waited for this day. But I can wait a while longer." He smiled through his tears. "You've grown, lad. And filled out, too."

"Uh, I guess so. But, Father, I'm not an officer yet and I might only be out of the *Rog* a few days. I mean, they sometimes bust 'em out pretty fast and—"

"Enough of that, young man!"

"Huh?"

"You'll make it. Let's have no more talk of 'busting out.'" Suddenly he smiled. "That's the first time I've been able to tell a sergeant to shut up."

"Well . . . I'll certainly try, Father. And if I do make it, I'll certainly put in for the old *Rog*. But—" I trailed off.

"Yes, I know. Your request won't mean anything unless there's a billet for you. Never mind. If this hour is all we have, we'll make the most of it—and I'm so proud of you I'm splitting my seams. How have you been, Johnnie?"

"Oh, fine, just fine." I was thinking that it wasn't all bad. He would be better off in the Roughnecks than in any other outfit. All my friends . . . they'd take care of him, keep him alive. I'd have to send a gram to Ace— Father like as not wouldn't even let them know he was related. "Father, how long have you been in?"

"A little over a year."

"And corporal already!"

Father smiled grimly. "They're making them fast these days."

I didn't have to ask what he meant. Casualties. There were always vacancies in the T.O.; you couldn't get enough trained soldiers to fill them. Instead I said, "Uh . . . but, Father, you're—Well, I mean, aren't you sort of old to be soldiering? I mean the Navy, or Logistics, or—"

"I wanted the M.I. and I got it!" he said emphatically. "And I'm no older than many sergeants—not as old, in fact. Son, the mere fact that I am twenty-two years older than you are doesn't put me in a wheel chair. And age has its advantages, too."

Well, there was something in that. I recalled how Sergeant Zim had always tried the older men first, when he was dealing out boot chevrons. And Father would never have goofed in Basic the way I had—no lashes for him. He was probably spotted as non-com material before he ever finished Basic. The Army needs a lot of really grown-up men in the middle grades; it's a paternalistic organization.

I didn't have to ask him why he had wanted M.I., nor why or how he had wound up in my ship—I just felt warm about it, more flattered by it than any praise he had ever given me in words. And I didn't want to ask him *why* he had joined up; I felt that I knew. Mother. Neither of us had mentioned her—too painful.

So I changed the subject abruptly. "Bring me up to date. Tell me where you've been and what you've done."

"Well, I trained at Camp San Martín—"

"Huh? Not Currie?"

"New one. But the same old lumps, I understand. Only

they rush you through two months faster, you don't get Sundays off. Then I requested the *Rodger Young*—and didn't get it—and wound up in McSlattery's Volunteers. A good outfit."

"Yes, I know." They had had a reputation for being rough, tough, and nasty—almost as good as the Roughnecks.

"I should say that it *was* a good outfit. I made several drops with them and some of the boys bought it and after a while I got these." He glanced at his chevrons. "I was a corporal when we dropped on Sheol—"

"You were *there*? So was I!" With a sudden warm flood of emotion I felt closer to my father than I ever had before in my life.

"I know. At least I knew your outfit was there. I was about fifty miles north of you, near as I can guess. We soaked up that counterattack when they came boiling up out of the ground like bats out of a cave." Father shrugged. "So when it was over I was a corporal without an outfit, not enough of us left to make a healthy cadre. So they sent me here. I could have gone with King's Kodiak Bears, but I had a word with the placement sergeant—and, sure as sunrise, the *Rodger Young* came back with a billet for a corporal. So here I am."

"And when did you join up?" I realized that it was the wrong remark as soon as I had made it—but I had to get the subject away from McSlattery's Volunteers; an orphan from a dead outfit wants to forget it.

Father said quietly, "Shortly after Buenos Aires."

"Oh. I see."

Father didn't say anything for several moments. Then he said softly, "I'm not sure that you do see, Son."

"Sir?"

"Mmm . . . it will not be easy to explain. Certainly, losing your mother had a great deal to do with it. But I didn't enroll to avenge her—even though I had that in mind, too. You had more to do with it—"

"*Me?*"

"Yes, you. Son, I always understood what you were doing better than your mother did—don't blame her; she never had a chance to know, any more than a bird can understand swimming. And perhaps I knew *why* you did it, even though I beg to doubt that you knew yourself, at the time. At least half of my anger at you was sheer resentment . . . that you had actually done something that I knew, buried deep in my heart, I should have done. But you weren't the cause of my joining up, either . . . you merely helped trigger it and you did control the service I chose."

He paused. "I wasn't in good shape at the time you enrolled. I was seeing my hypnotherapist pretty regularly—you never suspected that, did you?—but we had gotten no farther than a clear recognition that I was enormously dissatisfied. After you left, I took it out on you—but it was not you, and I knew it and my therapist knew it. I suppose I knew that there was real trouble brewing earlier than most; we were invited to bid on military components fully a month before the state of emergency was announced. We had converted almost entirely to war production while you were still in training.

"I felt better during that period, worked to death and too busy to see my therapist. Then I became more troubled than ever." He smiled. "Son, do you know about civilians?"

"Well . . . we don't talk the same language. I know that."

"Clearly enough put. Do you remember Madame Ruitman? I was on a few days leave after I finished Basic and I went home. I saw some of our friends, said good-by—she among them. She chattered away and said, 'So you're really going out? Well, if you reach Faraway, you really must look up my dear friends the Regatos.'

"I told her, as gently as I could, that it seemed unlikely, since the Arachnids had occupied Faraway.

"It didn't faze her in the least. She said, 'Oh, that's all right—they're civilians!'" Father smiled cynically.

"Yes, I know."

"But I'm getting ahead of my story. I told you that I was getting still more upset. Your mother's death released me for what I had to do . . . even though she and I were closer than most, nevertheless it set me free to do it. I turned the business over to Morales—"

"Old man Morales? Can he handle it?"

"Yes. Because he has to. A lot of us are doing things we didn't know we could. I gave him a nice chunk of stock—you know the old saying about the kine that tread the grain—and the rest I split two ways, in a trust: half to the Daughters of Charity, half to you whenever you want to go back and take it. If you do. Never mind. I had at last found out what was wrong with me." He stopped, then said very softly, "I had to perform an act of faith. I had to prove to myself that I was a man. Not just a producing-consuming economic animal . . . but a *man*."

At that moment, before I could answer anything, the wall speakers around us sang: "—*shines the name, shines the name of Rodger Young!*" and a girl's voice added, "Personnel for F.C.T. *Rodger Young*, stand to boat. Berth H. Nine minutes."

Father bounced to his feet, grabbed his kit roll. "That's mine! Take care of yourself, Son—and hit those exams. Or you'll find you're still not too big to paddle."

"I will, Father."

He embraced me hastily. "See you when we get back!" And he was gone, on the bounce.

In the Commandant's outer office I reported to a fleet sergeant who looked remarkably like Sergeant Ho, even to lacking an arm. However, he lacked Sergeant Ho's smile as well. I said, "Career Sergeant Juan Rico, to report to the Commandant pursuant to orders."

He glanced at the clock. "Your boat was down seventy-three minutes ago. Well?"

So I told him. He pulled his lip and looked at me meditatively. "I've heard every excuse in the book. But you've just added a new page. Your father, your own father, really was reporting to your old ship just as you were detached?"

"The bare truth, Sergeant. You can check it—Corporal Emilio Rico."

"We don't check the statements of the 'young gentlemen' around here. We simply cashier them if it ever turns out that they have not told the truth. Okay, a boy who wouldn't be late in order to see his old man off wouldn't be worth much in any case. Forget it."

"Thanks, Sergeant. Do I report to the Commandant now?"

"You've reported to him." He made a check mark on a list. "Maybe a month from now he'll send for you along with a couple of dozen others. Here's your room assignment, here's a checkoff list you start with—and you can start by cutting off those chevrons. But save them; you may need them later. But as of this moment you are 'Mister,' not 'Sergeant.'"

"Yes, sir."

"Don't call me 'sir.' I call *you* 'sir.' But you won't like it."

I am not going to describe Officer Candidates School. It's like Basic, but squared and cubed with books added. In the mornings we behaved like privates, doing the same old things we had done in Basic and in combat and being chewed out for the way we did them—by sergeants. In the afternoons we were cadets and "gentlemen," and recited on and were lectured concerning an endless list of subjects: math, science, galactography, xenology, hypnopedia, logistics, strategy and tactics, communications, military law, terrain reading, special weapons, psychology of leadership, anything from the care and feeding of privates to why Xerxes lost the big one. Most especially how to be a one-man catastrophe yourself while keeping track of fifty other men, nursing them, loving them, leading them, saving them—but *never* babying them.

We had beds, which we used all too little; we had rooms and showers and inside plumbing; and each four candidates had a civilian servant, to make our beds and clean our rooms and shine our shoes and lay out our uniforms and run errands. This service was not intended as a luxury and was not; its purpose was to give the student more time to accomplish the plainly impossible by relieving him of things any graduate of Basic can already do perfectly.

> *Six days shalt thou work and do all thou art able,*
> *The seventh the same and pound on the cable.*

Or the Army version ends:—*and clean out the stable*, which shows you how many centuries this sort of thing has been going on. I wish I could catch just one of those civilians who think we loaf and put them through one month of O.C.S.

In the evenings and all day Sundays we studied until our eyes burned and our ears ached—then slept (if we slept) with a hypnopedic speaker droning away under the pillow.

Our marching songs were appropriately downbeat: "No Army for mine, no Army for mine! I'd rather be behind the plow any old time!" and "Don't wanta study war no more," and "Don't make my boy a soldier, the weeping mother cried," and—favorite of all—the old classic "Gentlemen Rankers" with its chorus about the Little Lost Sheep: "—God ha' pity on such as we. Baa! Yah! Bah!"

Yet somehow I don't remember being unhappy. Too busy, I guess. There was never that psychological "hump" to get over, the one everybody hits in Basic; there was simply the ever-present fear of flunking out. My poor preparation in math bothered me especially. My roommate, a colonial from Hesperus with the oddly appropriate name of "Angel," sat up night after night, tutoring me.

Most of the instructors, especially the officers, were disabled. The only ones I can remember who had a full complement of arms, legs, eyesight, hearing, etc., were some of the non-commissioned combat instructors—and not all of those. Our coach in dirty fighting sat in a powered chair, wearing a plastic collar, and was completely paralyzed from the neck down. But his tongue wasn't paralyzed, his eye was photographic, and the savage way in which he could analyze and criticize what he had seen made up for his minor impediment.

At first I wondered why those obvious candidates for physical retirement and full-pay pension didn't take it and go home. Then I quit wondering.

I guess the high point in my whole cadet course was a visit from Ensign Ibañez, she of the dark eyes, junior watch officer and pilot-under-instruction of the Corvette Transport *Mannerheim*. Carmencita showed up, looking

incredibly pert in Navy dress whites and about the size of
a paperweight, while my class was lined up for evening
meal muster—walked down the line and you could hear
eyeballs click as she passed—walked straight up to the
duty officer and asked for me by name in a clear, penetrat-
ing voice.

The duty officer, Captain Chandar, was widely believed never to have smiled at his own mother, but he
smiled down at little Carmen, straining his face out of
shape, and admitted my existence . . . whereupon she
waved her long black lashes at him, explained that her
ship was about to boost and could she *please* take me out
to dinner?

And I found myself in possession of a highly irregular
and totally unprecedented three-hour pass. It may be
that the Navy has developed hypnosis techniques that
they have not yet gotten around to passing on to the
Army. Or her secret weapon may be older than that and
not usable by M.I. In any case I not only had a wonderful time but my prestige with my classmates, none too
high until then, climbed to amazing heights.

It was a glorious evening and well worth flunking two
classes the next day. It was somewhat dimmed by the fact
that we had each heard about Carl—killed when the
Bugs smashed our research station on Pluto—but only
somewhat, as we had each learned to live with such
things.

One thing did startle me. Carmen relaxed and took
off her hat while we were eating, and her blue-black hair
was all gone. I knew that a lot of the Navy girls shaved

their heads—after all, it's not practical to take care of long hair in a war ship and, most especially, a pilot can't risk having her hair floating around, getting in the way, in any free-fall maneuvers. Shucks, I shaved my own scalp, just for convenience and cleanliness. But my mental picture of little Carmen included this mane of thick, wavy hair.

But, do you know, once you get used to it, it's rather cute. I mean, if a girl looks all right to start with, she still looks all right with her head smooth. And it does serve to set a Navy girl apart from civilian chicks—sort of a lodge pin, like the gold skulls for combat drops. It made Carmen look distinguished, gave her dignity, and for the first time I fully realized that she really was an officer and a fighting man—as well as a very pretty girl.

I got back to barracks with stars in my eyes and whiffing slightly of perfume. Carmen had kissed me good-by.

The only O.C.S. classroom course the content of which I'm even going to mention was: History and Moral Philosophy.

I was surprised to find it in the curriculum. H. & M. P. has nothing to do with combat and how to lead a platoon; its connection with war (where it is connected) is in *why* to fight—a matter already settled for any candidate long before he reaches O.C.S. An M.I. fights because he is M.I.

I decided that the course must be a repeat for the benefit of those of us (maybe a third) who had never had it in school. Over 20 per cent of my cadet class were not from Terra (a much higher percentage of colonials sign up to serve than do people born on Earth—sometimes it makes you wonder) and of the three-quarters or so from Terra, some were from associated territories and other places where H. & M. P. might not be taught. So I figured it for a cinch course which would give me a little rest from tough courses, the ones with decimal points.

Wrong again. Unlike my high school course, you had to pass it. Not by examination, however. The course included examinations and prepared papers and quizzes and such—but no marks. What you had to have was the instructor's opinion that you were worthy of commission.

If he gave you a downcheck, a board sat on you, questioning not merely whether you could be an officer but whether you belonged in the Army at *any* rank, no matter how fast you might be with weapons—deciding whether to give you extra instruction . . . or just kick you out and let you be a civilian.

History and Moral Philosophy works like a delayed-action bomb. You wake up in the middle of the night and think: Now what did he mean by *that*? That had been true even with my high school course; I simply hadn't known what Colonel Dubois was talking about. When I was a kid I thought it was silly for the course to be in the science department. It was nothing like physics or chemistry; why wasn't it over in the fuzzy studies where it belonged? The only reason I paid attention was because there were such lovely arguments.

I had no idea that "Mr." Dubois was trying to teach

me *why* to fight until long after I had decided to fight anyhow.

Well, why *should* I fight? Wasn't it preposterous to expose my tender skin to the violence of unfriendly strangers? Especially as the pay at any rank was barely spending money, the hours terrible, and the working conditions worse? When I could be sitting at home while such matters were handled by thick-skulled characters who *enjoyed* such games? Particularly when the strangers against whom I fought never had done anything to me personally until I showed up and started kicking over their tea wagon—what sort of nonsense *is* this?

Fight because I'm an M.I.? Brother, you're drooling like Dr. Pavlov's dogs. Cut it out and start thinking.

Major Reid, our instructor, was a blind man with a disconcerting habit of looking straight at you and calling you by name. We were reviewing events after the war between the Russo-Anglo-American Alliance and the Chinese Hegemony, 1987 and following. But this was the day that we heard the news of the destruction of San Francisco and the San Joaquin Valley; I thought he would give us a pep talk. After all, even a civilian ought to be able to figure it out now—the Bugs or us. Fight or die.

Major Reid didn't mention San Francisco. He had one of us apes summarize the negotiated treaty of New Delhi, discuss how it ignored prisoners of war . . . and, by implication, dropped the subject forever; the armistice became a stalemate and prisoners stayed where they were—on one side; on the other side they were turned loose and, during the Disorders, made their way home—or not if they didn't want to.

Major Reid's victim summed up the unreleased prisoners: survivors of two divisions of British paratroopers, some thousands of civilians, captured mostly in Japan, the Philippines, and Russia and sentenced for "political" crimes.

"Besides that, there were many other military prisoners," Major Reid's victim went on, "captured during and before the war—there were rumors that some had been captured in an earlier war and never released. The total of unreleased prisoners was never known. The best estimates place the number around sixty-five thousand."

"Why the 'best'?"

"Uh, that's the estimate in the textbook, sir."

"Please be precise in your language. Was the number greater or less than one hundred thousand?"

"Uh, I don't know, sir."

"And nobody else knows. Was it greater than one thousand?"

"Probably, sir. Almost certainly."

"Utterly certain—because more than that eventually escaped, found their ways home, were tallied by name. I see you did not read your lesson carefully. *Mr. Rico!*"

Now I am the victim. "Yes, sir."

"Are a thousand unreleased prisoners sufficient reason to start or resume a war? Bear in mind that millions of innocent people may die, almost certainly *will* die, if war is started or resumed."

I didn't hesitate. "Yes, *sir*! More than enough reason."

"'More than enough.' Very well, is *one* prisoner, unreleased by the enemy, enough reason to start or resume a war?"

I hesitated. I knew the M.I. answer—but I didn't think that was the one he wanted. He said sharply, "Come, come, Mister! We have an upper limit of one thousand; I invited you to consider a lower limit of one. But you can't pay a promissory note which reads 'somewhere between one and one thousand pounds'—and starting a war is *much* more serious than paying a trifle of money. Wouldn't it be criminal to endanger a country—two countries in fact—to save one man? Especially as he may not deserve it? Or may die in the meantime? Thousands of people get killed every day in accidents . . . so why hesitate over one man? Answer! Answer yes, or answer no—you're holding up the class."

He got my goat. I gave him the cap trooper's answer. "Yes, sir!"

"'Yes' what?"

"It doesn't matter whether it's a thousand—or just one, sir. You fight."

"Aha! The number of prisoners is irrelevant. Good. Now prove your answer."

I was stuck. I *knew* it was the right answer. But I didn't know why. He kept hounding me. "Speak up, Mr. Rico. This is an exact science. You have made a mathematical statement; you must give proof. Someone may claim that you have asserted, by analogy, that one potato is worth

the same price, no more, no less, as one thousand potatoes. No?"

"No, sir!"

"Why not? Prove it."

"Men are not potatoes."

"Good, good, Mr. Rico! I think we have strained your
tired brain enough for one day. Bring to class tomorrow
a written proof, in symbolic logic, of your answer to my
original question. I'll give you a hint. See reference seven
in today's chapter. Mr. Salomon! How did the present
political organization evolve out of the Disorders? And
what is its moral justification?"

Sally stumbled through the first part. However, nobody can describe accurately how the Federation came
about; it just grew. With national governments in collapse at the end of the XXth century, something had to fill
the vacuum, and in many cases it was returned veterans.
They had lost a war, most of them had no jobs, many
were sore as could be over the terms of the Treaty of
New Delhi, especially the P.O.W. foul-up—and they
knew how to fight. But it wasn't revolution; it was more
like what happened in Russia in 1917—the system collapsed; somebody else moved in.

The first known case, in Aberdeen, Scotland, was
typical. Some veterans got together as vigilantes to stop
rioting and looting, hanged a few people (including
two veterans) and decided not to let anyone but veterans
on their committee. Just arbitrary at first—they trusted
each other a bit, they didn't trust anyone else. What

started as an emergency measure became constitutional practice . . . in a generation or two.

Probably those Scottish veterans, since they were finding it necessary to hang some veterans, decided that, if they had to do this, they weren't going to let any "bleedin', profiteering, black-market, double-time-for-overtime, army-dodging, unprintable" civilians have any say about it. They'd do what they were told, see?—while us apes straightened things out! That's my guess, because I might feel the same way . . . and historians agree that antagonism between civilians and returned soldiers was more intense than we can imagine today.

Sally didn't tell it by the book. Finally Major Reid cut him off. "Bring a summary to class tomorrow, three thousand words. Mr. Salomon, can you give me a reason—not historical nor theoretical but practical—why the franchise is today limited to discharged veterans?"

"Uh, because they are picked men, sir. Smarter."

"Pre*pos*terous!"

"Sir?"

"Is the word too long for you? I said it was a silly notion. Service men are not brighter than civilians. In many cases civilians are much more intelligent. That was the sliver of justification underlying the attempted *coup d'état* just before the Treaty of New Delhi, the so-called 'Revolt of the Scientists': let the intelligent elite run things and you'll have utopia. It fell flat on its foolish face of course. Because the pursuit of science, despite its social benefits, is itself not a social virtue; its practitioners can be men so self-centered as to be lacking in social responsibility. I've given you a hint, Mister; can you pick it up?"

Sally answered, "Uh, service men are disciplined, sir."

Major Reid was gentle with him. "Sorry. An appealing theory not backed up by facts. You and I are not permitted to vote as long as we remain in the Service, nor is it verifiable that military discipline makes a man self-disciplined once he is out; the crime rate of veterans is much like that of civilians. And you have forgotten that in peacetime most veterans come from non-combatant auxiliary services and have not been subjected to the full rigors of military discipline; they have merely been harried, overworked, and endangered—yet their votes count."

Major Reid smiled. "Mr. Salomon, I handed you a trick question. The practical reason for continuing our system is the same as the practical reason for continuing anything: It works satisfactorily.

"Nevertheless, it is instructive to observe the details. Throughout history men have labored to place the sovereign franchise in hands that would guard it well and use it wisely, for the benefit of all. An early attempt was absolute monarchy, passionately defended as the 'divine right of kings.'

"Sometimes attempts were made to select a wise monarch, rather than leave it up to God, as when the Swedes picked a Frenchman, General Bernadotte, to rule them. The objection to this is that the supply of Bernadottes is limited.

"Historic examples ranged from absolute monarch to utter anarch; mankind has tried thousands of ways and many more have been proposed, some weird in the extreme such as the antlike communism urged by Plato under the misleading title *The Republic*. But the intent has always been moralistic: to provide stable and benevolent government.

"All systems seek to achieve this by limiting franchise to those who are *believed* to have the wisdom to use it justly. I repeat '*all* systems'; even the so-called 'unlimited democracies' excluded from franchise not less than one-quarter of their populations by age, birth, poll tax, criminal record, or other."

Major Reid smiled cynically. "I have never been able to see how a thirty-year-old moron can vote more wisely than a fifteen-year-old genius . . . but that was the age of the 'divine right of the common man.' Never mind, they paid for their folly.

"The sovereign franchise has been bestowed by all sorts of rules—place of birth, family of birth, race, sex, property, education, age, religion, et cetera. All these systems worked and none of them well. All were regarded as tyrannical by many, all eventually collapsed or were overthrown.

"Now here are we with still another system . . . and our system works quite well. Many complain but none rebel; personal freedom for all is greatest in history, laws are few, taxes are low, living standards are as high as productivity permits, crime is at its lowest ebb. Why? Not because our voters are smarter than other people; we've disposed of that argument. Mr. Tammany—can you tell us why our system works better than any used by our ancestors?"

I don't know where Clyde Tammany got his name; I'd take him for a Hindu. He answered. "Uh, I'd venture to guess that it's because the electors are a small group who know that the decisions are up to them . . . so they study the issues."

"No guessing, please; this is exact science. And your guess is wrong. The ruling nobles of many another system were a small group fully aware of their grave power. Furthermore, our franchised citizens are not everywhere a small fraction; you know or should know that the percentage of citizens among adults ranges from over eighty per cent on Iskander to less than three per cent in some Terran nations—yet government is much the same everywhere. Nor are the voters picked men; they bring no special wisdom, talent, or training to their sovereign tasks. So what difference is there between our voters and wielders of franchise in the past? We have had enough guesses; I'll state the obvious: Under our system every voter and officeholder is a man who has demonstrated through voluntary and difficult service that he places the welfare of the group ahead of personal advantage.

"And that is the one practical difference.

"He may fail in wisdom, he may lapse in civic virtue. But his average performance is enormously better than that of any other class of rulers in history."

Major Reid paused to touch the face of an old-fashioned watch, "reading" its hands. "The period is almost over and we have yet to determine the moral reason for our success in governing ourselves. Now continued success is *never* a matter of chance. Bear in mind that this is science, not wishful thinking; the universe is what it *is*, not what we want it to be. To vote is to wield authority; it is the supreme authority from which all other authority derives—such as mine to make your lives miserable once a day. *Force*, if you will!—the franchise is force, naked and raw, the Power of the Rods and the Ax. Whether it is exerted by ten men or by ten billion, political authority is *force*.

"But this universe consists of paired dualities. What is the converse of authority? Mr. Rico."

He had picked one I could answer. "Responsibility, sir."

"Applause. Both for practical reasons and for mathematically verifiable moral reasons, authority and responsibility must be equal—else a balancing takes place as surely as current flows between points of unequal potential. To permit irresponsible authority is to sow disaster; to hold a man responsible for anything he does not control is to behave with blind idiocy. The unlimited democracies were unstable because their citizens were not responsible for the fashion in which they exerted their sovereign authority . . . other than through the tragic logic of history. The unique 'poll tax' that we must pay was unheard of. No attempt was made to determine whether a voter was socially responsible to the extent of his literally unlimited authority. If he voted the impossible, the disastrous possible happened instead—and responsibility was then forced on him willy-nilly and destroyed both him and his foundationless temple.

"Superficially, our system is only slightly different; we

have democracy unlimited by race, color, creed, birth, wealth, sex, or conviction, and anyone may win sovereign power by a usually short and not too arduous term of service—nothing more than a light workout to our cave-man ancestors. But that slight difference is one between a system that works, since it is constructed to match the facts, and one that is inherently unstable. Since sovereign franchise is the ultimate in human authority, we insure that all who wield it accept the ultimate in social responsibility—we require each person who wishes to exert control over the state to wager his own life—and lose it, if need be—to save the life of the state. The maximum responsibility a human can accept is thus equated to the ultimate authority a human can exert. Yin and yang, perfect and equal."

The Major added, "Can anyone define why there has never been revolution against our system? Despite the fact that every government in history has had such? Despite the notorious fact that complaints are loud and unceasing?"

One of the older cadets took a crack at it. "Sir, revolution is impossible."

"Yes. But why?"

"Because revolution—armed uprising—requires not only dissatisfaction but aggressiveness. A revolutionist has to be willing to fight and die—or he's just a parlor pink. If you separate out the aggressive ones and make them the sheep dogs, the sheep will never give you trouble."

"Nicely put! Analogy is always suspect, but that one is close to the facts. Bring me a mathematical proof tomorrow. Time for one more question—you ask it and I'll answer. Anyone?"

"Uh, sir, why not go—well, go the limit? Require everyone to serve and let everybody vote?"

"Young man, can you restore my eyesight?"

"Sir? Why, no, sir!"

"You would find it much easier than to instill moral virtue—social responsibility—into a person who doesn't have it, doesn't want it, and resents having the burden thrust on him. This is why we make it so hard to enroll, so easy to resign. Social responsibility above the level of family, or at most of tribe, requires imagination—devotion, loyalty, all the higher virtues—which a man must develop himself; if he has them forced down him, he will vomit them out. Conscript armies have been tried in the past. Look up in the library the psychiatric report on brainwashed prisoners in the so-called 'Korean War,' circa 1950—the Mayor Report. Bring an analysis to class." He touched his watch. "Dismissed."

Major Reid gave us a busy time.

But it was interesting. I caught one of those master's-thesis assignments he chucked around so casually; I had suggested that the Crusades were different from most wars. I got sawed off and handed this: *Required*: to prove that war and moral perfection derive from the same genetic inheritance. Briefly, thus: All wars arise from population pressure. (Yes, even the Crusades, though you have to dig into trade routes and birth rate and several other things to prove it.) Morals—*all* correct moral rules—derive from the instinct to survive; moral behavior is survival behavior above the individual level—as in a father who dies to save his children. But since population pressure results from the process of surviving through others, then war, because it results from population pressure, derives from the same inherited instinct which produces all moral rules suitable for human beings.

Check of proof: Is it possible to abolish war by relieving population pressure (and thus do away with the all-too-evident evils of war) through constructing a moral code under which population is limited to resources?

Without debating the usefulness or morality of planned parenthood, it may be verified by observation that any breed which stops its own increase gets crowded out by breeds which expand. Some human populations did so, in Terran history, and other breeds moved in and engulfed them.

Nevertheless, let's assume that the human race manages to balance birth and death, just right to fit its own planets, and thereby becomes peaceful. What happens?

Soon (about next Wednesday) the Bugs move in, kill off this breed which "ain'ta gonna study war no more" and the universe forgets us. Which still may happen. Either we spread and wipe out the Bugs, or they spread and wipe us out—because both races are tough and smart and want the same real estate.

Do you know how fast population pressure could cause us to fill the entire universe shoulder to shoulder? The answer will astound you, just the flicker of an eye in terms of the age of our race.

Try it—it's a compound-interest expansion.

But does Man have any "right" to spread through the universe?

Man is what he is, a wild animal with the will to survive, and (so far) the ability, against all competition. Unless one accepts that, anything one says about morals, war, politics—you name it—is nonsense. Correct morals arise from knowing what Man *is*—not what do-gooders and well-meaning old Aunt Nellies would like him to be.

The universe will let us know—later—whether or not Man has any "right" to expand through it.

In the meantime the M.I. will be in there, on the bounce and swinging, on the side of our own race.

Toward the end each of us was shipped out to serve under an experienced combat commander. This was a semifinal examination, your 'board-ship instructor could decide that you didn't have what it takes. You could demand a board but I never heard of anybody who did; they either came back with an upcheck—or we never saw them again.

Some hadn't failed; it was just that they were killed—because assignments were to ships about to go into action. We were required to keep kit bags packed—once at lunch, all the cadet officers of my company were tapped; they left without eating and I found myself cadet company commander.

Like boot chevrons, this is an uncomfortable honor, but in less than two days my own call came.

I bounced down to the Commandant's office, kit bag

over my shoulder and feeling grand. I was sick of late hours and burning eyes and never catching up, of looking stupid in class; a few weeks in the cheerful company of a combat team was just what Johnnie needed!

I passed some new cadets, trotting to class in close formation, each with the grim look that every O.C.S. candidate gets when he realizes that possibly he made a mistake in bucking for officer, and I found myself singing. I shut up when I was within earshot of the office.

Two others were there, Cadets Hassan and Byrd. Hassan the Assassin was the oldest man in our class and looked like something a fisherman had let out of a bottle, while Birdie wasn't much bigger than a sparrow and about as intimidating.

We were ushered into the Holy of Holies. The Commandant was in his wheel chair—we never saw him out of it except Saturday inspection and parade, I guess walking hurt. But that didn't mean you didn't see him—you could be working a prob at the board, turn around and find that wheel chair behind you, and Colonel Nielssen reading your mistakes.

He never interrupted—there was a standing order not to shout "Attention!" But it's disconcerting. There seemed to be about six of him.

The Commandant had a permanent rank of fleet general (yes, *that* Nielssen); his rank as colonel was temporary, pending second retirement, to permit him to be Commandant. I once questioned a paymaster about this and confirmed what the regulations seemed to say: The

Commandant got only the pay of a colonel—but would revert to the pay of a fleet general on the day he decided to retire again.

Well, as Ace says, it takes all sorts—I can't imagine choosing half pay for the privilege of riding herd on cadets.

Colonel Nielssen looked up and said, "Morning, gentlemen. Make yourselves comfortable." I sat down but wasn't comfortable. He glided over to a coffee machine, drew four cups, and Hassan helped him deal them out. I didn't want coffee but a cadet doesn't refuse the Commandant's hospitality.

He took a sip. "I have your orders, gentlemen," he announced, "and your temporary commissions." He went on, "But I want to be sure you understand your status."

We had already been lectured about this. We were going to be officers just enough for instruction and testing—"supernumerary, probationary, and temporary." Very junior, quite superfluous, on good behavior, and extremely temporary; we would revert to cadet when we got back and could be busted at any time by the officers examining us.

We would be "temporary third lieutenants"—a rank as necessary as feet on a fish, wedged into the hairline between fleet sergeants and real officers. It is as low as you can get and still be called an "officer." If anybody ever saluted a third lieutenant, the light must have been bad.

"Your commission reads 'third lieutenant,'" he went on, "but your pay stays the same, you continue to be addressed as 'Mister,' the only change in uniform is a shoulder pip even smaller than cadet insignia. You continue under instruction since it has not yet been settled that you are fit to be officers." The Colonel smiled. "So why call you a 'third lieutenant'?"

I had wondered about that. Why this whoopty-do of "commissions" that weren't real commissions?

Of course I knew the textbook answer.

"Mr. Byrd?" the Commandant said.

"Uh . . . to place us in the line of command, sir."

"Exactly!" Colonel glided to a T.O. on one wall. It was the usual pyramid, with chain of command defined all the way down. "Look at this—" He pointed to a box connected to his own by a horizontal line; it read: ASSISTANT TO COMMANDANT (Miss Kendrick).

"Gentlemen," he went on, "I would have trouble running this place without Miss Kendrick. Her head is a rapid-access file to everything that happens around here." He touched a control on his chair and spoke to the air. "Miss Kendrick, what mark did Cadet Byrd receive in military law last term?"

Her answer came back at once: "Ninety-three per cent, Commandant."

"Thank you." He continued, "You see? I sign anything if Miss Kendrick has initialed it. I would hate to have an investigating committee find out how often she signs my name and I don't even see it. Tell me, Mr. Byrd . . . if I drop dead, does Miss Kendrick carry on to keep things moving?"

"Why, uh—" Birdie looked puzzled. "I suppose, with routine matters, she would do what was necess—"

"She wouldn't do a blessed thing!" the Colonel thundered. "Until Colonel Chauncey *told* her what to do—*his* way. She is a very smart woman and understands what you apparently do not, namely, that she is *not* in the line of command and has no authority."

He went on, "'Line of command' isn't just a phrase; it's as real as a slap in the face. If I ordered you to combat *as a cadet* the most you could do would be to pass along somebody else's orders. If your platoon leader bought out and you then gave an order to a private—a good order, sensible and wise—you would be wrong and he would be just as wrong if he obeyed it. Because a cadet cannot be in the line of command. A cadet has no military existence, no rank, and is not a soldier. He is a student who will become a soldier—either an officer, or at his former rank. While he is under Army discipline, he is not *in* the Army. That is why—"

A zero. A nought with no rim. If a cadet wasn't even in the Army—"Colonel!"

"Eh? Speak up, young man. Mr. Rico."

I had startled myself but I had to say it. "But . . . if we aren't in the Army . . . then we aren't M.I. Sir?"

He blinked at me. "This worries you?"

"I, uh, don't believe I like it much, sir." I didn't like it at all. I felt naked.

"I see." He didn't seem displeased. "You let me worry about the space-lawyer aspects of it, son."

"But—"

"That's an order. You are technically not an M.I. But the M.I. hasn't forgotten you; the M.I. *never* forgets its own no matter where they are. If you are struck dead this instant, you will be cremated as Second Lieutenant Juan Rico, Mobile Infantry, of—" Colonel Nielssen stopped. "Miss Kendrick, what was Mr. Rico's ship?"

"The *Rodger Young*."

"Thank you." He added, "—in and of TFCT *Rodger Young*, assigned to mobile combat team Second Platoon of George Company, Third Regiment, First Division, M.I.—the 'Roughnecks,'" he recited with relish, not consulting anything once he had been reminded of my ship. "A good outfit, Mr. Rico—proud and nasty. Your Final Orders go back to them for Taps and that's the way your name would read in Memorial Hall. That's why we always commission a dead cadet, son—so we can send him home to his mates."

I felt a surge of relief and homesickness and missed a few words. ". . . lip buttoned while I talk, we'll have you back in the M.I. where you belong. You must be temporary officers for your 'prentice cruise because there is no room for deadheads in a combat drop. You'll fight—and take orders—and *give* orders. *Legal* orders, because you will hold rank and be ordered to serve in that team; that makes any order you give in carrying out your assigned duties as binding as one signed by the C-in-C.

"Even more," the Commandant went on, "once you are in line of command, you must be ready instantly to assume higher command. If you are in a one-platoon team—quite likely in the present state of the war—and you are assistant platoon leader when your platoon leader buys it . . . then . . . *you* . . . are . . . *It!*"

He shook his head. "Not 'acting platoon leader.' Not a cadet leading a drill. Not a 'junior officer under instruction.' Suddenly you are the Old Man, the Boss, Commanding Officer Present—and you discover with a sickening shock that fellow human beings are depending on *you alone* to tell them what to do, how to fight, how to complete the mission and get out alive. They wait for the sure voice of command—while seconds trickle away—and it's up to you to *be* that voice, make decisions, give the right orders . . . and not only the right ones but in a calm, unworried tone. Because it's a cinch, gentlemen, that your team is in trouble—*bad* trouble!— and a strange voice with panic in it can turn the best combat team in the Galaxy into a leaderless, lawless, fear-crazed mob.

"The whole merciless load will land without warning. You must act at once and you'll have only God over you. Don't expect Him to fill in tactical details; that's *your* job. He'll be doing all that a soldier has a right to expect if He helps you keep the panic you are sure to feel out of your voice."

The Colonel paused. I was sobered and Birdie was looking terribly serious and awfully young and Hassan was scowling. I wished that I were back in the drop room of the *Rog*, with not too many chevrons and an after-chow bull session in full swing. There was a lot to be said for the job of assistant section leader—when you come right to it, it's a lot easier to *die* than it is to use your head.

The Commandant continued: "That's the Moment

of Truth, gentlemen. Regrettably there is no method known to military science to tell a real officer from a glib imitation with pips on his shoulders, other than through ordeal by fire. Real ones come through—or die gallantly; imitations crack up.

"Sometimes, in cracking up, the misfits die. But the tragedy lies in the loss of others . . . good men, sergeants and corporals and privates, whose only lack is fatal bad fortune in finding themselves under the command of an incompetent.

"We try to avoid this. First is our unbreakable rule that every candidate must be a trained trooper, blooded under fire, a veteran of combat drops. No other army in history has stuck to this rule, although some came close. Most great military schools of the past—Saint Cyr, West Point, Sandhurst, Colorado Springs—didn't even pretend to follow it; they accepted civilian boys, trained them, commissioned them, sent them out with no battle experience to command men . . . and sometimes discovered too late that this smart young 'officer' was a fool, a poltroon, or a hysteric.

"At least we have no misfits of those sorts. We know you are good soldiers—brave and skilled, proved in battle—else you would not be here. We know that your intelligence and education meet acceptable minimums. With this to start on, we eliminate as many as possible of the not-quite-competent—get them quickly back in ranks before we spoil good cap troopers by forcing them beyond their abilities. The course is very hard—because what will be expected of you later is still harder.

"In time we have a small group whose chances look fairly good. The major criterion left untested is one we *cannot* test here; that undefinable something which is the difference between a leader in battle . . . and one who merely has the earmarks but not the vocation. So we field-test for it.

"Gentlemen!—you have reached that point. Are you ready to take the oath?"

There was an instant of silence, then Hassan the Assassin answered firmly, "Yes, Colonel," and Birdie and I echoed.

The Colonel frowned. "I have been telling you how wonderful you are—physically perfect, mentally alert, trained, disciplined, blooded. The very model of the smart young officer—" He snorted. "Nonsense! You *may* become officers someday. I hope so . . . we not only hate to waste money and time and effort, but also, and *much* more important, I shiver in my boots every time I send one of you half-baked not-quite-officers up to the Fleet, knowing what a Frankensteinian monster I may be turning loose on a good combat team. If you understood what you are up against, you wouldn't be so all-fired ready to take the oath the second the question is put to you. You may turn it down and force me to let you go back to your permanent ranks. But you *don't* know.

"So I'll try once more. Mr. Rico! Have you ever thought how it would feel to be court-martialed for losing a regiment?"

I was startled silly. "Why—No, sir, I never have." To be court-martialed—for *any* reason—is eight times as bad for an officer as for an enlisted man. Offenses which will get privates kicked out (maybe with lashes, possibly without) rate death in an officer. Better never to have been born!

"Think about it," he said grimly. "When I suggested that your platoon leader might be killed, I was by no means citing the ultimate in military disaster. Mr. Hassan! What is the largest number of command levels ever knocked out in a single battle?"

The Assassin scowled harder than ever. "I'm not sure, sir. Wasn't there a while during Operation Bughouse when a major commanded a brigade, before the Soveki-poo?"

"There was and his name was Fredericks. He got a decoration and a promotion. If you go back to the Second Global War, you can find a case in which a naval junior officer took command of a major ship and not only fought it but sent signals as if he were admiral. He was vindicated even though there were officers senior to him in line of command who were not even wounded. Special circumstances—a breakdown in communications. But I am thinking of a case in which four levels were wiped out in six minutes—as if a platoon leader were to blink his eyes and find himself commanding a brigade. Any of you heard of it?"

Dead silence.

"Very well. It was one of those bush wars that flared up on the edges of the Napoleonic wars. This young officer was the most junior in a naval vessel—wet navy, of course—wind-powered, in fact. This youngster was about the age of most of your class and was not commissioned. He carried the title of 'temporary third lieutenant'—note that this is the title you are about to carry. He had no combat experience; there were four officers in the chain of command above him. When the battle started his commanding officer was wounded. The kid picked him up and carried him out of the line of fire. That's all—make a pickup on a comrade. But he did it without being ordered to leave his post. The other officers all bought it while he was doing this and he was tried for 'deserting his post of duty as *commanding officer* in the presence of the enemy.' Convicted. Cashiered."

I gasped. "For *that*? Sir."

"Why not? True, we make pickup. But we do it under different circumstances from a wet-navy battle, and by orders to the man making pickup. But pickup is never an excuse for breaking off battle in the presence of the enemy. This boy's family tried for a century and a half to get his conviction reversed. No luck, of course. There was doubt about some circumstances but no doubt that he had left his post during battle without orders. True, he was green as grass—but he was lucky not to be hanged." Colonel Nielssen fixed me with a cold eye. "Mr. Rico— could this happen to *you*?"

I gulped. "I hope not, sir."

"Let me tell you how it could on this very 'prentice cruise. Suppose you are in a multiple-ship operation, with a full regiment in the drop. Officers drop first, of course. There are advantages to this and disadvantages, but we do it for reasons of morale; no trooper ever hits the ground on a hostile planet without an officer. Assume the Bugs know this—and they may. Suppose they work up some trick to wipe out those who hit the ground first . . . but not good enough to wipe out the whole drop. Now suppose, since you are a supernumerary, you have to take any vacant capsule instead of being fired with the first wave. Where does that leave you?"

"Uh, I'm not sure, sir."

"You have just inherited command of a regiment. *What are you going to do with your command, Mister?* Talk fast—the Bugs won't wait!"

"Uh . . ." I caught an answer right out of the book and parroted it. "I'll take command and act as circumstances permit, sir, according to the tactical situation as I see it."

"You will, eh?" The Colonel grunted. "And you'll buy a farm too—that's all anybody can do with a foul-up like that. But I hope you'll go down swinging—and shouting orders to somebody, whether they make sense or not. We don't expect kittens to fight wildcats and win—we merely expect them to try. All right, stand up. Put up your right hands."

He struggled to his feet. Thirty seconds later we were officers—"temporary, probationary, and supernumerary."

I thought he would give us our shoulder pips and let us go. We aren't supposed to buy them—they're a loan, like the temporary commission they represent. Instead he lounged back and looked almost human.

"See here, lads—I gave you a talk on how rough it's

going to be. I want you to worry about it, doing it in advance, planning what steps you might take against any combination of bad news that can come your way, keenly aware that your life belongs to your men and is not yours to throw away in a suicidal reach for glory . . . and that your life isn't yours to save, either, if the situation requires that you expend it. I want you to worry yourself sick *before* a drop, so that you can be unruffled when the trouble starts.

"Impossible, of course. Except for one thing. What is the *only* factor that can save you when the load is too heavy? Anyone?"

Nobody answered.

"Oh, come now!" Colonel Nielssen said scornfully. "You aren't recruits. Mr. Hassan!"

"Your leading sergeant, sir," the Assassin said slowly.

"Obviously. He's probably older than you are, more drops under his belt, and he certainly knows his team better than you do. Since he isn't carrying that dreadful, numbing load of top command, he may be thinking more clearly than you are. Ask his advice. You've got one circuit just for that.

"It won't decrease his confidence in you; he's used to being consulted. If you don't, he'll decide you are a fool, a cocksure know-it-all—and he'll be right.

"But you don't have to *take* his advice. Whether you use his ideas, or whether they spark some different plan—make your decision and snap out orders. The one thing—the *only* thing!—that can strike terror in the heart of a good platoon sergeant is to find that he's working for a boss who can't make up his mind.

"There never has been an outfit in which officers and men were more dependent on each other than they are in the M.I., and sergeants are the glue that holds us together. Never forget it."

The Commandant whipped his chair around to a cabinet near his desk. It contained row on row of pigeonholes, each with a little box. He pulled out one and opened it. "Mr. Hassan—"

"Sir?"

"These pips were worn by Captain Terrence O'Kelly on his 'prentice cruise. Does it suit you to wear them?"

"Sir?" The Assassin's voice squeaked and I thought the big lunk was going to break into tears. "Yes, sir!"

"Come here." Colonel Nielssen pinned them on, then said, "Wear them as gallantly as he did . . . but *bring them back*. Understand me?"

"Yes, sir. I'll do my best."

"I'm sure you will. There's an air car waiting on the roof and your boat boosts in twenty-eight minutes. Carry out your orders, sir!"

The Assassin saluted and left; the Commandant turned and picked out another box. "Mr. Byrd, are you superstitious?"

"No, sir."

"Really? I am, quite. I take it you would not object to wearing pips which have been worn by five officers, all of whom were killed in action?"

Birdie barely hesitated. "No, sir."

"Good. Because these five officers accumulated seventeen citations, from the Terran Medal to the Wounded Lion. Come here. The pip with the brown discoloration must always be worn on your left shoulder—and don't try to buff it off! Just try not to get the other one marked in the same fashion. Unless necessary, and you'll know when it is necessary. Here is a list of former wearers. You have thirty minutes until your transportation leaves. Bounce up to Memorial Hall and look up the record of each."

"Yes, sir."

"Carry out your orders, sir!"

He turned to me, looked at my face and said sharply, "Something on your mind, son? Speak up!"

"Uh—" I blurted it out. "Sir, that temporary third lieutenant—the one that got cashiered. How could I find out what happened?"

"Oh. Young man, I didn't mean to scare the daylights out of you; I simply intended to wake you up. The battle was on one June 1813 old style between USF *Chesapeake* and HMF *Shannon*. Try the *Naval Encyclopedia*; your ship will have it." He turned back to the case of pips and frowned.

Then he said, "Mr. Rico, I have a letter from one of your high school teachers, a retired officer, requesting that you be issued the pips he wore as a third lieutenant. I am sorry to say that I must tell him 'No.'"

"Sir?" I was delighted to hear that Colonel Dubois was still keeping track of me—and very disappointed, too.

"Because I *can't*. I issued those pips two years ago—and they never came back. Real estate deal. Hmm—" He took a box, looked at me. "You could start a new pair. The metal isn't important; the importance of the request lies in the fact that your teacher wanted you to have them."

"Whatever you say, sir."

"Or"—he cradled the box in his hands—"you could wear these. They have been worn five times . . . and the last four candidates to wear them have all failed of commission—nothing dishonorable but pesky bad luck. Are you willing to take a swing at breaking the hoodoo? Turn them into good-luck pips instead?"

I would rather have petted a snark. But I answered, "All right, sir. I'll take a swing at it."

"Good." He pinned them on me. "Thank you, Mr. Rico. You see, these were mine, I wore them first . . . and it would please me mightily to have them brought back to me with that streak of bad luck broken, have you go on and graduate."

I felt ten feet tall. "I'll try, sir!"

"I know you will. You may now carry out your orders, sir. The same air car will take both you and Byrd. Just a moment—Are your mathematics textbooks in your bag?"

"Sir? No, sir."

"Get them. The Weightmaster of your ship has been advised of your extra baggage allowance."

I saluted and left, on the bounce. He had me shrunk down to size as soon as he mentioned math.

My math books were on my study desk, tied into a

package with a daily assignment sheet tucked under the cord. I gathered the impression that Colonel Nielssen never left anything unplanned—but everybody knew that.

Birdie was waiting on the roof by the air car. He glanced at my books and grinned. "Too bad. Well, if we're in the same ship, I'll coach you. What ship?"

"*Tours*."

"Sorry, I'm for the *Moskva*." We got in, I checked the pilot, saw that it had been pre-set for the field, closed the door and the car took off. Birdie added, "You could be worse off. The Assassin took not only his math books but two other subjects."

Birdie undoubtedly knew and he had not been showing off when he offered to coach me; he was a professor type except that his ribbons proved that he was a soldier too.

Instead of studying math Birdie taught it. One period each day he was a faculty member, the way little Shujumi taught judo at Camp Currie. The M.I. doesn't waste anything; we can't afford to. Birdie had a B.S. in math on his eighteenth birthday, so naturally he was assigned extra duty as instructor—which didn't keep him from being chewed out at other hours.

Not that he got chewed out much. Birdie had that rare combo of brilliant intellect, solid education, common sense, and guts, which gets a cadet marked as a potential general. We figured he was a cinch to command a brigade by the time he was thirty, what with the war.

But my ambitions didn't soar that high. "It would be a dirty, rotten shame," I said, "if the Assassin flunked out," while thinking that it would be a dirty, rotten shame if *I* flunked out.

"He won't," Birdie answered cheerfully. "They'll sweat him through the rest if they have to put him in a hypno booth and feed him through a tube. Anyhow," he added, "Hassan could flunk out and get promoted for it."

"Huh?"

"Didn't you know? The Assassin's permanent rank is first lieutenant—field commission, naturally. He reverts to it if he flunks out. See the regs."

I knew the regs. If I flunked math, I'd revert to buck sergeant, which is better than being slapped in the face with a wet fish any way you think about it . . . and I'd thought about it, lying awake nights after busting a quiz.

But this was different. "Hold it," I protested. "He gave up *first* lieutenant, permanent grade . . . and has just made temporary *third* lieutenant . . . in order to become a *second* lieutenant? Are you crazy? Or is he?"

Birdie grinned. "Just enough to make us both M.I."

"But—I don't get it."

"Sure you do. The Assassin has no education that he didn't pick up in the M.I. So how high can he go? I'm sure he could command a regiment in battle and do a real swingin' job—provided somebody else planned the operation. But commanding in battle is only a fraction of what an officer does, especially a senior officer. To direct a war, or even to plan a single battle and mount the operation, you have to have theory of games, operational analysis, symbolic logic, pessimistic synthesis, and a dozen other skull subjects. You can sweat them out on your own if you've got the grounding. But have them you must, or you'll never get past captain, or possibly major. The Assassin knows what he is doing."

"I suppose so," I said slowly. "Birdie, Colonel Nielssen must know that Hassan was an officer—is an officer, really."

"Huh? Of course."

"He didn't talk as if he knew. We all got the same lecture."

"Not quite. Did you notice that when the Commandant wanted a question answered a particular way he always asked the Assassin?"

I decided it was true. "Birdie, what is your permanent rank?"

The car was just landing; he paused with a hand on the latch and grinned. "PFC—I don't dare flunk out!"

I snorted. "You won't. You can't!" I was surprised that he wasn't even a corporal, but a kid as smart and well educated as Birdie would go to O.C.S. just as quickly as he proved himself in combat . . . which, with the war on, could be only months after his eighteenth birthday.

Birdie grinned still wider. "We'll see."

"You'll graduate. Hassan and I have to worry, but not you."

"So? Suppose Miss Kendrick takes a dislike to me." He opened the door and looked startled. "Hey! They're sounding my call. So long!"

"See you, Birdie."

But I did not see him and he did not graduate. He was commissioned two weeks later and his pips came back with their eighteenth decoration—the Wounded Lion, posthumous.

CH: 13

Youse guys think this deleted outfit is a blankety-blank nursery.
Well, it ain't! See?

—Remark attributed to a Hellenic corporal
before the walls of Troy, 1194 B.C.

The *Rodger Young* carries one platoon and is crowded; the *Tours* carries six—and is roomy. She has the tubes to drop them all at once and enough spare room to carry twice that number and make a second drop. This would make her very crowded, with eating in shifts, hammocks in passageways and drop rooms, rationed water, inhale when your mate exhales, and get your elbow out of my eye! I'm glad they didn't double up while I was in her.

But she has the speed and lift to deliver such crowded troops still in fighting condition to any point in Federation space and much of Bug space; under Cherenkov drive she cranks Mike 400 or better—say Sol to Capella, forty-six light-years, in under six weeks.

Of course, a six-platoon transport is not big compared with a battle wagon or passenger liner; these things are compromises. The M.I. prefers speedy little one-platoon corvettes which give flexibility for any operation, while if it was left up to the Navy we would have nothing but regimental transports. It takes almost as many Navy files to run a corvette as it does to run a monster big enough for a regiment—more maintenance and housekeeping, of course, but soldiers can do that. After all, those lazy troopers do nothing but sleep and eat and polish buttons—do 'em good to have a little regular work. So says the Navy.

The real Navy opinion is even more extreme: The Army is obsolete and should be abolished.

The Navy doesn't say this officially—but talk to a Naval officer who is on R&R and feeling his oats; you'll get an earful. They think they can fight any war, win it, send a few of their own people down to hold the conquered planet until the Diplomatic Corps takes charge.

I admit that their newest toys can blow any planet right out of the sky—I've never seen it but I believe it. Maybe I'm as obsolete as *Tyrannosaurus rex*. I don't feel obsolete and us apes can do things that the fanciest ship cannot. If the government doesn't want those things done, no doubt they'll tell us.

Maybe it's just as well that neither the Navy nor the M.I. has the final word. A man can't buck for Sky Marshal unless he has commanded both a regiment and a capital ship—go through M.I. and take his lumps and then become a Naval officer (I think little Birdie had that in mind), or first become an astrogator-pilot and follow it with Camp Currie, etc.

I'll listen respectfully to any man who has done both.

Like most transports, the *Tours* is a mixed ship; the

most amazing change for me was to be allowed "North of Thirty." The bulkhead that separates ladies' country from the rough characters who shave is not necessarily No. 30 but, by tradition, it is called "bulkhead thirty" in any mixed ship. The wardroom is just beyond it and the rest of ladies' country is farther forward. In the *Tours* the wardroom also served as messroom for enlisted women, who ate just before we did, and it was partitioned between meals into a recreation room for them and a lounge for their officers. Male officers had a lounge called the cardroom just abaft thirty.

Besides the obvious fact that drop & retrieval require the best pilots (i.e., female), there is very strong reason why female Naval officers are assigned to transports: It is good for trooper morale.

Let's skip M.I. traditions for a moment. Can you think of anything sillier than letting yourself be fired out of a spaceship with nothing but mayhem and sudden death at the other end? However, if someone must do this idiotic stunt, do you know of a surer way to keep a man keyed up to the point where he is willing than by keeping him constantly reminded that the only good reason why men fight is a living, breathing reality?

In a mixed ship, the last thing a trooper hears before a drop (maybe the last word he ever hears) is a woman's voice, wishing him luck. If you don't think this is important, you've probably resigned from the human race.

The *Tours* had fifteen Naval officers, eight ladies and seven men; there were eight M.I. officers including (I am happy to say) myself. I won't say "bulkhead thirty" caused me to buck for O.C.S. but the privilege of eating with the ladies is more incentive than any increase in pay. The Skipper was president of the mess, my boss Captain Blackstone was vice-president—not because of rank; three Naval officers ranked him; but as C.O. of the strike force he was de facto senior to everybody but the Skipper.

Every meal was formal. We would wait in the cardroom until the hour struck, follow Captain Blackstone in and stand behind our chairs; the Skipper would come in followed by her ladies and, as she reached the head of the table, Captain Blackstone would bow and say, "Madam President . . . ladies," and she would answer, "Mr. Vice . . . gentlemen," and the man on each lady's right would seat her.

This ritual established that it was a social event, not an officers' conference; thereafter ranks or titles were used, except that junior Naval officers and myself alone among the M.I. were called "Mister" or "Miss"—with one exception which fooled me.

My first meal aboard I heard Captain Blackstone called "Major," although his shoulder pips plainly read "captain." I got straightened out later. There can't be two captains in a Naval vessel so an Army captain is bumped one rank socially rather than commit the unthinkable of calling him by the title reserved for the one and only monarch. If a Naval captain is aboard as anything but skipper, he or she is called "Commodore" even if the skipper is a lowly lieutenant.

The M.I. observes this by avoiding the necessity in the wardroom and paying no attention to the silly custom in our own part of the ship.

Seniority ran downhill from each end of the table, with the Skipper at the head and the strike force C.O. at the foot, the junior midshipmen at his right and myself at the Skipper's right. I would most happily have sat by the junior midshipman; she was awfully pretty—but the arrangement is planned chaperonage; I never even learned her first name.

I knew that I, as the lowliest male, sat on the Skipper's right—but I didn't know that I was supposed to seat her. At my first meal she waited and nobody sat down—until the third assistant engineer jogged my elbow. I haven't been so embarrassed since a very unfortunate incident in kindergarten, even though Captain Jorgenson acted as if nothing had happened.

When the Skipper stands up the meal is over. She was pretty good about this but once she stayed seated only a few minutes and Captain Blackstone got annoyed. He stood up but called out, "Captain—"

She stopped. "Yes, Major?"

"Will the Captain please give orders that my officers and myself be served in the cardroom?"

She answered coldly, "Certainly, sir." And we were. But no Naval officer joined us.

The following Saturday she exercised her privilege of inspecting the M.I. aboard—which transport skippers almost never do. However, she simply walked down the ranks without commenting. She was not really a martinet and she had a nice smile when she wasn't being stern. Captain Blackstone assigned Second Lieutenant "Rusty" Graham to crack the whip over me about math; she found out about it, somehow, and told Captain Blackstone to have me report to her office for one hour after lunch each day, whereupon she tutored me in math and bawled me out when my "homework" wasn't perfect.

Our six platoons were two companies as a rump battalion; Captain Blackstone commanded Company D, Blackie's Blackguards, and also commanded the rump battalion. Our battalion commander by the T.O., Major Xera, was with A and B companies in the *Tours'* sister ship *Normandy Beach*—maybe half a sky away; he commanded us only when the full battalion dropped together—except that Cap'n Blackie routed certain reports and letters through him. Other matters went directly to Fleet, Division, or Base, and Blackie had a truly wizard fleet sergeant to keep such things straight and to help him handle both a company and a rump battalion in combat.

Administrative details are not simple in an army spread through many light-years in hundreds of ships. In the old *Valley Forge*, in the *Rodger Young*, and now in the *Tours* I was in the same regiment, the Third ("Pampered Pets") Regiment of the First ("Polaris") M.I. Division. Two battalions formed from available units had been called the "Third Regiment" in Operation Bughouse but I did not see "my" regiment; all I saw was PFC Bamburger and a lot of Bugs.

I might be commissioned in the Pampered Pets,

grow old and retire in it—and never even see my regimental commander. The Roughnecks had a company commander but he also commanded the first platoon ("Hornets") in another corvette; I didn't know his name until I saw it on my orders to O.C.S. There is a legend about a "lost platoon" that went on R&R as its corvette was decommissioned. Its company commander had just been promoted and the other platoons had been attached tactically elsewhere. I've forgotten what happened to the platoon's lieutenant but R&R is a routine time to detach an officer—theoretically after a relief has been sent to understudy him, but reliefs are always scarce.

They say this platoon enjoyed a local year of the fleshpots along Churchill Road before anybody missed them.

I don't believe it. But it could happen.

The chronic scarcity of officers strongly affected my duties in Blackie's Blackguards. The M.I. has the lowest percentage of officers in any army of record and this factor is just part of the M.I.'s unique "divisional wedge." "D.W." is military jargon but the idea is simple: If you have 10,000 soldiers, how many fight? And how many just peel potatoes, drive lorries, count graves, and shuffle papers?

In the M.I., 10,000 men fight.

In the mass wars of the XXth century it sometimes took 70,000 men (fact!) to enable 10,000 to fight.

I admit it takes the Navy to place us where we fight; however, an M.I. strike force, even in a corvette, is at least three times as large as the transport's Navy crew. It also takes civilians to supply and service us; about 10 per cent of us are on R&R at any time; and a few of the very best of us are rotated to instruct at boot camps.

While a few M.I. are on desk jobs you will always find that they are shy an arm or leg, or some such. These are the ones—the Sergeant Hos and the Colonel Nielssens—who refuse to retire, and they really ought to count twice since they release able-bodied M.I. by filling jobs which require fighting spirit but not physical perfection. They do work that civilians can't do—or we would hire civilians. Civilians are like beans; you buy 'em as needed for any job which merely requires skill and savvy.

But you can't buy fighting spirit.

It's scarce. We use all of it, waste none. The M.I. is the smallest army in history for the size of the population it guards. You can't buy an M.I., you can't conscript him, you can't coerce him—you can't even keep him if he wants to leave. He can quit thirty seconds before a drop, lose his nerve and not get into his capsule and all that happens is that he is paid off and can never vote.

At O.C.S. we studied armies in history that were driven like galley slaves. But the M.I. is a free man; all that drives him comes from inside—that self-respect and need for the respect of his mates and his pride in being one of them called morale, or *esprit de corps*.

The root of our morale is: "Everybody works, everybody fights." An M.I. doesn't pull strings to get a soft, safe job; there aren't any. Oh, a trooper will get away with what he can; any private with enough savvy to mark time to music can think up reasons why he should not clean compartments or break out stores; this is a soldier's ancient right.

But *all* "soft, safe" jobs are filled by civilians; that goldbricking private climbs into his capsule certain that *everybody*, from general to private, is doing it with him. Light-years away and on a different day, or maybe an hour or so later—no matter. What does matter is that *everybody* drops. This is why he enters the capsule, even though he may not be conscious of it.

If we ever deviate from this, the M.I. will go to pieces. All that holds us together is an idea—one that binds more strongly than steel but its magic power depends on keeping it intact.

It is this "everybody fights" rule that lets the M.I. get by with so few officers.

I know more about this than I want to, because I asked a foolish question in Military History and got stuck with an assignment which forced me to dig up stuff ranging from *De Bello Gallico* to Tsing's classic *Collapse of the Golden Hegemony*. Consider an ideal M.I. division—on paper, because you won't find one elsewhere. How many officers does it require? Never mind units attached from other corps; they may not be present during a ruckus and they are not like M.I.—the special talents attached to Logistics & Communications are all ranked as officers. If it will make a memory man, a telepath, a senser, or a lucky man happy to have me salute him, I'm glad to oblige; he is more valuable than I am and I could not replace him if I lived to be two hundred. Or take the

K-9 Corps, which is 50 per cent "officers" but whose other 50 per cent are neodogs.

None of these is in line of command, so let's consider only us apes and what it takes to lead us.

This imaginary division has 10,800 men in 216 platoons, each with a lieutenant. Three platoons to a company calls for 72 captains; four companies to a battalion calls for 18 majors or lieutenant colonels. Six regiments with six colonels can form two or three brigades, each with a short general, plus a medium-tall general as top boss.

You wind up with 317 officers out of a total, all ranks, of 11,117.

There are no blank files and every officer commands a team. Officers total 3 per cent—which is what the M.I. does have, but arranged somewhat differently. In fact a good many platoons are commanded by sergeants and many officers "wear more than one hat" in order to fill some utterly necessary staff jobs.

Even a platoon leader should have "staff"—his platoon sergeant.

But he can get by without one and his sergeant can get by without him. But a general *must* have staff; the job is too big to carry in his hat. He needs a big planning staff and a small combat staff. Since there are never enough officers, the team commanders in his flag transport double as his planning staff and are picked from the M.I.'s best mathematical logicians—then they drop with their own teams. The general drops with a small combat staff, plus a small team of the roughest, on-the-bounce troopers in the M.I. Their job is to keep the general from being bothered by rude strangers while he is managing the battle. Sometimes they succeed.

Besides necessary staff billets, any team larger than a platoon ought to have a deputy commander. But there are never enough officers so we make do with what we've got. To fill each necessary combat billet, one job to one officer, would call for a 5 per cent ratio of officers—but 3 per cent is all we've got.

In place of that optimax of 5 per cent that the M.I. never can reach, many armies in the past commissioned 10 per cent of their number, or even 15 per cent—and sometimes a preposterous 20 per cent! This sounds like a fairy tale but it was a fact, especially during the XXth century. What kind of an army has more "officers" than corporals? (And more non-coms than privates!)

An army organized to lose wars—if history means anything. An army that is mostly organization, red tape, and overhead, most of whose "soldiers" never fight.

But what do "officers" *do* who do not command fighting men?

Fiddlework, apparently—officers' club officer, morale officer, athletics officer, public information officer, recreation officer, PX officer, transportation officer, legal officer, chaplain, assistant chaplain, junior assistant chaplain, officer-in-charge of anything anybody can think of—even *nursery* officer!

In the M.I., such things are extra duty for combat officers or, if they are *real* jobs, they are done better and cheaper and without demoralizing a fighting outfit by hiring civilians. But the situation got so smelly in one of the XXth century major powers that *real* officers, ones who commanded fighting men, were given special insignia to distinguish them from the swarms of swivel-chair hussars.

The scarcity of officers got steadily worse as the war wore on, because the casualty rate is always highest among officers . . . and the M.I. *never* commissions a man simply to fill a vacancy. In the long run, each boot regiment must supply its own share of officers and the percentage can't be raised without lowering the standards—The strike force in the *Tours* needed thirteen officers—six platoon leaders, two company commanders and two deputies, and a strike force commander staffed by a deputy and an adjutant.

What it had was six . . . and me.

TABLE OF ORGANIZATION

"Rump Battalion" Strike Force—

Cpt. Blackstone

("first hat")

Fleet Sergeant

C Company—	D Company—
"Warren's Wolverines"	"Blackie's Blackguards"
1st Lt. Warren	Cpt. Blackstone
	("second hat")
1st plat.—	1st plat.—
1st Lt. Bayonne	(1st Lt. Silva, Hosp.)

C Company—	D Company—
2nd plat.—	2nd plat.—
2nd Lt. Sukarno	2nd Lt. Khoroshen
3rd plat.—	3rd plat.—
2nd Lt. N'gam	2nd Lt. Graham

I would have been under Lieutenant Silva, but he left for hospital the day I reported, ill with some sort of twitching awfuls. But this did not necessarily mean that I would get his platoon. A temporary third lieutenant is not considered an asset; Captain Blackstone could place me under Lieutenant Bayonne and put a sergeant in charge of his own first platoon, or even "put on a third hat" and take the platoon himself.

In fact, he did both and nevertheless assigned me as platoon leader of the first platoon of the Blackguards. He did this by borrowing the Wolverine's best buck sergeant to act as his battalion staffer, then he placed his fleet sergeant as platoon sergeant of his first platoon—a job two grades below his chevrons. Then Captain Blackstone spelled it out for me in a head-shrinking lecture: I would appear on the T.O. as platoon leader, but Blackie himself and the fleet sergeant would run the platoon.

As long as I behaved myself, I could go through the motions. I would even be allowed to drop as platoon leader—but one word from my platoon sergeant to my company commander and the jaws of the nutcracker would close.

It suited me. It was my platoon as long as I could

swing it—and if I couldn't, the sooner I was shoved aside the better for everybody. Besides, it was a lot less nerve-racking to get a platoon that way than by sudden catastrophe in battle.

I took my job very seriously, for it was *my* platoon—the T.O. said so. But I had not yet learned to delegate authority and, for about a week, I was around troopers' country much more than is good for a team. Blackie called me into his stateroom. "Son, what in Ned do you think you are doing?"

I answered stiffly that I was trying to get my platoon ready for action.

"So? Well, that's not what you are accomplishing. You are stirring them like a nest of wild bees. Why the deuce do you think I turned over to you the best sergeant in the Fleet? If you will go to your stateroom, hang yourself on a hook, and *stay* there! . . . until 'Prepare for Action' is sounded, he'll hand that platoon over to you tuned like a violin."

"As the Captain pleases, sir," I agreed glumly.

"And that's another thing—I can't stand an officer who acts like a confounded *kay*det. Forget that silly third-person talk around me—save it for generals and the Skipper. Quit bracing your shoulders and clicking your heels. Officers are supposed to look *relaxed*, son."

"Yes, sir."

"And let that be the last time you say 'sir' to me for one solid week. Same for saluting. Get that grim *kay*det look off your face and hang a smile on it."

"Yes, s— Okay."

"That's better. Lean against the bulkhead. Scratch yourself. Yawn. Anything but that tin-soldier act."

I tried . . . and grinned sheepishly as I discovered that breaking a habit is not easy. Leaning was harder work than standing at attention. Captain Blackstone studied me. "Practice it," he said. "An officer can't look scared or tense; it's contagious. Now tell me, Johnnie, what your platoon needs. Never mind the piddlin' stuff; I'm not interested in whether a man has the regulation number of socks in his locker."

I thought rapidly. "Uh . . . do you happen to know if Lieutenant Silva intended to put Brumby up for sergeant?"

"I do happen to know. What's *your* opinion?"

"Well . . . the record shows that he has been acting section leader the past two months. His efficiency marks are good."

"I asked for your recommendation, Mister."

"Well, s—Sorry. I've never seen him work on the ground, so I can't have a real opinion; anybody can soldier in the drop room. But the way I see it, he's been acting sergeant too long to bust him back to chaser and promote a squad leader over him. He ought to get that third chevron before we drop—or he ought to be transferred when we get back. Sooner, if there's a chance for a spaceside transfer."

Blackie grunted. "You're pretty generous in giving away my Blackguards—for a third lieutenant."

I turned red. "Just the same, it's a soft spot in my platoon. Brumby ought to be promoted, or transferred. I don't want him back in his old job with somebody promoted over his head; he'd likely turn sour and I'd have an even worse soft spot. If he can't have another chevron, he ought to go to repple-depple for cadre. Then he won't be humiliated and he gets a fair shake to make sergeant in another team—instead of a dead end here."

"Really?" Blackie did not quite sneer. "After that masterly analysis, apply your powers of deduction and tell me why Lieutenant Silva failed to transfer him three weeks ago when we arrived around Sanctuary."

I had wondered about that. The time to transfer a man is the earliest possible instant after you decide to let him go—and without warning; it's better for the man and the team—so says the book. I said slowly, "Was Lieutenant Silva already ill at that time, Captain?"

"No."

The pieces matched. "Captain, I recommend Brumby for immediate promotion."

His eyebrows shot up. "A minute ago you were about to dump him as useless."

"Uh, not quite. I said it had to be one or the other—but I didn't know which. Now I know."

"Continue."

"Uh, this assumes that Lieutenant Silva is an efficient officer—"

"*Hummmph!* Mister, for your information, 'Quick' Silva has an unbroken string of 'Excellent—Recommended for Promotion' on his Form Thirty-One."

"But I knew that he was good," I plowed on, "because

I inherited a good platoon. A good officer might not promote a man for—oh, for many reasons—and still not put his misgivings in writing. But in this case, if he could not recommend him for sergeant, then he wouldn't keep him with the team—so he would get him out of the ship at the first opportunity. But he didn't. Therefore I know he intended to promote Brumby." I added, "But I can't see why he didn't push it through three weeks ago, so that Brumby could have worn his third chevron on R&R."

Captain Blackstone grinned. "That's because you don't credit *me* with being efficient."

"S—I beg pardon?"

"Never mind. You've proved who killed Cock Robin and I don't expect a still-moist *kay*det to know all the tricks. But listen and learn, son. As long as this war goes on, don't *ever* promote a man just before you return to Base."

"Uh . . . why not, Captain?"

"You mentioned sending Brumby to Replacement Depot if he was not to be promoted. But that's just where he would have gone if we *had* promoted him three weeks ago. You don't know how hungry that noncom desk at repple-depple is. Paw through the dispatch file and you'll find a demand that we supply two sergeants for cadre. With a platoon sergeant being detached for O.C.S. and a buck sergeant spot vacant, I was under complement and able to refuse." He grinned savagely. "It's a rough war, son, and your own people will steal your best men if you don't watch 'em." He took two sheets of paper out of a drawer. "There—"

One was a letter from Silva to Cap'n Blackie, recommending Brumby for sergeant; it was dated over a month ago.

The other was Brumby's warrant for sergeant—dated the day *after* we left Sanctuary.

"That suit you?" he asked.

"Huh? Oh, yes indeed!"

"I've been waiting for you to spot the weak place in your team, and tell me what had to be done. I'm pleased that you figured it out—but only middlin' pleased because an experienced officer would have analyzed it at once from the T.O. and the service records. Never mind, that's how you gain experience. Now here's what you do. Write me a letter like Silva's, date it yesterday. Tell your platoon sergeant to tell Brumby that you have put him up for a third stripe—and don't mention that Silva did so. You didn't know that when you made the recommendation, so we'll keep it that way. When I swear Brumby in, I'll let him know that both his officers recommended him independently—which will make him feel good. Okay, anything more?"

"Uh . . . not in organization—unless Lieutenant Silva planned to promote Naidi, vice Brumby. In which case we could promote one PFC to lance . . . and that would allow us to promote four privates to PFC, including three vacancies now existing. I don't know whether it's your policy to keep the T.O. filled up tight or not?"

"Might as well," Blackie said gently, "as you and I know that some of those lads aren't going to have many days in which to enjoy it. Just remember that we don't make a man a PFC until after he has been in combat—not in Blackie's Blackguards we don't. Figure it out with your platoon sergeant and let me know. No hurry . . . any time before bedtime tonight. Now . . . anything else?"

"Well— Captain, I'm worried about the suits."

"So am I. All platoons."

"I don't know all the other platoons, but with five recruits to fit, plus four suits damaged and exchanged, and two more downchecked this past week and replaced from stores—well, I don't see how Cunha and Navarre can warm up that many and run routine tests on forty-one others and get it all done by our calculated date. Even if no trouble develops—"

"Trouble always develops."

"Yes, Captain. But that's two hundred and eighty-six man-hours just for warm & fit, and plus a hundred and twenty-three hours of routine checks. And it always takes longer."

"Well, what do *you* think can be done? The other platoons will lend you help if they finish their suits ahead of time. Which I doubt. Don't ask to borrow help from the Wolverines; we're more likely to lend them help."

"Uh . . . Captain, I don't know what you'll think of this, since you told me to stay out of troopers' country. But when I was a corporal, I was assistant to the Ordnance & Armor sergeant."

"Keep talking."

"Well, right at the last I was the O&A sergeant. But I was just standing in another man's shoes—I'm not a finished O&A mechanic. But I'm a pretty darn good assistant and if I was allowed to, well, I can either warm new suits, or run routine checks—and give Cunha and Navarre that much more time for trouble."

Blackie leaned back and grinned. "Mister, I have searched the regs carefully . . . and I can't find the one that says an officer mustn't get his hands dirty." He added, "I mention that because some 'young gentlemen' who have been assigned to me apparently had read such a regulation. All right, draw some dungarees—no need to get your uniform dirty along with your hands. Go aft and find your platoon sergeant, tell him about Brumby and order him to prepare recommendations to close the gaps in the T.O. in case I should decide to confirm your recommendation for Brumby. Then tell him that you are going to put in all your time on ordnance and armor—and that you want him to handle everything else. Tell him that if he has any problems to look you up in the armory. Don't tell him you consulted me—just give him orders. Follow me?"

"Yes, s— Yes, I do."

"Okay, get on it. As you pass through the cardroom, please give my compliments to Rusty and tell him to drag his lazy carcass in here."

For the next two weeks I was never so busy—not even in boot camp. Working as an ordnance & armor mech about ten hours a day was not all that I did. Math, of course—and no way to duck it with the Skipper tutoring me. Meals—say an hour and a half a day. Plus the mechanics of staying alive—shaving, showering, putting buttons in uniforms and trying to chase down the Navy master-at-arms, get him to unlock the laundry to locate clean uniforms ten minutes before inspection. (It is an unwritten law of the Navy that facilities must *always* be locked when they are most needed.)

Guard mount, parade, inspections, a minimum of platoon routine, took another hour a day. But besides, I was "George." Every outfit has a "George." He's the most junior officer and has the extra jobs—athletics officer, mail censor, referee for competitions, school officer, correspondence courses officer, prosecutor courts-martial, treasurer of the welfare mutual loan fund, custodian of registered publications, stores officer, troopers' mess officer, et cetera ad endless nauseam.

Rusty Graham had been "George" until he happily turned it over to me. He wasn't so happy when I insisted on a sight inventory on everything for which I had to sign. He suggested that if I didn't have sense enough to accept a commissioned officer's signed inventory then perhaps a direct order would change my tune. So I got sullen and told him to put his orders in writing—with a certified copy so that I could keep the original and endorse the copy over to the team commander.

Rusty angrily backed down—even a second lieutenant isn't stupid enough to put such orders in writing. I wasn't happy either as Rusty was my roommate and was then still my tutor in math, but we held the sight inventory. I got chewed out by Lieutenant Warren for being stupidly officious but he opened his safe and let me check his registered publications. Captain Blackstone opened his with no comment and I couldn't tell whether he approved of my sight inventory or not.

Publications were okay but accountable property was not. Poor Rusty! He had accepted his predecessor's count and now the count was short—and the other officer was not merely gone, he was dead. Rusty spent a restless night (and so did I!), then went to Blackie and told him the truth.

Blackie chewed him out, then went over the missing items, found ways to expend most of them as "lost in combat." It reduced Rusty's shortages to a few days' pay—but Blackie had him keep the job, thereby postponing the cash reckoning indefinitely.

Not all "George" jobs caused that much headache. There were no courts-martial; good combat teams don't have them. There was no mail to censor as the ship was in Cherenkov drive. Same for welfare loans for similar reasons. Athletics I delegated to Brumby; referee was "if and when." The troopers' mess was excellent; I initialed menus and sometimes inspected the galley, i.e., I scrounged a sandwich without getting out of dungarees when working late in the armory. Correspondence courses meant a lot of paperwork since quite a few were continuing their educations, war or no war—but I delegated my platoon sergeant and the records were kept by the PFC who was his clerk.

Nevertheless "George" jobs soaked up about two hours every day—there were so many.

You see where this left me—ten hours O&A, three hours math, meals an hour and a half, personal one hour, military fiddlework one hour, "George" two hours, sleep eight hours; total, twenty-six and a half hours. The ship wasn't even on the twenty-five-hour Sanctuary day; once we left we went on Greenwich standard and the universal calendar.

The only slack was in my sleeping time.

I was sitting in the cardroom about one o'clock one morning, plugging away at math, when Captain Blackstone came in. I said, "Good evening, Captain."

"Morning, you mean. What the deuce ails you, son? Insomnia?"

"Uh, not exactly."

He picked up a stack of sheets, remarking, "Can't your sergeant handle your paperwork? Oh, I see. Go to bed."

"But, Captain—"

"Sit back down. Johnnie, I've been meaning to talk to you. I never see you here in the cardroom, evenings. I walk past your room, you're at your desk. When your bunkie goes to bed, you move out here. What's the trouble?"

"Well . . . I just never seem to get caught up."

"Nobody ever does. How's the work going in the armory?"

"Pretty well. I think we'll make it."

"I think so, too. Look, son, you've got to keep a sense of proportion. You have two prime duties. First is to see that your platoon's equipment is ready—you're doing that. You don't have to worry about the platoon itself, I told you that. The second—and just as important—you've got to be ready to fight. You're muffing that."

"I'll be ready, Captain."

"Nonsense and other comments. You're getting no exercise and losing sleep. Is that how to train for a drop? When you lead a platoon, son, you've got to be on the bounce. From here on you will exercise from sixteen-thirty to eighteen hundred each day. You will be in your sack with lights out at twenty-three hundred—and if you lie awake fifteen minutes two nights in a row, you will report to the Surgeon for treatment. Orders."

"Yes, sir." I felt the bulkheads closing in on me and added desperately, "Captain, I don't see *how* I can get to bed by twenty-three—and still get everything *done*."

"Then you won't. As I said, son, you must have a sense of proportion. Tell me how you spend your time."

So I did. He nodded. "Just as I thought." He picked up my math "homework," tossed it in front of me. "Take this. Sure, you want to work on it. But why work so hard before we go into action?"

"Well, I thought—"

"'Think' is what you didn't do. There are four possibilities, and only one calls for finishing these assignments. First, you might buy a farm. Second, you might buy a small piece and be retired with an honorary commission. Third, you might come through all right . . . but get a downcheck on your Form Thirty-One from your examiner, namely me. Which is just what you're aching for at the present time—why, son, I won't even let you drop if you show up with eyes red from no sleep and muscles flabby from too much chair parade. The fourth possibility is that you take a grip on yourself . . . in which case I might let you take a swing at leading a platoon. So let's assume that you do and put on the finest show since Achilles slew Hector and I pass you. In that case only— you'll need to finish these math assignments. So do them on the trip back.

"That takes care of that—I'll tell the Skipper. The rest of those jobs you are relieved of, right now. On our way home you can spend your time on math. If we get home. But you'll never get anywhere if you don't learn to keep first things first. Go to bed!"

A week later we made rendezvous, coming out of drive and coasting short of the speed of light while the fleet exchanged signals. We were sent Briefing, Battle Plan, our Mission & Orders—a stack of words as long as a novel— and were told not to drop.

Oh, we were to be in the operation but we would ride down like gentlemen, cushioned in retrieval boats. This we could do because the Federation already held the surface; Second, Third, and Fifth M.I. Divisions had taken it—and paid cash.

The described real estate didn't seem worth the price. Planet P is smaller than Terra, with a surface gravity of 0.7, is mostly arctic-cold ocean and rock, with lichenous flora and no fauna of interest. Its air is not breathable for long, being contaminated with nitrous oxide and too much ozone. Its one continent is about half the size of Australia, plus many worthless islands; it would probably require as much terra-forming as Venus before we could use it.

However, we were not buying real estate to live on; we went there because Bugs were there—and they were there on our account, so Staff thought. Staff told us that Planet P was an uncompleted advance base (prob. 87 ± 6 per cent) to be used against us.

Since the planet was no prize, the routine way to get rid of this Bug base would be for the Navy to stand off at a safe distance and render this ugly spheroid uninhabitable by Man or Bug. But the C-in-C had other ideas.

The operation was a raid. It sounds incredible to call a battle involving hundreds of ships and thousands of casualties a "raid," especially as, in the meantime, the Navy and a lot of other cap troopers were keeping things stirred up many light-years into Bug space in order to divert them from reinforcing Planet P.

But the C-in-C was not wasting men; this giant raid could determine who won the war, whether next year or thirty years hence. We needed to learn more about Bug psychology. Must we wipe out every Bug in the Galaxy? Or was it possible to trounce them and impose a peace? We did not know; we understood them as little as we understand termites.

To learn their psychology we had to communicate with them, learn their motivations, find out why they fought and under what conditions they would stop; for these, the Psychological Warfare Corps needed prisoners.

Workers are easy to capture. But a Bug worker is hardly more than animate machinery. Warriors can be captured by burning off enough limbs to make them helpless—but they are almost as stupid without a director as workers. From such prisoners our own professor types had learned important matters—the development of that oily gas that killed them but not us came from analyzing the biochemistries of workers and warriors, and we had had other new weapons from such research even in the short time I had been a cap trooper. But to discover why Bugs fight we needed to study members of their brain caste. Also, we hoped to exchange prisoners.

So far, we had never taken a brain Bug alive. We had either cleaned out colonies from the surface, as on Sheol, or (as had too often been the case) raiders had gone down their holes and not come back. A lot of brave men had been lost this way.

Still more had been lost through retrieval failure. Sometimes a team on the ground had its ship or ships knocked out of the sky. What happens to such a team? Possibly it dies to the last man. More probably it fights until power and ammo are gone, then survivors are captured as easily as so many beetles on their backs.

From our co-belligerents the Skinnies we knew that many missing troopers were alive as prisoners—thousands we hoped, hundreds we were sure. Intelligence believed that prisoners were always taken to Klendathu; the Bugs are as curious about us as we are about them—a race of individuals able to build cities, starships, armies, may be even more mysterious to a hive entity than a hive entity is to us.

As may be, we wanted those prisoners back!

In the grim logic of the universe this may be a weakness. Perhaps some race that never bothers to rescue an individual may exploit this human trait to wipe us out. The Skinnies have such a trait only slightly and the Bugs don't seem to have it at all—nobody *ever* saw a Bug come to the aid of another because he was wounded; they co-operate perfectly in fighting but units are abandoned the instant they are no longer useful.

Our behavior is different. How often have you seen a headline like this?—TWO DIE ATTEMPTING RESCUE OF DROWNING CHILD. If a man gets lost in the mountains, hundreds will search and often two or three searchers are killed. But the next time somebody gets lost just as many volunteers turn out.

Poor arithmetic . . . but very human. It runs through all our folklore, all human religions, all our literature—a racial conviction that when one human needs rescue, others should not count the price.

Weakness? It might be the unique strength that wins us a Galaxy.

Weakness or strength, Bugs don't have it; there was no prospect of trading fighters for fighters.

But in a hive polyarchy, some castes are valuable—or so our Psych Warfare people hoped. If we could capture brain Bugs, alive and undamaged, we might be able to trade on good terms.

And suppose we captured a queen!

What is a queen's trading value? A regiment of troopers? Nobody knew, but Battle Plan ordered us to capture Bug "royalty," brains and queens, *at any cost*, on the gamble that we could trade them for human beings.

The third purpose of Operation Royalty was to develop methods: how to go down, how to dig them out, how to win with less than total weapons. Trooper for warrior, we could now defeat them above ground; ship for ship, our Navy was better; but, so far, we had had no luck when we tried to go down their holes.

If we failed to exchange prisoners on any terms, then we still had to: (a) win the war, (b) do so in a way that gave us a fighting chance to rescue our own people, or (c)—might as well admit it—die trying and lose. Planet P was a field test to determine whether we could learn how to root them out.

Briefing was read to every trooper and he heard it again in his sleep during hypno preparation. So, while we all knew that Operation Royalty was laying the ground-work toward eventual rescue of our mates, we also knew that Planet P held no human prisoners—it had never been raided. So there was no reason to buck for medals in a wild hope of being personally in on a rescue; it was just another Bug hunt, but conducted with massive force and new techniques. We were going to peel that planet like an onion, until we *knew* that every Bug had been dug out.

The Navy had plastered the islands and that unoccupied part of the continent until they were radioactive glaze; we could tackle Bugs with no worries about our rear. The Navy also maintained a ball-of-yarn patrol in tight orbits around the planet, guarding us, escorting transports, keeping a spy watch on the surface to make sure that Bugs did not break out behind us despite that plastering.

Under the Battle Plan, the orders for Blackie's Black-guards charged us with supporting the prime Mission when ordered or as opportunity presented, relieving another company in a captured area, protecting units of other corps in that area, maintaining contact with M.I. units around us—and smacking down any Bugs that showed their ugly heads.

So we rode down in comfort to an unopposed landing. I took my platoon out at a powered-armor trot. Blackie went ahead to meet the company commander he was relieving, get the situation and size up the terrain. He headed for the horizon like a scared jack rabbit.

I had Cunha send his first sections' scouts out to locate the forward corners of my patrol area and I sent my platoon sergeant off to my left to make contact with a patrol from the Fifth Regiment. We, the Third Regiment, had a grid three hundred miles wide and eighty miles deep to hold; my piece was a rectangle forty miles deep and seventeen wide in the extreme left flank forward corner. The Wolverines were behind us, Lieutenant Khoroshen's platoon on the right and Rusty beyond him.

Our First Regiment had already relieved a Vth Div. regiment ahead of us, with a "brick wall" overlap which placed them on my corner as well as ahead. "Ahead" and "rear," "right flank" and "left," referred to orientation set up in dead-reckoning tracers in each command suit to match the grid of the Battle Plan. We had no true front, simply an area, and the only fighting at the moment was going on several hundred miles away, to our arbitrary right and rear.

Somewhere off that way, probably two hundred miles, should be 2nd platoon, G Co, 2nd Batt, 3rd Reg—commonly known as "The Roughnecks."

Or the Roughnecks might be forty light-years away. Tactical organization never matches the Table of Organization; all I knew from Plan was that something called the "2nd Batt" was on our right flank beyond the boys from the *Normandy Beach*. But that battalion could have been borrowed from another division. The Sky Marshal plays his chess without consulting the pieces.

Anyhow, I should not be thinking about the Roughnecks; I had all I could do as a Blackguard. My platoon was okay for the moment—safe as you can be on a hostile planet—but I had plenty to do before Cunha's first squad reached the far corner. I needed to:

1. Locate the platoon leader who had been holding my area.

2. Establish corners and identify them to section and squad leaders.

3. Make contact liaison with eight platoon leaders on my sides and corners, five of whom should already be in position (those from Fifth and First Regiments) and three (Khoroshen of the Blackguards and Bayonne and Sukarno of the Wolverines) who were now moving into position.

4. Get my own boys spread out to their initial points as fast as possible by shortest routes.

The last had to be set up first, as the open column in which we disembarked would not do it. Brumby's last squad needed to deploy to the left flank; Cunha's leading squad needed to spread from dead ahead to left oblique; the other four squads must fan out in between.

This is a standard square deployment and we had simulated how to reach it quickly in the drop room; I called out: "Cunha! Brumby! Time to spread 'em out," using the non-com circuit.

"Roger sec one!"—"Roger sec two!"

"Section leaders take charge . . . and caution each recruit. You'll be passing a lot of Cherubs. I don't want 'em shot at by mistake!" I bit down for my private circuit and said, "Sarge, you got contact on the left?"

"Yes, sir. They see me, they see you."

"Good. I don't see a beacon on our anchor corner—"

"Missing."

"—so you coach Cunha by D.R. Same for the lead scout—that's Hughes—and have Hughes set a new beacon." I wondered why the Third or Fifth hadn't replaced that anchor beacon—my forward left corner where three regiments came together.

No use talking. I went on: "D.R. check. You bear two seven five, miles twelve."

"Sir, reverse is nine six, miles twelve scant."

"Close enough. I haven't found my opposite number yet, so I'm cutting out forward at max. Mind the shop."

"Got 'em, Mr. Rico."

I advanced at max speed while clicking over to officers' circuit: "Square Black One, answer. Black One, Chang's

Cherubs—do you read me? Answer." I wanted to talk with the leader of the platoon we were relieving—and not for any perfunctory I-relieve-you-sir: I wanted the ungarnished word.

I didn't like what I had seen.

Either the top brass had been optimistic in believing that we had mounted overwhelming force against a small, not fully developed Bug base—or the Blackguards had been awarded the spot where the roof fell in. In the few moments I had been out of the boat I had spotted half a dozen armored suits on the ground—empty I hoped, dead men possibly, but 'way too many any way you looked at it.

Besides that, my tactical radar display showed a full platoon (my own) moving into position but only a scattering moving back toward retrieval or still on station. Nor could I see any system to their movements.

I was responsible for 680 square miles of hostile terrain and I wanted very badly to find out all I could *before* my own squads were deep into it. Battle Plan had ordered a new tactical doctrine which I found dismaying: Do not close the Bugs' tunnels. Blackie had explained this as if it had been his own happy thought, but I doubt if he liked it.

The strategy was simple, and, I guess, logical . . . if we could afford the losses. Let the Bugs come up. Meet them and kill them on the surface. Let them keep on coming up. Don't bomb their holes, don't gas their holes—let them out. After a while—a day, two days, a week—if we really did have overwhelming force, they would stop coming up. Planning Staff estimated (don't ask me how!) that the Bugs would expend 70 per cent to 90 per cent of their warriors before they stopped trying to drive us off the surface.

Then we would start the unpeeling, killing surviving warriors as we went down and trying to capture "royalty" alive. We knew what the brain caste looked like; we had seen them dead (in photographs) and we knew they could not run—barely functional legs, bloated bodies that were mostly nervous system. Queens no human had ever seen, but Bio War Corps had prepared sketches of what they should look like—obscene monsters larger than a horse and utterly immobile.

Besides brains and queens there might be other "royalty" castes. As might be—encourage their warriors to come out and die, then capture alive anything but warriors and workers.

A necessary plan and very pretty, on paper. What it meant to me was that I had an area 17 × 40 miles which might be riddled with unstopped Bug holes. I wanted co-ordinates on each one.

If there were too many . . . well, I might accidentally plug a few and let my boys concentrate on watching the rest. A private in a marauder suit can cover a lot of terrain, but he can look at only one thing at a time; he is not superhuman.

I bounced several miles ahead of the first squad, still calling the Cherub platoon leader, varying it by calling *any* Cherub officer and describing the pattern of my transponder beacon (dah-di-dah-dah).

No answer—

At last I got a reply from my boss: "Johnnie! Knock off the noise. Answer me on conference circuit."

So I did, and Blackie told me crisply to quit trying to find the Cherub leader for Square Black One; there wasn't one. Oh, there might be a non-com alive somewhere but the chain of command had broken.

By the book, somebody always moves up. But it *does* happen if too many links are knocked out. As Colonel Nielssen had once warned me, in the dim past . . . almost a month ago.

Captain Chang had gone into action with three officers besides himself; there was one left now (my classmate, Abe Moise) and Blackie was trying to find out from him the situation. Abe wasn't much help. When I joined the conference and identified myself, Abe thought I was his battalion commander and made a report almost heart-breakingly precise, especially as it made no sense at all.

Blackie interrupted and told me to carry on. "Forget about a relief briefing. The situation is whatever you see that it is—so stir around and *see*."

"Right, Boss!" I slashed across my own area toward the far corner, the anchor corner, as fast as I could move, switching circuits on my first bounce. "Sarge! How about that beacon?"

"No place on that corner to put it, sir. A fresh crater there, about scale six."

I whistled to myself. You could drop the *Tours* into a size six crater. One of the dodges the Bugs used on us when we were sparring, ourselves on the surface, Bugs underground, was land mines. (They never seemed to use missiles, except from ships in space.) If you were near the spot, the ground shock got you; if you were in the air when one went off, the concussion wave could tumble your gyros and throw your suit out of control.

I had never seen larger than a scale-four crater. The theory was that they didn't dare use too big an explosion because of damage to their troglodyte habitats, even if they cofferdammed around it.

"Place an offset beacon," I told him. "Tell section and squad leaders."

"I have, sir. Angle one one oh, miles one point three. Da-di-dit. You should be able to read it, bearing about three three five from where you are." He sounded as calm as a sergeant-instructor at drill and I wondered if I were letting my voice get shrill.

I found it in my display, above my left eyebrow—long and two shorts. "Okay. I see Cunha's first squad is nearly in position. Break off that squad, have it patrol the crater. Equalize the areas—Brumby will have to take four more miles of depth." I thought with annoyance that each man already had to patrol fourteen square miles; spreading the butter so thin meant seventeen square miles per man—and a Bug can come out of a hole less than five feet wide.

I added, "How 'hot' is that crater?"

"Amber-red at the edge. I haven't been in it, sir."

"Stay out of it. I'll check it later." Amber-red would kill an unprotected human but a trooper in armor can take it for quite a time. If there was that much radiation at the edge, the bottom would no doubt fry your eyeballs. "Tell

Naidi to pull Malan and Bjork back to amber zone, and have them set up ground listeners." Two of my five recruits were in that first squad—and recruits are like puppies; they stick their noses into things.

"Tell Naidi that I am interested in two things: movement inside the crater . . . and noises in the ground around it." *We* wouldn't send troopers out through a hole so radioactive that mere exit would kill them. But Bugs would, if they could reach us that way. "Have Naidi report to me. To you and me, I mean."

"Yes, sir." My platoon sergeant added, "May I make a suggestion?"

"Of course. And don't stop to ask permission next time."

"Navarre can handle the rest of the first section. Sergeant Cunha could take the squad at the crater and leave Naidi free to supervise the ground-listening watch."

I knew what he was thinking. Naidi, so newly a corporal that he had never before had a squad on the ground, was hardly the man to cover what looked like the worst danger point in Square Black One; he wanted to pull Naidi back for the same reasons I had pulled the recruits back.

I wondered if he knew what I was thinking? That "nutcracker"—he was using the suit he had worn as Blackie's battalion staffer, he had one more circuit than I had, a private one to Captain Blackstone.

Blackie was probably patched in and listening via that extra circuit. Obviously my platoon sergeant did not agree with my disposition of the platoon. If I didn't take his advice, the next thing I heard might be Blackie's voice cutting in: "Sergeant, take charge. Mr. Rico, you're relieved."

But— Confound it, a corporal who wasn't allowed to boss his squad wasn't a corporal . . . and a platoon leader who was just a ventriloquist's dummy for his platoon sergeant was an empty suit!

I didn't mull this. It flashed through my head and I answered at once. "I can't spare a corporal to baby-sit with two recruits. Nor a sergeant to boss four privates and a lance."

"But—"

"Hold it. I want the crater watch relieved every hour. I want our first patrol sweep made rapidly. Squad leaders will check any hole reported and get beacon bearings so that section leaders, platoon sergeant and platoon leader can check them as they reach them. If there aren't too many, we'll put a watch on each—I'll decide later."

"Yes, sir."

"Second time around. I want a slow patrol, as tight as possible, to catch holes we miss on the first sweep. Assistant squad leaders will use snoopers on that pass. Squad leaders will get bearings on any troopers—or suits—on the ground; the Cherubs may have left some live wounded. But no one is to stop even to check physicals until I order it. We've got to know the Bug situation first."

"Yes, sir."

"Suggestions?"

"Just one," he answered. "I think the squad chasers should use their snoopers on that first fast pass."

"Very well, do it that way." His suggestion made sense as the surface air temperature was much lower than the Bugs use in their tunnels; a camouflaged vent hole should show a plume like a geyser by infrared vision. I glanced at my display. "Cunha's boys are almost at limit. Start your parade."

"Very well, sir!"

"Off." I clicked over to the wide circuit and continued to make tracks for the crater while I listened to everybody at once as my platoon sergeant revised the pre-plan—cutting out one squad, heading it for the crater, starting the rest of the first section in a two-squad countermarch while keeping the second section in a rotational sweep as pre-planned but with four miles increased depth; got the sections moving, dropped them and caught the first squad as it converged on the anchor crater, gave it its instructions; cut back to the section leaders in plenty of time to give them new beacon bearings at which to make their turns.

He did it with the smart precision of a drum major on parade and he did it faster and in fewer words than I could have done it. Extended-order powered suit drill, with a platoon spread over many miles of countryside, is much more difficult than the strutting precision of parade—but it has to be exact, or you'll blow the head off your mate in action . . . or, as in this case, you sweep part of the terrain twice and miss another part.

But the drillmaster has only a radar display of his formation; he can see with his eyes only those near him. While I listened I watched it in my own display—glowworms crawling past my face in precise lines, "crawling" because even forty miles an hour is a slow crawl when you compress a formation twenty miles across into a display a man can see.

I listened to everybody at once because I wanted to hear the chatter inside the squads.

There wasn't any. Cunha and Brumby gave their secondary commands—and shut up. The corporals sang out only as squad changes were necessary; section and squad chasers called out occasional corrections of interval or alignment—and privates said nothing at all.

I heard the breathing of fifty men like muted sibilance of surf, broken only by necessary orders in the fewest possible words. Blackie had been right; the platoon had been handed over to me "tuned like a violin."

They didn't need *me*! I could go home and my platoon would get along just as well.

Maybe better—

I wasn't sure I had been right in refusing to cut Cunha out to guard the crater; if trouble broke there and those boys couldn't be reached in time, the excuse that I had done it "by the book" was worthless. If you get killed, or let someone else get killed, "by the book" it's just as permanent as any other way.

I wondered if the Roughnecks had a spot open for a buck sergeant.

Most of Square Black One was as flat as the prairie around Camp Currie and much more barren. For this I was thankful; it gave us our only chance of spotting a

Bug coming up from below and getting him first. We were spread so widely that four-mile intervals between men and about six minutes between waves of a fast sweep was as tight a patrol as we could manage. This isn't tight enough; any one spot would remain free of observation for at least three or four minutes between patrol waves—and a lot of Bugs can come out of a very small hole in three to four minutes.

Radar can see farther than the eye, of course, but it cannot see as accurately.

In addition we did not dare use anything but short-range selective weapons—our own mates were spread around us in all directions. If a Bug popped up and you let fly with something lethal, it was certain that not too far beyond that Bug was a cap trooper; this sharply limits the range and force of the frightfulness you dare use. On this operation only officers and platoon sergeants were armed with rockets and, even so, we did not expect to use them. If a rocket fails to find its target, it has a nasty habit of continuing to search until it finds one . . . and it cannot tell a friend from foe; a brain that can be stuffed into a small rocket is fairly stupid.

I would happily have swapped that area patrol with thousands of M.I. around us, for a simple one-platoon strike in which you know where your own people are and anything else is an enemy target.

I didn't waste time moaning; I never stopped bouncing toward that anchor-corner crater while watching the ground and trying to watch the radar picture as well. I didn't find any Bug holes but I did jump over a dry wash, almost a canyon, which could conceal quite a few. I didn't stop to see; I simply gave its co-ordinates to my platoon sergeant and told him to have somebody check it.

That crater was even bigger than I had visualized; the *Tours* would have been lost in it. I shifted my radiation counter to directional cascade, took readings on floor and sides—red to multiple red right off the scale, very unhealthy for long exposure even to a man in armor; I estimated its width and depth by helmet range finder, then prowled around and tried to spot openings leading underground.

I did not find any but I did run into crater watches set out by adjacent platoons of the Fifth and First Regiments, so I arranged to split up the watch by sectors such that the combined watch could yell for help from all three platoons, the patch-in to do this being made through First Lieutenant Do Campo of the "Head Hunters" on our left. Then I pulled out Naidi's lance and half his squad (including the recruits) and sent them back to platoon, reporting all this to my boss, and to my platoon sergeant.

"Captain," I told Blackie, "we aren't getting any ground vibrations. I'm going down inside and check for holes. The readings show that I won't get too much dosage if I—"

"Youngster, stay out of that crater."

"But Captain, I just meant to—"

"Shut up. You can't learn anything useful. Stay out."

"Yes, sir."

The next nine hours were tedious. We had been preconditioned for forty hours of duty (two revolutions of

Planet P) through forced sleep, elevated blood sugar count, and hypno indoctrination, and of course the suits are self-contained for personal needs. The suits can't last that long, but each man was carrying extra power units and super H.P. air cartridges for recharging. But a patrol with no action is dull, it is easy to goof off.

I did what I could think of, having Cunha and Brumby take turns as drill sergeant (thus leaving platoon sergeant and leader free to rove around): I gave orders that no sweeps were to repeat in pattern so that each man would always check terrain that was new to him. There are endless patterns to cover a given area, by combining the combinations. Besides that, I consulted my platoon sergeant and announced bonus points toward honor squad for first verified hole, first Bug destroyed, etc.—boot camp tricks, but staying alert means staying alive, so anything to avoid boredom.

Finally we had a visit from a special unit: three combat engineers in a utility air car, escorting a talent—a spatial senser. Blackie warned me to expect them. "Protect them and give them what they want."

"Yes, sir. What will they need?"

"How should I know? If Major Landry wants you to take off your skin and dance in your bones, do it!"

"Yes, sir. Major Landry."

I relayed the word and set up a bodyguard by subareas. Then I met them as they arrived because I was curious; I had never seen a special talent at work. They landed beside my right flank and got out. Major Landry and two officers were wearing armor and hand flamers but the talent had no armor and no weapons—just an oxygen mask. He was dressed in a fatigue uniform without insignia and he seemed terribly bored by everything. I was not introduced to him. He looked like a sixteen-year-old boy . . . until I got close and saw a network of wrinkles around his weary eyes.

As he got out he took off his breathing mask. I was horrified, so I spoke to Major Landry, helmet to helmet without radio. "Major—the air around here is 'hot.' Besides that, we've been warned that—"

"Pipe down," said the Major. "He knows it."

I shut up. The talent strolled a short distance, turned and pulled his lower lip. His eyes were closed and he seemed lost in thought.

He opened them and said fretfully, "How can one be expected to work with all those silly people jumping around?"

Major Landry said crisply, "Ground your platoon."

I gulped and started to argue—then cut in the all-hands circuit: "First Platoon Blackguards—*ground and freeze!*"

It speaks well for Lieutenant Silva that all I heard was a double echo of my order, as it was repeated down to squad. I said, "Major, can I let them move around on the ground?"

"No. And shut up."

Presently the senser got back in the car, put his mask on. There wasn't room for me, but I was allowed—ordered, really—to grab on and be towed; we shifted a couple of miles. Again the senser took off his mask and walked around. This time he spoke to one of the other combat engineers, who kept nodding and sketching on a pad.

The special-mission unit landed about a dozen times in my area, each time going through the same apparently pointless routine; then they moved on into the Fifth Regiment's grid. Just before they left, the officer who had been sketching pulled a sheet out of the bottom of his sketch box and handed it to me. "Here's your sub map. The wide red band is the only Bug boulevard in your area. It is nearly a thousand feet down where it enters but it climbs steadily toward your left rear and leaves at about minus four hundred fifty. The light blue network joining it is a big Bug colony; the only places where it comes within a hundred feet of the surface I have marked. You might put some listeners there until we can get over there and handle it."

I stared at it. "Is this map reliable?"

The engineer officer glanced at the senser, then said very quietly to me, "Of course it is, you idiot! What are you trying to do? Upset him?"

They left while I was studying it. The artist-engineer had done double sketching and the box had combined them into a stereo picture of the first thousand feet under the surface. I was so bemused by it that I had to be reminded to take the platoon out of "freeze"—then I withdrew the ground listeners from the crater, pulled two men from each squad and gave them bearings from that infernal map to have them listen along the Bug highway and over the town.

I reported it to Blackie. He cut me off as I started to describe the Bug tunnels by co-ordinates. "Major Landry relayed a facsimile to me. Just give me co-ordinates of your listening posts."

I did so. He said, "Not bad, Johnnie. But not quite what I want, either. You've placed more listeners than you need over their mapped tunnels. String four of them along that Bug race track, place four more in a diamond around their town. That leaves you four. Place one in the triangle formed by your right rear corner and the main tunnel; the other three go in the larger area on the other side of the tunnel."

"Yes, sir." I added, "Captain, can we depend on this map?"

"What's troubling you?"

"Well . . . it seems like magic. Uh, black magic."

"Oh. Look, son, I've got a special message from the Sky Marshal to you. He says to tell you that map is official . . . and that he will worry about everything else so that you can give full time to your platoon. Follow me?"

"Uh, yes, Captain."

"But the Bugs can burrow mighty fast, so you give special attention to the listening posts *outside* the area of the tunnels. Any noise from those four outside posts louder than a butterfly's roar is to be reported at once, re-gardless of its nature."

"Yes, sir."

"When they burrow, it makes a noise like frying bacon—in case you've never heard it. Stop your patrol sweeps. Leave one man on visual observation of the crater. Let half your platoon sleep for two hours, while the other half pairs off to take turns listening."

"Yes, sir."

"You may see some more combat engineers. Here's the revised plan. A sapper company will blast down and cork that main tunnel where it comes nearest the surface, either at your left flank, or beyond in 'Head Hunter' territory. At the same time another engineer company will do the same where that tunnel branches about thirty miles off to your right in the First Regiment's bailiwick. When the corks are in, a long chunk of their main street and a biggish settlement will be cut off. Meanwhile, the same sort of thing will be going on a lot of other places. Thereafter—we'll see. Either the Bugs break through to the surface and we have a pitched battle, or they sit tight and we go down after them, a sector at a time."

"I see." I wasn't sure that I did, but I understood my part: rearrange my listening posts; let half my platoon sleep. Then a Bug hunt—on the surface if we were lucky, underground if we had to.

"Have your flank make contact with that sapper company when it arrives. Help 'em if they want help."

"Right, Cap'n," I agreed heartily. Combat engineers are almost as good an outfit as the infantry; it's a pleasure to work with them. In a pinch they fight, maybe not expertly but bravely. Or they go ahead with their work, not even lifting their heads while a battle rages around them. They have an unofficial, very cynical and very ancient motto: "First we dig 'em, then we die in 'em," to supplement their official motto: "Can do!" Both mottoes are literal truth.

"Get on it, son."

Twelve listening posts meant that I could put a half squad at each post, either a corporal or his lance, plus three privates, then allow two of each group of four to sleep while the other two took turns listening. Navarre and the other section chaser could watch the crater and sleep, turn about, while section sergeants could take turns in charge of the platoon. The redisposition took no more than ten minutes once I had detailed the plan and given out bearings to the sergeants: nobody had to move very far. I warned everybody to keep eyes open for a company of engineers. As soon as each section reported its listening posts in operation I clicked to the wide circuit: "Odd numbers! Lie down, prepare to sleep . . . one . . . two . . . three . . . four . . . five—sleep!"

A suit is not a bed, but it will do. One good thing about hypno preparation for combat is that, in the un-likely event of a chance to rest, a man can be put to sleep instantly by post-hypnotic command triggered by some-one who is not a hypnotist—and awakened just as in-stantly, alert and ready to fight. It is a life-saver, because a man can get so exhausted in battle that he shoots at things that aren't there and can't see what he should be fighting.

But I had no intention of sleeping. I had not been told to—and I had not asked. The very thought of sleep-ing when I knew that perhaps many thousands of Bugs were only a few hundred feet away made my stomach jump. Maybe that senser was infallible, perhaps the Bugs could not reach us without alerting our listening posts.

Maybe—But I didn't want to chance it.

I clicked to my private circuit. "Sarge—"

"Yes, sir."

"You might as well get a nap. I'll be on watch. Lie down and prepare to sleep . . . one . . . two—"

"Excuse me, sir. I have a suggestion."

"Yes?"

"If I understand the revised plan, no action is expected for the next four hours. You could take a nap now, and then—"

"Forget it, Sarge! I am not going to sleep. I am going to make the rounds of the listening posts and watch for that sapper company."

"Very well, sir."

"I'll check number three while I'm here. You stay here with Brumby and catch some rest while I—"

"*Johnnie!*"

I broke off. "Yes, Captain?" Had the Old Man been listening?

"Are your posts all set?"

"Yes, Captain, and my odd numbers are sleeping. I am about to inspect each post. Then—"

"Let your sergeant do it. I want you to rest."

"But, Captain—"

"Lie down. That's a direct order. Prepare to sleep . . . one . . . two . . . three—*Johnnie!*"

"Captain, with your permission, I would like to inspect my posts first. Then I'll rest, if you say so, but I would rather remain awake. I—"

Blackie guffawed in my ear. "Look, son, you've slept for an hour and ten minutes."

"*Sir?*"

"Check the time." I did so—and felt foolish. "You wide-awake, son?"

"Yes, sir. I think so."

"Things have speeded up. Call your odd numbers and put your even numbers to sleep. With luck, they may get an hour. So swap 'em around, inspect your posts, and call me back."

I did so and started my rounds without a word to my platoon sergeant. I was annoyed at both him and Blackie—at my company commander because I resented being put to sleep against my wishes; and as for my platoon sergeant, I had a dirty hunch that it wouldn't have been done if he weren't the real boss and myself just a figurehead.

But after I had checked posts number three and one (no sounds of any sort, both were forward of the Bug area), I cooled down. After all, blaming a sergeant, even a fleet sergeant, for something a captain did was silly. "Sarge—"

"Yes, Mr. Rico?"

"Do you want to catch a nap with the even numbers? I'll wake you a minute or two before I wake them."

He hesitated slightly. "Sir, I'd like to inspect the listening posts myself."

"Haven't you already?"

"No, sir. I've been asleep the past hour."

"*Huh?*"

He sounded embarrassed. "The Captain required me to do so. He placed Brumby temporarily in charge and put me to sleep immediately after he relieved you."

I started to answer, then laughed helplessly. "Sarge? Let's you and I go off somewhere and go back to sleep. We're wasting our time; Cap'n Blackie is running this platoon."

"I have found, sir," he answered stiffly, "that Captain Blackstone invariably has a reason for anything he does."

I nodded thoughtfully, forgetting that I was ten miles from my listener. "Yes. You're right, he always has a reason. Mmm . . . since he had us both sleep, he must want us both awake and alert now."

"I think that must be true."

"Mmm . . . any idea why?"

He was rather long in answering. "Mr. Rico," he said slowly, "if the Captain knew he would tell us; I've never known him to hold back information. But sometimes he does things a certain way without being able to explain why. The Captain's hunches—well, I've learned to respect them."

"So? Squad leaders are all even numbers; they're asleep."

"Yes, sir."

"Alert the lance of each squad. We won't wake anybody . . . but when we do, seconds may be important."

"Right away."

I checked the remaining forward post, then covered the four posts bracketing the Bug village, jacking my phones in parallel with each listener. I had to force myself to listen, because you could *hear* them, down there below, chittering to each other. I wanted to run and it was all I could do not to let it show.

I wondered if that "special talent" was simply a man with incredibly acute hearing.

Well, no matter how he did it, the Bugs were where he said they were. Back at O.C.S. we had received demonstrations of recorded Bug noises; these four posts were picking up typical nest noises of a large Bug town—that chittering which may be their speech (though why should they need to talk if they are all remotely controlled by the brain caste?), a rustling like sticks and dry leaves, a high background whine which is always heard at a settlement and which had to be machinery—their air conditioning perhaps.

I did not hear the hissing, cracking noise they make in cutting through rock.

The sounds along the Bug boulevard were unlike the settlement sounds—a low background rumble which increased to a roar every few moments, as if heavy traffic were passing. I listened at post number five, then got an idea—checked it by having the stand-by man at each of the four posts along the tunnel call out "*Mark!*" to me each time the roaring got loudest.

Presently I reported. "Captain—"

"Yeah, Johnnie?"

"The traffic along this Bug race is all moving one way, from me toward you. Speed is approximately a hundred and ten miles per hour, a load goes past about once a minute."

"Close enough," he agreed. "I make it one-oh-eight with a headway of fifty-eight seconds."

"Oh." I felt dashed, and changed the subject. "I haven't seen that sapper company."

"You won't. They picked a spot in the middle rear of 'Head Hunter' area: Sorry, I should have told you. Anything more?"

"No, sir." We clicked off and I felt better. Even Blackie could forget . . . and there hadn't been anything *wrong* with my idea. I left the tunnel zone to inspect the listening post to right and rear of the Bug area, post twelve.

As with the others, there were two men asleep, one listening, one stand-by, I said to the stand-by, "Getting anything?"

"No, sir."

The man listening, one of my five recruits, looked up and said, "Mr. Rico, I think this pickup has just gone sour."

"I'll check it," I said. He moved to let me jack in with him.

"Frying bacon" so loud you could smell it!

I hit the all-hands circuit. "First platoon *up*! Wake up, call off, and report!"

—And clicked over to officers' circuit. "Captain! Captain Blackstone! *Urgent!*"

"Slow down, Johnnie. Report."

"'Frying bacon' sounds, sir," I answered, trying desperately to keep my voice steady. "Post twelve at co-ordinates Easter Nine, Square Black One."

"Easter Nine," he agreed. "Decibels?"

I looked hastily at the meter on the pickup. "I don't know, Captain. Off the scale at the max end. It sounds like they're right under my feet!"

"Good!" He applauded—and I wondered how he could feel that way. "Best news we've had today! Now listen, son. Get your lads awake—"

"They are, sir!"

"Very well. Pull back two listeners, have them spot-check around post twelve. Try to figure where the Bugs are going to break out. *And stay away from that spot!* Understand me?"

"I hear you, sir," I said carefully. "But I do not understand."

He sighed. "Johnnie, you'll turn my hair gray yet. Look, son, we *want* them to come out, the more the better. You don't have the firepower to handle them other than by blowing up their tunnel as they reach the surface—and that is the one thing *you must not do*! If they come out in force, a regiment can't handle them. But that's just what the General wants, and he's got a brigade of heavy weapons in orbit, waiting for it. So you spot that breakthrough, fall back and keep it under observation. If you are lucky enough to have a major breakthrough in your area, your reconnaissance will be patched through all the way to the top. So stay lucky and stay alive! Got it?"

"Yes, sir. Spot the breakthrough. Fall back and avoid contact. Observe and report."

"Get on it!"

I pulled back listeners nine and ten from the middle

stretch of "Bug Boulevard" and had them close in on co-ordinates Easter Nine from right and left, stopping every half mile to listen for "frying bacon." At the same time I lifted post twelve and moved it toward our rear, while checking for a dying away of the sound.

In the meantime my platoon sergeant was regrouping the platoon in the forward area between the Bug settlement and the crater—all but twelve men who were ground-listening. Since we were under orders not to attack, we both worried over the prospect of having the platoon spread too widely for mutual support. So he rearranged them in a compact line five miles long, with Brumby's section on the left, nearer the Bug settlement. This placed the men less than three hundred yards apart (almost shoulder to shoulder for cap troopers), and put nine of the men still on listening stations within support distance of one flank or the other. Only the three listeners working with me were out of reach of ready help.

I told Bayonne of the Wolverines and Do Campo of the Head Hunters that I was no longer patrolling and why, and I reported our regrouping to Captain Blackstone.

He grunted. "Suit yourself. Got a prediction on that breakthrough?"

"It seems to center about Easter Ten, Captain, but it is hard to pin down. The sounds are very loud in an area about three miles across—and it seems to get wider. I'm trying to circle it at an intensity level just barely on scale." I added, "Could they be driving a new horizontal tunnel just under the surface?"

He seemed surprised. "That's possible. I hope not—we want them to come up." He added, "Let me know if the center of the noise moves. Check on it."

"Yes, sir. Captain—"

"Huh? Speak up."

"You told us not to attack when they break out. If they break out. What *are* we to do? Are we just spectators?"

There was a longish delay, fifteen or twenty seconds, and he may have consulted "upstairs." At last he said. "Mr. Rico, you are not to attack at or near Easter Ten. Anywhere else—the idea is to hunt Bugs."

"Yes, sir," I agreed happily. "We hunt Bugs."

"Johnnie!" he said sharply. "If you go hunting medals instead of Bugs—and I find out—you're going to have a mighty sad-looking Form Thirty-One!"

"Captain," I said earnestly. "I don't *ever* want to win a medal. The idea is to hunt Bugs."

"Right. Now quit bothering me."

I called my platoon sergeant, explained the new limits under which we would work, told him to pass the word along and to make sure that each man's suit was freshly charged, air and power.

"We've just finished that, sir. I suggest that we relieve the men with you." He named three reliefs.

That was reasonable, as my ground listeners had had no time to recharge. But the reliefs he named were all scouts.

Silently I cussed myself for utter stupidity. A scout's suit is as fast as a command suit, twice the speed of a marauder. I had been having a nagging feeling of something left undone, and had checked it off to the nervousness I always feel around Bugs.

Now I knew. Here I was, ten miles away from my platoon with a party of three men—each in a marauder suit. When the Bugs broke through, I was going to be faced with an impossible decision . . . unless the men with me could rejoin as fast as I could. "That's good," I agreed, "but I no longer need three men. Send Hughes, right away. Have him relieve Nyberg. Use the other three scouts to relieve the listening posts farthest forward."

"Just Hughes?" he said doubtfully.

"Hughes is enough. I'm going to man one listener myself. Two of us can straddle the area; we know where they are now." I added, "Get Hughes down here on the bounce."

For the next thirty-seven minutes nothing happened. Hughes and I swung back and forth along the forward and rear arcs of the area around Easter Ten, listening five seconds at a time, then moving on. It was no longer necessary to seat the microphone in rock; it was enough to touch it to the ground to get the sound of "frying bacon" strong and clear. The noise area expanded but its center did not change. Once I called Captain Blackstone to tell him the sound had abruptly stopped, and again three minutes later to tell him it had resumed; otherwise I used the scouts' circuit and let my platoon sergeant take care of the platoon and the listening posts near the platoon.

At the end of this time everything happened at once.

*   *   *

A voice called out on the scouts' circuit, "'Bacon Fry'! Albert Two!"

I clicked over and called out, "Captain! 'Bacon Fry' at Albert Two, Black One!"—clicked over to liaison with the platoons surrounding me: "Liaison flash! 'Bacon frying' at Albert Two, Square Black One"—and immediately heard Do Campo reporting: "'Frying bacon' sounds at Adolf Three, Green Twelve."

I relayed that to Blackie and cut back to my own scouts' circuit, heard: "Bugs! *Bugs! HELP!*"

"Where?"

No answer. I clicked over. "Sarge! Who reported Bugs?"

He rapped back, "Coming up out of their town—about Bangkok Six."

"*Hit 'em!*" I clicked over to Blackie. "Bugs at Bangkok Six, Black One—I am attacking!"

"I heard you order it," he answered calmly. "How about Easter Ten?"

"Easter Ten is—" The ground fell away under me and I was engulfed in Bugs.

I didn't know what had happened to me. I wasn't hurt; it was a bit like falling into the branches of a tree—but those branches were alive and kept jostling me while my gyros complained and tried to keep me upright. I fell ten or fifteen feet, deep enough to be out of the daylight.

Then a surge of living monsters carried me back up into the light—and training paid off: I landed on my feet, talking and fighting: "Breakthrough at Easter Ten—no, Easter Eleven, where I am now. Big hole and they're pouring up.

Hundreds. More than that." I had a hand flamer in each hand and was burning them down as I reported.

"Get out of there, Johnnie!"

"Wilco!"—and I started to jump.

And stopped. Checked the jump in time, stopped flaming, and really looked—for I suddenly realized that I ought to be dead. "Correction," I said, looking and hardly believing. "Breakthrough at Easter Eleven is a feint. No warriors."

"Repeat."

"Easter Eleven, Black One. Breakthrough here is entirely by workers so far. No warriors. I am surrounded by Bugs and they are still pouring out, but not a one of them is armed and those nearest me all have typical worker features. I have not been attacked." I added, "Captain, do you think this could be just a diversion? With their real breakthrough to come somewhere else?"

"Could be," he admitted. "Your report is patched through right to Division, so let them do the thinking. Stir around and check what you've reported. Don't assume that they are all workers—you may find out the hard way."

"Right, Captain." I jumped high and wide, intending to get outside that mass of harmless but loathsome monsters.

That rocky plain was covered with crawly black shapes in all directions. I overrode my jet controls and increased the jump, calling out, "Hughes! Report!"

"Bugs, Mr. Rico! Zillions of 'em! I'm a-burnin' 'em down!"

"Hughes, take a close look at those Bugs. Any of them fighting back? Aren't they all workers?"

"Uh—" I hit the ground and bounced again. He went on, "Hey! You're right, sir! How did you know?"

"Rejoin your squad, Hughes." I clicked over. "Captain, several thousand Bugs have exited near here from an undetermined number of holes. I have not been attacked. Repeat, I have not been attacked at all. If there are any warriors among them, they must be holding their fire and using workers as camouflage."

He did not answer.

There was an extremely brilliant flash far off to my left, followed at once by one just like it but farther away to my right front; automatically I noted time and bearings. "Captain Blackstone—answer!" At the top of my jump I tried to pick out his beacon, but that horizon was cluttered by low hills in Square Black Two.

I clicked over and called out, "Sarge! Can you relay to the Captain for me?"

At that very instant my platoon sergeant's beacon blinked out.

I headed on that bearing as fast as I could push my suit. I had not been watching my display closely, my platoon sergeant had the platoon and I had been busy, first with ground-listening and, most lately, with a few hundred Bugs. I had suppressed all but the non-com's beacons to allow me to see better.

I studied the skeleton display, picked out Brumby and Cunha, their squad leaders and section chasers. "Cunha!

Where's the platoon sergeant?" "He's reconnoitering a hole, sir."

"Tell him I'm on my way, rejoining." I shifted circuits without waiting. "First Platoon Blackguards to second platoon—answer!"

"What do you want?" Lieutenant Khoroshen growled.

"I can't raise the Captain."

"You won't, he's out."

"Dead?"

"No. But he's lost power—so he's out."

"Oh. Then you're company commander?"

"All right, all right, so what? Do you want help?"

"Uh . . . no. No, sir."

"Then shut up," Khoroshen told me, "until you do need help. We've got more than we can handle here."

"Okay." I suddenly found that *I* had more than I could handle. While reporting to Khoroshen, I shifted to full display and short range, as I was almost closed with my platoon—and now I saw my first section disappear one by one, Brumby's beacon disappearing first.

"Cunha! What's happening to the first section?"

His voice sounded strained. "They are following the platoon sergeant down."

If there's anything in the book that covers this, I don't know what it is. Had Brumby acted without orders? Or had he been given orders I hadn't heard? Look, the man was already down a Bug hole, out of sight and hearing—is this a time to go legal? We would sort such things out tomorrow. If any of us had a tomorrow—

"Very well," I said. "I'm back now. Report." My last

jump brought me among them; I saw a Bug off to my right and I got him before I hit. No worker, this—it had been firing as it moved.

"I've lost three men," Cunha answered, gasping. "I don't know what Brumby lost. They broke out three places at once—that's when we took the casualties. But we're mopping them—"

A tremendous shock wave slammed me just as I bounced again, slapped me sideways. Three minutes thirty-seven seconds—call it thirty miles. Was that our sappers "putting down their corks"? "First section! Brace yourselves for another shock wave!" I landed sloppily, almost on top of a group of three or four Bugs. They weren't dead but they weren't fighting; they just twitched. I donated them a grenade and bounced again. "Hit 'em *now*!" I called out. "They're groggy. And mind that next—"

The second blast hit as I was saying it. It wasn't as violent. "Cunha! Call off your section. And everybody stay on the bounce and mop up."

The call-off was ragged and slow—too many missing files as I could see from my physicals display. But the mop-up was precise and fast. I ranged around the edge and got half a dozen Bugs myself—the last of them suddenly became active just before I flamed it. Why did concussion daze them more than it did us? Because they were unarmored? Or was it their brain Bug, somewhere down below, that was dazed?

The call-off showed nineteen effectives, plus two dead, two hurt, and three out of action through suit failure—and two of these latter Navarre was repairing by vandalizing power units from suits of dead and wounded. The third suit failure was in radio & radar and could not be repaired, so Navarre assigned the man to guard the wounded, the nearest thing to pickup we could manage until we were relieved.

In the meantime I was inspecting, with Sergeant Cunha, the three places where the Bugs had broken through from their nest below. Comparison with the sub map showed, as one could have guessed, that they had cut exits at the places where their tunnels were closest to the surface.

One hole had closed; it was a heap of loose rock. The second one did not show Bug activity; I told Cunha to post a lance and a private there with orders to kill single Bugs, close the hole with a bomb if they started to pour out—it's all very well for the Sky Marshal to sit up there and decide that holes must not be closed, but I had a situation, not a theory.

Then I looked at the third hole, the one that had swallowed up my platoon sergeant and half my platoon.

Here a Bug corridor came within twenty feet of the surface and they had simply removed the roof for about fifty feet. Where the rock went, what caused that "frying bacon" noise while they did it, I could not say. The rocky roof was gone and the sides of the hole were sloped and grooved. The map showed what must have happened; the other two holes came up from small side tunnels, this tunnel was part of their main labyrinth—so the other two had been diversions and their main attack had come from here.

Can those Bugs see through solid rock?

Nothing was in sight down that hole, neither Bug nor human. Cunha pointed out the direction the second section had gone. It had been seven minutes and forty seconds since the platoon sergeant had gone down, slightly over seven since Brumby had gone after him. I peered into the darkness, gulped and swallowed my stomach. "Sergeant, take charge of your section," I said, trying to make it sound cheerful. "If you need help, call Lieutenant Khoroshen."

"Orders, sir?"

"None. Unless some come down from above. I'm going down and find the second section—so I may be out of touch for a while." Then I jumped down in the hole at once, because my nerve was slipping.

Behind me I heard: "Sec*tion!*"

"First squad!"—"Second squad!"—"Third squad!"

"By squads! *Follow me!*"—and Cunha jumped down, too.

It's not nearly so lonely that way.

I had Cunha leave two men at the hole to cover our rear, one on the floor of the tunnel, one at surface level. Then I led them down the tunnel the second section had followed, moving as fast as possible—which wasn't fast as the roof of the tunnel was right over our heads. A man can move in sort of a skating motion in a powered suit without lifting his feet, but it is neither easy nor natural; we could have trotted without armor faster.

Snoopers were needed at once—whereupon we confirmed something that had been theorized: Bugs see by infrared. That dark tunnel was well lighted when seen by snoopers. So far it had no special features, simply glazed rock walls arching over a smooth, level floor.

We came to a tunnel crossing the one we were in and I stopped short of it. There are doctrines for how you should dispose a strike force underground—but what good are they? The only certainty was that the man who had written the doctrines had never himself tried them . . . because, before Operation Royalty, nobody had come back up to tell what had worked and what had not.

One doctrine called for guarding every intersection such as this one. But I had already used two men to guard our escape hole; if I left 10 per cent of my force at each intersection, mighty soon I would be ten-percented to death.

I decided to keep us together . . . decided, too, that none of us would be captured. Not by Bugs. Far better a nice, clean real estate deal . . . and with that decision a load was lifted from my mind and I was no longer worried.

I peered cautiously into the intersection, looked both ways. No Bugs. So I called out over the non-coms' circuit: "Brumby!"

The result was startling. You hardly hear your own voice when using suit radio, as you are shielded from your output. But here, underground in a network of smooth corridors, my output came back to me as if the whole complex were one enormous wave guide:

"BRRRRUMMBY!"

My ears rang with it.

And then rang again: "MR. RRRICCCO!"

"Not so loud," I said, trying to talk very softly myself. "Where are you?"

Brumby answered, not quite so deafeningly, "Sir, I don't know. We're lost."

"Well, take it easy. We're coming to get you. You can't be far away. Is the platoon sergeant with you?"

"No, sir. We never—"

"Hold it." I clicked in my private circuit. "Sarge—"

"I read you, sir." His voice sounded calm and he was holding the volume down. "Brumby and I are in radio contact but we have not been able to make rendezvous."

"Where are you?"

He hesitated slightly. "Sir, my advice is to make rendezvous with Brumby's section—then return to the surface."

"Answer my question."

"Mr. Rico, you could spend a week down here and not find me . . . and I am not able to move. You must—"

"Cut it, Sarge! Are you wounded?"

"No, sir, but—"

"Then why can't you move? Bug trouble?"

"Lots of it. They can't reach me now . . . but I can't come out. So I think you had better—"

"Sarge, you're wasting time! I am certain you know exactly what turns you took. Now tell me, while I look at the map. And give me a vernier reading on your D.R. tracer. That's a direct order. Report."

He did so, precisely and concisely. I switched on my

head lamp, flipped up the snoopers, and followed it on the map. "All right," I said presently. "You're almost directly under us and two levels down—and I know what turns to take. We'll be there as soon as we pick up the second section. Hang on." I clicked over. "Brumby—"

"Here, sir."

"When you came to the first tunnel intersection, did you go right, left, or straight ahead?"

"Straight ahead, sir."

"Okay. Cunha, bring 'em along. Brumby, have you got Bug trouble?"

"Not now, sir. But that's how we got lost. We tangled with a bunch of them . . . and when it was over, we were turned around."

I started to ask about casualties, then decided that bad news could wait; I wanted to get my platoon together and get out of there. A Bug town with no bugs in sight was somehow more upsetting than the Bugs we had expected to encounter. Brumby coached us through the next two choices and I tossed tanglefoot bombs down each corridor we did not use. "Tanglefoot" is a derivative of the nerve gas we had been using on Bugs in the past—instead of killing, it gives any Bug that trots through it a sort of shaking palsy. We had been equipped with it for this one operation, and I would have swapped a ton of it for a few pounds of the real stuff. Still, it might protect our flanks.

In one long stretch of tunnel I lost touch with Brumby—some oddity in reflection of radio waves, I guess, for I picked him up at the next intersection.

But there he could not tell me which way to turn. This was the place, or near the place, where the Bugs had hit them.

And here the Bugs hit us.

I don't know where they came from. One instant everything was quiet. Then I heard the cry of "Bugs! *Bugs!*" from back of me in the column, I turned—and suddenly Bugs were everywhere. I suspect that those smooth walls are not as solid as they look; that's the only way I can account for the way they were suddenly all around us and among us.

We couldn't use flamers, we couldn't use bombs; we were too likely to hit each other. But the Bugs didn't have any such compunctions among themselves if they could get one of us. But we had hands and we had feet—

It couldn't have lasted more than a minute, then there were no more Bugs, just broken pieces of them on the floor . . . and four cap troopers down.

One was Sergeant Brumby, dead. During the ruckus the second section had rejoined. They had been not far away, sticking together to keep from getting further lost in that maze, and had heard the fight. Hearing it, they had been able to trace it by sound, where they had not been able to locate us by radio.

Cunha and I made certain that our casualties were actually dead, then consolidated the two sections into one of four squads and down we went—and found the Bugs that had our platoon sergeant besieged.

That fight didn't last any time at all, because he had warned me what to expect. He had captured a brain Bug and was using its bloated body as a shield. He could not get out, but they could not attack him without (quite literally) committing suicide by hitting their own brain.

We were under no such handicap; we hit them from behind.

Then I was looking at the horrid thing he was holding and I was feeling exultant despite our losses, when suddenly I heard close up that "frying bacon" noise. A big piece of roof fell on me and Operation Royalty was over as far as I was concerned.

I woke up in bed and thought that I was back at O.C.S. and had just had a particularly long and complicated Bug nightmare. But I was not at O.C.S.; I was in a temporary sick bay of the transport *Argonne*, and I really had had a platoon of my own for nearly twelve hours.

But now I was just one more patient, suffering from nitrous oxide poisoning and overexposure to radiation through being out of armor for over an hour before being retrieved, plus broken ribs and a knock in the head which had put me out of action.

It was a long time before I got everything straight about Operation Royalty and some of it I'll never know. Why Brumby took his section underground, for example. Brumby is dead and Naidi bought the farm next to his and I'm simply glad that they both got their chevrons and were wearing them that day on Planet P when nothing went according to plan.

I did learn, eventually, why my platoon sergeant decided to go down into that Bug town. He had heard my report to Captain Blackstone that the "major breakthrough" was actually a feint, made with workers sent up to be slaughtered. When real warrior Bugs broke out where he was, he had concluded (correctly and minutes sooner than Staff reached the same conclusion) that the Bugs were making a desperation push, or they would not expend their workers simply to draw our fire.

He saw that their counterattack made from Bug town was not in sufficient force, and concluded that the enemy did not have many reserves—and decided that, at this one golden moment, one man acting alone might have a chance of raiding, finding "royalty" and capturing it. Remember, that was the whole purpose of the operation; we had plenty of force simply to sterilize Planet P, but our object was to capture royalty castes and to learn how to go down in. So he tried it, snatched that one moment— and succeeded on both counts.

It made it "mission accomplished" for the First Platoon of the Blackguards. Not very many platoons, out of many, many hundreds, could say that; no queens were captured (the Bugs killed them first) and only six brains. None of the six were ever exchanged, they didn't live long enough. But the Psych Warfare boys did get live specimens, so I suppose Operation Royalty was a success.

My platoon sergeant got a field commission. I was not offered one (and would not have accepted)—but I was not surprised when I learned that he had been commissioned. Cap'n Blackie had told me that I was getting "the best sergeant in the fleet" and I had never had any doubt that Blackie's opinion was correct. I had met my platoon sergeant before. I don't think any other Black-

328 ROBERT A. HEINLEIN

guard knew this—not from me and certainly not from him. I doubt if Blackie himself knew it. But I had known my platoon sergeant since my first day as a boot.

His name is Zim.

My part in Operation Royalty did not seem a success to me. I was in the *Argonne* more than a month, first as a patient, then as an unattached casual, before they got around to delivering me and a few dozen others to Sanctuary; it gave me too much time to think—mostly about casualties, and what a generally messed-up job I had made out of my one short time on the ground as platoon leader. I knew I hadn't kept everything juggled the way the Lieutenant used to—why, I hadn't even managed to get wounded still swinging; I had let a chunk of rock fall on me.

And casualties—I didn't know how many there were; I just knew that when I closed ranks there were only four squads where I had started with six. I didn't know how many more there might have been before Zim got them to the surface, before the Blackguards were relieved and retrieved.

I didn't even know whether Captain Blackstone was still alive (he was—in fact he was back in command about the time I went underground) and I had no idea what the procedure was if a candidate was alive and his examiner was dead. But I felt that my Form Thirty-One was sure to make me a buck sergeant again. It really didn't seem important that my math books were in another ship.

Nevertheless, when I was let out of bed the first week I was in the *Argonne*, after loafing and brooding a day I borrowed some books from one of the junior officers and got to work. Math is hard work and it occupies your mind—and it doesn't hurt to learn all you can of it, no matter what rank you are; everything of any importance is founded on mathematics.

When I finally checked in at O.C.S. and turned in my pips, I learned that I was a cadet again instead of a sergeant. I guess Blackie gave me the benefit of the doubt.

My roommate, Angel, was in our room with his feet on the desk—and in front of his feet was a package, my math books. He looked up and looked surprised. "Hi, Juan! We thought you had bought it!"

"Me? The Bugs don't like me that well. When do you go out?"

"Why, I've been out," Angel protested. "Left the day after you did, made three drops and been back a week. What took you so long?"

"Took the long way home. Spent a month as a passenger."

"Some people are lucky. What drops did you make?"

"Didn't make any," I admitted.

He stared. "Some people have *all* the luck!"

Perhaps Angel was right; eventually I graduated. But he supplied some of the luck himself, in patient tutoring. I guess my "luck" has usually been people—Angel and Jelly and the Lieutenant and Carl and Lieutenant Colonel

Dubois, yes and my father, and Blackie . . . and Brumby . . . and Ace—and always Sergeant Zim. Brevet Captain Zim, now, with permanent rank of First Lieutenant. It wouldn't have been *right* for me to have wound up senior to him.

Bennie Montez, a classmate of mine, and I were at the Fleet landing field the day after graduation, waiting to go up to our ships. We were still such brand-new second lieutenants that being saluted made us nervous and I was covering it by reading the list of ships in orbit around Sanctuary—a list so long that it was clear that something big was stirring, even though they hadn't seen fit to mention it to me. I felt excited. I had my two dearest wishes, in one package—posted to my old outfit and while my father was still there, too. And now this, whatever it was, meant that I was about to have the polish put on me by "makee-learnee" under Lieutenant Jelal, with some important drop coming up.

I was so full of it all that I couldn't talk about it, so I studied the lists. Whew, what a lot of ships! They were posted by types, too many to locate otherwise. I started reading off the troop carriers, the only ones that matter to an M.I.

There was the *Mannerheim*! Any chance of seeing Carmen? Probably not, but I could send a dispatch and find out.

Big ships—the new *Valley Forge* and the new *Ypres*, *Marathon*, *El Alamein*, *Iwo*, *Gallipoli*, *Leyte*, *Marne*, *Tours*, *Gettysburg*, *Hastings*, *Alamo*, *Waterloo*—all places where mud feet had made their names to shine.

Little ships, the ones named for foot sloggers: *Hora-tius*, *Alvin York*, *Swamp Fox*, the *Rog* herself, bless her heart, *Colonel Bowie*, *Devereux*, *Vercingetorix*, *Sandino*, *Aubrey Cousens*, *Kamehameha*, *Audie Murphy*, *Xeno-phon*, *Aguinaldo*—

I said, "There ought to be one named Magsaysay."

Bennie said, "What?"

"Ramón Magsaysay," I explained. "Great man, great soldier—probably be chief of psychological warfare if he were alive today. Didn't you ever study any history?"

"Well," admitted Bennie, "I learned that Simón Bolívar built the Pyramids, licked the Armada, and made the first trip to the moon."

"You left out marrying Cleopatra."

"Oh, that. Yup. Well, I guess every country has its own version of history."

"I'm sure of it." I added something to myself and Bennie said, "What did you say?"

"Sorry, Bernardo. Just an old saying in my own lan-guage. I suppose you could translate it, more or less, as: 'Home is where the heart is.'"

"But what language was it?"

"Tagalog. My native language."

"Don't they talk Standard English where you come from?"

"Oh, certainly. For business and school and so forth. We just talk the old speech around home a little. Traditions. You know."

"Yeah, I know. My folks chatter in Español the same

way. But where do you—" The speaker started playing "Meadowland"; Bennie broke into a grin. "Got a date with a ship! Watch yourself, fellow! See you."

"Mind the Bugs." I turned back and went on reading ships' names: *Pal Maleter*, *Montgomery*, *Tchaka*, *Geronimo*—

Then came the sweetest sound in the world: "—*shines the name, shines the name of Rodger Young!*"

I grabbed my kit and hurried. "Home is where the heart is"—I was going home.

CH:14

Am I my brother's keeper?

—Genesis IV:9

How think ye? If a man have an hundred sheep, and one of them be gone astray, doth he not leave the ninety and nine, and goeth into the mountains, and seeketh that which is gone astray?

—Matthew XII:12

How much then is a man better than a sheep?

—Matthew XVIII:12

In the Name of God, the Beneficent, the Merciful . . . whoso saveth the life of one, it shall be as if he had saved the life of all mankind.

—The Koran, Sûrah V, 32

Each year we gain a little. You have to keep a sense of proportion.

"Time, sir." My j.o. under instruction, Candidate or "Third Lieutenant" Bearpaw, stood just outside my door. He looked and sounded awfully young, and was about as harmless as one of his scalp-hunting ancestors.

"Right, Jimmie." I was already in armor. We walked aft to the drop room. I said, as we went, "One word, Jimmie. Stick with me and keep out of my way. Have fun and use up your ammo. If by any chance I buy it, you're the boss—but if you're smart, you'll let your platoon sergeant call the signals."

"Yes, sir."

As we came in, the platoon sergeant called them to attention and saluted. I returned it, said, "At ease," and started down the first section while Jimmie looked over the second.

Then I inspected the second section, too, checking everything on every man. My platoon sergeant is much more careful than I am, so I didn't find anything, I never do. But it makes the men feel better if their Old Man scrutinizes everything—besides, it's my job.

Then I stepped out in the middle. "Another Bug hunt, boys. This one is a little different, as you know. Since they still hold prisoners of ours, we can't use a nova bomb on Klendathu—so this time we go down, stand on it, hold it, take it away from them. The boat won't be down to retrieve us; instead it'll fetch more ammo and rations. If you're taken prisoner, keep your chin up and follow the rules—because you've got the whole outfit behind you, you've got the whole Federation behind you; we'll come and get you. That's what the boys from the *Swamp Fox* and the *Montgomery* have been depending on. Those who are still alive are waiting, knowing that we will show up. And here we are. Now we go get 'em.

"Don't forget that we'll have help all around us, lots of help above us. All we have to worry about is our one little piece, just the way we rehearsed it.

"One last thing. I had a letter from Captain Jelal just

before we left. He says that his new legs work fine. But he also told me to tell *you* that he's got you in mind . . . and he expects your names to *shine!*

"And so do I. Five minutes for the Padre."

I felt myself beginning to shake. It was a relief when I could call them to attention again and add: "By sections . . . port and starboard . . . prepare for drop!"

I was all right then while I inspected each man into his cocoon down one side, with Jimmie and the platoon sergeant taking the other. Then we buttoned Jimmie into the No. 3 center-line capsule. Once his face was covered up, the shakes really hit me.

My platoon sergeant put his arm around my armored shoulders. "Just like a drill, Son."

"I know it, Father." I stopped shaking at once. "It's the waiting, that's all."

"I know. Four minutes. Shall we get buttoned up, sir?"

"Right away, Father." I gave him a quick hug, let the Navy drop crew seal us in. The shakes didn't start up again. Shortly I was able to report: "Bridge! Rico's Roughnecks . . . ready for drop!"

"Thirty-one seconds, Lieutenant." She added, "Good luck, boys! This time we take 'em!"

"Right, Captain."

"Check. Now some music while you wait?" She switched it on:

"To the everlasting glory of the Infantry—"

Historical Note

YOUNG, RODGER W., Private, 148th Infantry, 37th Infantry Division (the Ohio Buckeyes); born Tiffin, Ohio, 28 April 1918; died 31 July 1943, on the island New Georgia, Solomons, South Pacific, while single-handedly attacking and destroying an enemy machine-gun pillbox. His platoon had been pinned down by intense fire from this pillbox; Private Young was wounded in the first burst. He crawled toward the pillbox, was wounded a second time but continued to advance, firing his rifle as he did so. He closed on the pillbox, attacked and destroyed it with hand grenades, but in so doing he was wounded a third time and killed.

His bold and gallant action in the face of overwhelming odds enabled his teammates to escape without loss; he was awarded posthumously the Medal of Honor.

Robert A. Heinlein

Robert Anson Heinlein was born in Missouri in 1907, and was raised there. He graduated from the U.S. Naval Academy in 1929, but was forced by illness to retire from the Navy in 1934. He settled in California and over the next five years held a variety of jobs while doing postgraduate work in mathematics and physics at the University of California. In 1939 he sold his first science fiction story to *Astounding* magazine and soon devoted himself to the genre.

He was a four-time winner of the Hugo Award for his novels *Stranger in a Strange Land* (1961), *Starship Troopers* (1959), *Double Star* (1956), and *The Moon Is a Harsh Mistress* (1966). His Future History series, incorporating both short stories and novels, was first mapped out in 1941. The series charts the social, political, and technological changes shaping human society from the present through several centuries into the future.

Robert A. Heinlein's books were among the first

works of science fiction to reach bestseller status in both hardcover and paperback. He continued working into his eighties, and his work never ceased to amaze, to entertain, and to generate controversy. By the time he died, in 1988, it was evident that he was one of the formative talents of science fiction: a writer whose unique vision, unflagging energy, and persistence, over the course of five decades, made a great impact on the American mind.

Want to connect with fellow science fiction and fantasy fans?

For news on all your favorite Ace and Roc authors, sneak peeks into the newest releases, book giveaways, and much more—

"Like" Ace and Roc Books on Facebook!

facebook.com/AceRocBooks

M988JV1011

Can't get enough paranormal romance?

Looking for a place to get the latest information and connect with fellow fans?

"Like" Project Paranormal on Facebook!

- Participate in author chats
- Enter book giveaways
- Learn about the latest releases
- Get book recommendations and more!

facebook.com/ProjectParanormalBooks

Penguin Berkley Jove ACE NAL Signet Obsidian Signet Eclipse RoC W

M883G1011

Step into another
world with

DESTINATION
ELSEWHERE

A new
community
dedicated
to the best
Science Fiction
and Fantasy.

Visit
 /DestinationElsewhere

M1160G0712